This book is a gift

made possible by a grant from

The Indianapolis Foundation

—————THE—————
INDIANAPOLIS
F·O·U·N·D·A·T·I·O·N

119 English Foundation Building
615 North Alabama Street
Indianapolis, Indiana 46204
(317) 634-7497

The Facts On File
DICTIONARY
OF
MARINE SCIENCE

The Facts On File
DICTIONARY
OF
MARINE SCIENCE

BARBARA CHARTON

John H. Tietjen, Ph.D.
Scientific Adviser

Edmund H. Immergut, Ph.D.
Series Editor

Facts On File
New York • Oxford • Sydney

The Facts On File Dictionary of Marine Science

copyright © 1988 by Barbara Charton

Library of Congress Card Number: 82-15715
British CIP data available on request

Printed in the United States
ISBN 0-8160-1031-5 (HC)
ISBN 0-8160-2369-7 (PB)
10 9 8 7 6 5 4 3 2

ACKNOWLEDGMENTS

No long project goes forward without aid, encouragement, support, and helpful criticism. Many people have contributed to this work and I am very grateful to them all. I would like to thank most particularly Dr. E. H. Immergut, my editor, and Dr. J. H. Tietjen, of City College (CUNY). Both contributed much time, effort, and care, all of which were essential in moving this work toward fruition. In addition to Dr. Tietjen's technical support, I would like to thank Dr. N. Vincente of the Fondation Ricard (France) and Dr. R. Gallo of IPSOI (Université d'Aix-Marseille, France). In addition, I have depended on the indulgence of many friends: D. F. Lapidus, Dr. and Mrs. J. B. F. N. Engberts, Dr. and Mme. R. Gallo, Dr. and Mrs. A. H. C. Jakobs, and my long-suffering family. There is no way to thank Marvin who lived with it all, read it all, and stayed objective. I can only promise him a smoother path for the next effort.

BC
Brooklyn, NY

A

abalone Snails, class Gastropoda, subclass Prosobranchia, genus *Heliotis.* These animals do not look like snails, but superficially resemble clamlike mollusks. The shell exhibits a characteristic row of holes.

Abalone are found in the Pacific Ocean and off the Australian and South African coasts. The red abalone *(H. rufescens),* the largest species, is taken commercially along the western coast of North America. This animal usually has a shell about 30 cm (12 inches) long. The large muscle, the foot, is sliced and broiled as abalone "steak." With the reappearance of a sizable number of sea otters off the coasts of Washington, Oregon, and British Columbia, the abalone is now scarcer than it was in the 1930s and 1940s. See *sea otter.*

absolute humidity The maximum amount of water vapor that a unit volume of air can contain at saturation. Absolute humidity is a function of temperature because of the gaseous state of air: as the temperature rises the vapor pressure of the water rises, which means that more molecules of liquid water have sufficient energy to make the transition from the liquid to the gaseous state. The higher the temperature, the higher the vapor pressure of the water and therefore the higher the absolute humidity.

abyssal environment The environment at the bottom of deep ocean, one of the most constant marine biomes. The temperature is between 0 and 2°C. The salinity is also relatively constant, hovering at about 35%, as are the concentrations of various ions (phosphate, silicate, hydroxyl, and carbonate) are also fairly uniform. There is sufficient oxygen to support life. The life forms that exist at great depths depend for food on terrestrial or upper water layers; they are for the most part suspension feeders, scavengers, or carnivores, but their greatest difference from animals found at higher levels is their ability to live at a constant high pressure (200 to 700 atm). The number of animals in the abyssal environment is low, and their rate of reproduction is also thought to be low. See *abyssal floor, hadal zone.*

abyssal floor Also called abyssal plain. The flat, relatively featureless bottom of deep ocean, i.e., at a depth greater than 2000 m. The average depth of the abyssal floor is about 4,000 m (13,000 feet). There are very few animal species that live in the inhospitable cold, dark regions of the ocean bottom. There is little or no seismic activity, and the major geologic event is sedimentation. Particulate matter coming from a continent, or bits of calcareous material of marine origin, will finally filter down onto the ocean floor. However, the ocean floor is not without activity. Storms and strong currents carry sediment about, scouring the bottom. Also, the Atlantic Ocean receives rushes of very dense, cold water from both polar regions, which carries sediment with it. The great currents of very cold water are more prevalent in the Atlantic than in the Pacific Ocean since the shallow Bering Sea and the Aleutian island chain effectively bar the cold Arctic waters from entering the Pacific.

abyssopelagic Referring to that part of the water column at great depths, usually below 4,000 m (13,500 feet). The term refers to any organism or condition found in these hadal regions. *See ocean bottom.*

acantharia A group of marine protists. These tiny organisms comprise less than 10% of the plankton even in periods of planktonic bloom and in the Gulf Stream, where they are most numerous. Acantharia have internal rigid support systems made of celestite (strontium sulfate-$SrSO_4$). This compound is soluble at depths greater than 200 m (650 feet), and as a result, the skeletons of Acantharia do not settle to the ocean bottom when the organisms die.

Acantharia migrate vertically each day because they harbor algae or dinoflagellates within their cytoplasm, and this periodic rise allows these photosynthesizing symbionts to move into the photosphere during daylight hours. At night the Acantharia sink once again to lower levels where they find food and, in turn, provide food for other organisms. *See migration.*

acanthocephala Small (about 1 cm or 0.4-inch-long), spiny-headed, sausage-shaped worms, all of which are parasitic and distantly related to nematodes. These animals are named for the impressive, hook-covered proboscis that attaches them to the gut of their host.

The body cavity of the acanthocephalans is filled almost totally with gonads. The life cycle is a complex one in which larval stages infest intermediate hosts. Isopods are frequently the first or intermediate hosts. The *Corynosoma* is a parasite of several species as it works its way up the food chain. Its first host is an amphipod in which it changes into a more adult form. Then, upon ingestion by a fish, it becomes a parasiste of that fish. If this fish is eaten by a bird or seal, yet another change of the worm renders it a parasite in that host. In vast numbers, acanthocephalans can starve their hosts or create fatal gut blockages. *See food chain, Nematoda, parasite, proboscis.*

acidity (basicity) A measure of the concentration of acid (base) in water or in a solution. The acidity or basicity of seawater is described by the pH scale. Pure water has a pH of 7; a strong acid would have a pH of 1 or 2 and a strong base would have a pH of 12 or 13; seawater is slightly alkaline at 7.2. The acidity or alkalinity (basicity) of seawater is a function of depth, the quantity and nature of the gases dissolved in the water, local conditions, and the presence (or absence) of plant and animal life. *See acids and bases, buffer, minerals, pH, water.*

acids and bases An acid is a chemical entity that can transfer a proton (i.e., a hydrogen ion or a hydrogen atom nucleus) to a base. A base is a chemical entity that can accept a proton. This definition of acids and bases, known as the Brønsted definition, is the most appropriate for an aquatic environment.

Acipenseriformes The order of fish that includes the sturgeons (Acipenseridae). As adults, Acipenseriformes have a cartilaginous skeleton; when immature they have a bony skeleton and have scutes (armor plates) instead of scales. *See sturgeon.*

acorn worm A type of hemichordate whose anterior (head) end resembles an acorn. Not worms at all, these animals burrow in the sand or mud of the sea bottom down to depths exceeding 3000 m (10,000 feet). Most species are about 5 cm (2 inches) long, but the species *Balanoglossus gigas* is over 180 cm (6 feet) long.

There is considerable variation between species of acorn worms, with some of the deep-water specimens possibly being free swimming. Feeding modes also differ; some species filter seawater, extracting debris from it, while others feed by trapping food particles in a mucus film and conveying it toward their mouth by cilia. The sexes are separate. Some species produce large numbers of eggs that have almost no food reserve; others have few eggs with a relatively large yolk.

acoustics The study of the production, transmission, and detection of sound.

Any disturbance in a medium—gaseous, liquid, or solid—will produce a sound wave, which the medium will disperse.

Sound waves can be used to plumb the depths of the ocean and to map bottom terrain. A ship or buoy can generate a sound wave that will travel through the water at a known rate, strike the bottom, and rebound. Receiving equipment can then measure the interval from the generation to the return of the wave, from which the distance to the bottom may be calculated.

Plankton swarms or large schools of fish also reflect sound waves, because they are so tightly packed that their mass is echo-producing. Commercial fishermen capitalize on this fact and use SONAR to find fish. See *echo, LORAN, SONAR, sounding.*

Acropora   Representative reef-building corals. See *atoll, coral, Cnidaria, reef.*

activation analysis   The bombardment of samples of rock, water, or fossils by slow (thermal) neutrons, protons, or other nuclear fragments of known energy in order to date them. The radiation coming off the sample is then compared to established standards, which allows the analyst to identify the elements in the sample and determine their ages. The advantage of this type of analysis is that, unlike carbon dating, it does not destroy the sample. See *radioactivity*

Aden, Gulf of   A deepwater area between the Red and Arabian seas and between the African and Arabian coasts. The Gulf of Aden is a part of the East African Rift Valley system. Its most important undersea feature is the Sheba Ridge, a continuation of the ridge system in the Indian Ocean, which is also the site of seismic and volcanic activity.

The deepest part of the Gulf is the Alula-Fartak Trench, with a recorded low point of 5,360 m (17,566 feet). The bottom also exhibits a series of fault lines running from northeast to southwest, caused by plate movement as the African and Arabian crustal fragments move apart. The sediment of the bottom is terriginous and is carried into the water from the surrounding landforms—mainly by the wind. The water of the Gulf is warmer than normal ocean water because there is little water entering the system in the form of rain or river runoff, and the Gulf is unusually saline because there is considerable evaporation in the hot climate. The surrounding land is largely desert.

The Strait of Bab-el-Mandeb connects the Gulf to the Red Sea. There is no large-scale fishing in the Gulf, despite the large variety and considerable numbers of fish that feed on the upwelling of nutrients along the coasts. See *Indian Ocean, Rift Valley.*

Adriatic Plate   The leading edge of the Eurasian Plate, that abuts the Hellenic and African plates. See *plate tectonics.*

Adriatic Sea   An arm of the Mediterranean separating the eastern Italian coast from the Balkan and Hellenic peninsulas. The Sea is about 131,000 km² (50,000 square miles). The maximum depth is 1,324 m (4035 feet) west of Corfu, and the average depth is about 450 m (1,450 feet). The sea has one relatively smooth coast on the west, the Italian one, and a very indented one with many islands, the Balkan coast, on the east. The latter is a continuation of the mountain structure of the East European landmass.

The Adriatic is rather unusual for the Mediterranean because it has a greater tidal range (the difference between the high and low water marks) than that of the larger body itself. The average tide range is less than 30 cm (1 foot) in the Mediterranean, whereas the tide ranges in the Adriatic are three times larger. The nutrient and oxygen loads in this sea are relatively low, like those in the rest of the Mediterranean. See *Mediterranean.*

Aegean Sea   The easternmost embayment of the Mediterranean, bounded by Greece on the west, Turkey on the east, and Crete on the south. The Dardanelles are the strait that connects the Aegean Sea with the Sea of Marmora to the northeast. This is in turn connected to the Black Sea by the Bosporus. The Aegean is dotted with small islands that rise from a relatively deep floor. The deepest point, 3,543 m (11,627 feet), is found near Crete. The total area of the sea is about 214,000 km² (83,000 square miles).

Like most of the Mediterranean, the Aegean is relatively tideless except for the strait between continental Greece and the island of Euboea, where an unpredictable and often violent current runs. This anomaly has been the subject of study for millenia; Aristotle commented on it.

The geology of the Aegean is, like that of the islands in it, largely limestone. The water is both nutrient and oxygen poor even when compared to other Mediterranean water. See *archipelago, Mediterranean.*

aerosol   Particulate matter dispersed in air. The particles may be either liquid or solid. The major constituent of maritime aerosols is common salt, which in the form of very finely divided particles has been carried aloft either by the evaporation of water or by wave action. Sand and other terriginous solids are other common solid components of aerosols. Liquid aerosols, such as fogs made up of water particles, are also generated by the spray resulting from wave action.

The solid particles in some aerosols are windblown pollutants that can be carried for great distances, depending on wind velocity and particulate size. See *atmosphere, fog, pollution, sediment.*

African Plate   The tectonic plate extending under most of the African continent. At its northeastern edge, the Eastern Rift Valley is a subduction zone (the descending plate). Eventually the rift will widen and be invaded by ocean water as the African and Arabian plates continue to separate. See *plate tectonics.*

Agassiz,   Alexander   (1835–1910)   An authority on coral formations, he was born in Switzerland and came to the United States as a boy. His professional career was shaped largely by that of his father, natural historian

Jean-Louis Agassiz. Alexander Agassiz's publications include a major scientific study of starfish and a popular work on oceanography, co-authored with his stepmother, Elizabeth Cady Agassiz.

Agassiz made several voyages, most of them on the *Albatross.* In 1875 his discovery of a coral reef at a depth of 1,000 m (3,300 feet) led to a contradiction of Darwinian theories of coral reef construction and to Agassiz's continuing study of these formations.

He made other trips, at the behest of the United States Navy, including explorations of the Caribbean Sea, the path of the Gulf Stream, and The Great Barrier Reef.

In 1899 Agassiz sailed from San Francisco to the Marquesas Basin, where he dredged and found many sharks' teeth at great depths, although sharks are not hadal creatures. His explanation for this find (one that later evidence supported) was that in Tertiary times the Marquesas Basin was closer to the photic zone than it is now, with greater light and warmer water, and thus sufficient prey for sharks. Subsequently the area subsided.

On Agassiz's last voyage, again on the *Albatross,* he studied the Humboldt Current. See *Agassiz, Jean-Louis;* Albatross; *Caribbean; Gulf Stream; Pacific.*

Agassiz,   Jean-Louis   Rodolphe (1807–1873)   A natural historian who made major contributions to the study of fish fossils and glaciers. Jean-Louis Agassiz was born in the Swiss canton of Fribourg, was educated in Zurich and Heidelberg, and received a medical degree from the University of Munich in 1830. Agassiz went to France to work with Cuvier, who was studying the classification of fossil fish. While in Paris he met geographer and naturalist Friedrich Humboldt, whose recommendation helped Agassiz obtain his first academic position, as Professor of Natural History in Neuchâtel (Switzerland).

While in Switzerland Agassiz continued the work he had begun in Paris on the fishes of Brazil. The result was a superb five-volume treatise on fossil fish, published in 1834.

Agassiz belonged to a group, centered in Neuchâtel, that seriously studied natural history, although most of its members were trained in other disciplines. The group made extensive explorations of the mountains around them, and their careful observations were helpful to Agassiz in his own work on glaciers, in which he used existing glaciers to explain phenomena such as scouring scars, transported rocks, and moraines (the mixture of rocks left at the leading edge of a glacier).

Agassiz opposed the Darwinian theory of the evolution of species as the result of natural selection. He believed that changes in and on the Earth created diversification and change in living organisms and fossil remains. Until the work of Mendel on genetics, there was no mechanism for the theory of natural selection, so Agassiz's theory, which was logical and appeared to explain the evidence, found some acceptance.

In 1846, at the suggestion of Charles Lyell, the noted British geologist, Agassiz left Switzerland for an American lecture tour. He was well received, and at the end of the tour accepted a professorship in natural history at Harvard University. He began a new collection of natural history artifacts, which eventually grew to become the Harvard Museum of Comparative Zoology. This institution is now named for Agassiz.

Agnatha A class of primitive vertebrate fishes. The name means jawless and as a result, members of the class are all suction or filter feeders. See *hagfish, lamprey*.

Agulhas Current A southerly moving oceanic current along the East African coast. The Agulhas Current begins as an east-to-west flow but is deflected by the coastline. The water temperature varies with depth, although the surface water is warm year-round. The portion of the Agulhas Current between Madagascar and the African continent is called the Mozambique Current.

air A mixture of gases, predominantly nitrogen and oxygen. Air also contains

Constituents of Air

Substance	% by volume
Nitrogen (N_2)	78.11
Oxygen (O_2)	20.95
Water vapor (H_2O)	0–7
Argon (A)	0.93
Carbon dioxide (CO_2)	0.01–0.1 (0.03 is average)
Ozone (O_3)	0–0.1 (greatest concentration in the stratosphere)
Neon (Ne)	1.8×10^{-3}
Sulfur dioxide (SO_2)	$0 - 1 \times 10^{-4}$
Krypton (Kr)	1.1×10^{-4}
Methane (CH_4)	2×10^{-4}
Xenon (Xe)	5×10^{-5}
Nitrogen oxides (NO_x)*	$0 - 3 \times 10^{-5}$
Hydrogen (H_2)	5×10^{-5} (at higher altitudes)

*This, in turn, is a mixture of several gases, N_2O, NO, NO_2, N_2O_4, and N_2O_6; most of these nitrogen oxides are classified as pollutants. They are produced in gasoline burning engines, where the temperature is high enough to cause normally unreactive atmospheric nitrogen to react with oxygen.

carbon dioxide, water vapor, and trace amounts of some twenty other gases. It is the gaseous vehicle in aerosols and suspensions of larger particles, which it disperses. See *pollution.*

air mass   The moving, homogeneous aggregate of air as it pertains to weather. The leading *edge*, or *front*, is the site of rapid change as one mass interacts with another. The character of an air mass depends on the land or water it traverses. The meteorologic abbreviation cPK, for example, means a continental and polar or cold, dry air mass, mPK indicates cold polar air of maritime origin. See *meteorology.*

Alaska Current   The final branch of the Aleutian Current moving northeast into the Gulf of Alaska. See *Aleutian Current.*

albatross (bird)   A large, web-footed seabird; a member of the genus *Phoebetria* or *Diomedia.* There are about fifteen species of albatrosses. The largest and best known is *Diomedea exulans,* a white bird with a 3.5-m (12-foot) wingspan and a heavy, powerful bill. The usual food of this particular species of albatross is squid.

These large birds live for long periods well beyond the continental shelves in temperate to subpolar latitudes. Like their relatives the petrels, the albatrosses stay on the water for months, coming to land only to nest and raise their young. Most albatrosses are sighted in the Southern Hemisphere. See *petrel.*

Albatross (research vessel)   The first ship to be built as a research vessel. The *Albatross* was designed to explore the environments of commercial fishing grounds, and its construction was underwritten by the U.S. Department of Fisheries. Naturalist Alexander Agassiz used the ship from the 1870s through the 1890s for deep sea sampling. See *Agassiz, Alexander.*

Albert I, Prince of Monaco (1848–1922)   Albert-Honoré Charles Grimaldi founded two outstanding institutions of oceanographic research: the Musée Oceanographique de Monaco (Oceanographic Museum of Monaco) and the Institut Oceanographique de Paris (Oceanographic Institution of Paris).

After service in the French navy during the Franco-Prussian War, Albert devoted his life to oceanography. In 1873 he bought and refitted a 200-ton schooner and renamed it the *Hirondelle.* It made four cruises and was followed by the *Princess Alice* (there were two of that name) and the *Hirondelle II.*

Albert went to sea almost every summer for forty years. In 1885 his major study of the Gulf Stream appeared. A long study of drift currents in the North Atlantic followed, in conjunction with mapping of water temperatures and movements of isothermal bodies of water. He established three observatories in the Azores as meteorological research stations.

Albert's work with animals included experiments with *Physalia* (Portuguese man of war)—a coelenterate whose venom results in deep anesthesia. He designed and used plankton nets and with them brought to the surface animals from depths exceeding 6,000 m (19,000 feet).

Albert was an early expounder of the prudent and ecologically harmless use of the sea. He warned of overfishing in the early years of the 20th century, and in addition to the oceanographic institutes in Monaco and Paris, he founded the Musée Anthropologique and the Institut de Paleontologie Humaine for the study of human evolution. His wonderful museum and aquarium in Monaco are open to the public.

Aleutian Current   An ocean current that moves along the Aleutian island chain from west to east. This subarctic water divides, and part turns north into the Bering Sea and moves along the north coast of some of the Aleutian islands, another branch forms the Alaska Current. The surface water of the Alaska Current is colder and less saline than the water at the depth of the current. See *Alaska Current.*

Aleutian Trench   An ocean deep that skirts the southern border of the Aleutian Island chain. The deepest portions of the Aleutian Trench are more than 7300 m (24,000 feet) deep.

algae   An ancient group of primitive plants. Algae range from unicellular organisms to large, complex colonies of kelp, and from the blue-green algae to yellow, green, brown, and red organisms. All species are autotrophic (photosynthesizing). Some algae, such as dinoflagellates and euglenoids, have flagellae and are motile. These life forms support major marine food webs. See *blue-green algae, diatoms, kelp.*

alkalinity and acidity   The hydrogen ion (H^+) concentration of a liquid or solution, usually expressed in logarithmic form as pH. The pH of seawater varies with local conditions. Both alkalinity and acidity are functions of dissolved minerals and gases. See *acids and bases, buffer, pH.*

alpha ray   Alpha rays are produced by radioactive decay. The rate of decay is used in dating marine fossils and other remains or radioactive elements. Alpha rays consist of the nuclei of helium, atoms which are composed of two protons and two neutrons. These particles emanate from a substance as part of its radioactive decay. See *atom, radioactivity.*

Alps   The European portion of the great east-west mountain system stretching from the Pyrenees to Malaysia. The Alps extend from the Gulf of Genoa to the Vienna Woods. The massif is divided into a western section, comprising southeastern France and northwestern Italy; an eastern section comprising Germany, Switzerland, Austria, and Yugoslavia; and a central portion extending from north-central Italy to southern Switzerland.

According to the theory of plate tectonics, the alpine upheaval is the result of the collision of the African and Eurasian plates, each of which has a landmass at its leading edge. The Alps are much deformed and folded, the result of glacial movement. The mountains are relatively young, dating from the Tertiary Period. This is notable in light of the number of very sharp, comparatively uneroded peaks, such as the Matterhorn and Mont Blanc, the latter being the highest peak in the range at 4,800 m (15,800 feet) above sea level. Glaciers cover a significant portion of the Alps. Major river systems that rise in the Alps are the Rhine, Rhône, Po, and Danube. See *fossil, Mediterranean, plate tectonics, Tertiary Period.*

Alvin (research vessel)   A submersible built in the United States in 1964, the *Alvin* is operated by the Woods Hole Oceanographic Institution (on Cape Cod, Massachusetts). This light-weight, manned vessel has descended to depths of 3,000 m (10,000 feet) in the Mid-Atlantic Rift Valley, where it was used to investigate the undersea mountain range. The *Alvin* has also been used to investigate vent communities in the Pacific Ocean and other deep sea areas, and was involved in the exploration of the wreck of the *Titanic.* See Archimede, *submersible, vent communities.*

Amazon River   The world's largest river in volume of water. The Amazon delivers an immense quantity of water into the South Atlantic, draining most of north-central South America.

The river narrows, near Obidos, Brazil, about 320 km (200 miles) from its outlet in the Atlantic Ocean. At this point the river is 2.5 km (1.5 miles) wide and has a channel about 70 m (200 feet) deep. The water surging into the Atlantic creates a 4-m (13-foot)-high tidal bore called the Pororoca.

The Amazon is 6,275 km (about 3,900 miles) long, making it one of the world's longest rivers. It runs along an almost straight west-to-east line across tropical South America from its headwaters in Peru to its outlet in the Atlantic. Water in the river moves relatively slowly, carrying loads of various sediments. The sediment load is a function of the regions through which the Amazon's tributaries move; some are milky and carry clay; others are black from suspended humus.

Navigation upriver is possible on many of the tributaries of the Amazon, as well as on the main river, into Peru.

ambergris   A waxy substance found floating in the ocean. It is produced by sperm whales, and found in the digestive tracts of these mammals where it is thought to protect the whale's digestive system against the hard beaks of squid, which form the major part of the sperm whale's diet. When ambergris is fresh, it is a black, greasy, smelly substance. On drying it becomes grayish-black and smells faintly of musk. Ambergris was at one time used as a pharmaceutical and a fixative, which is the part of a perfume that keeps the volatile aromatic materials dissolved in the perfume medium (usually alcohol). For these uses ambergris has been largely replaced by synthetic materials. See *sperm whale*.

amino acid   A chemical component of all proteins, and found in all living organisms. Autotrophic or self-feeding organisms—usually single celled organisms such as bacteria—manufacture amino acids from simpler compounds such as water (H_2O), carbon dioxide (CO_2), and the nitrate ion (NO_3^-). Amino acids are characterized by an acid moiety and an amino group:

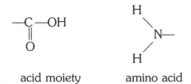

acid moiety   amino acid

There are theoretically thousands of possible amino acids, but only about twenty commonly occur in proteins. These relatively few amino acids, when linked together in specific sequences form the different proteins. The sequences in which they are linked together are species-spe-

cific. Thus, sharks of one specific genus and species have identical proteins made up of identical sequences of amino acids, and all cod of a certain species have identical proteins, but the shark proteins are slightly different in amino acid composition from the cod proteins. See *autotroph, peptide, protein*.

ammonia (NH_3)   A basic chemical compound, the end-product of both protein metabolism and the breakdown of nucleic acids in marine organisms. Bacteria that break down living or formerly living matter liberate ammonia as part of their metabolic processes. Ammonia is usually a gas at normal temperatures, and dissolves in water. See *nitrogen cycle*.

ammonites   Fossilized chambered cephalopods, like the nautiloids. Ammonites have been extinct since the Cretaceous Period; Ammonites are in the order Ammonoidea along with clymenias, goniatites and ceratites. Before their extinction, ammonites were very widely distributed, possibly because their larvae were free swimming or drifting (meroplanktonic). Ammonites were characterized by a tightly coiled planar shell with a bulbous, calcareous protoconch. This protoconch was continuous with the shell, enlarging in a logarithmic spiral. This process by which an organism such as an ammonite moves from a smaller chamber such as the protoconch to a larger one such as the shell is analogous to the molting (ecdysis) of arthropods. The size of the average ammonite was 10 to 30 cm (4 to 12 inches), although adult individuals with a shell size ranging from 1 cm to 3 m have been found. If, like the present nautilus, ammonites contained nitrogen bubbles, they could not have lived at great depths. Their food and method of locomotion are unknown. For unknown reasons also, ammonites managed to survive two near-extinctions and then reevolved many new species, finally dying out at the end of the Cretaceous Period. See *Cephalopoda, Cretaceous Period, fossils, nautilus*.

Ampere Seamount An undersea peak off the North African Coast northeast of the Madeira islands. See *Atlantis Seamount.*

Amphibia Cold-blooded vertebrates closely related to the fishes from which they evolved. Unlike fish, most amphibians (frogs, toads, salamanders) do not tolerate salt water, although the frog *Rana cancrivora* is an exception. It is a large crab-eating frog inhabiting Indonesia. The general characteristics of amphibians are a moist, absorbing skin, small lungs, four limbs, and gills at some stage of these animals' development. They deposit their eggs, which lack amnions, in water.

amphidromic point The center of zero tide. The tide crest in any tidal period rotates around this point. See *tide.*

Amphineura The chitons, marine mollusks of the order Placophora. The Amphineura are represented by both living and fossil species. The mantle secretes a segmented shell, giving the animal the appearance of a segmented worm. All amphineurans have a ventral foot, anterior mouth with a radula, no eye stalks or tentacles, and breathe through gills in the mantle cavity. The sexes are usually separate.

amphioxus A member of the subphylum Chordata, the genera *Branchiostoma* and *Asymetron* are the only genera of amphioxus. These animals have a notochord and an elongate, essentially headless, tailless, finless body 4 to 7 cm long. They live in semitropical or temperate coastal waters, where they burrow tail-first into the sand. They are filter feeders with primitive breathing systems. The amphioxus excretory system resembles that of the annelids, and its circulation that of fish. However, the amphioxus has colorless blood and no real heart. The sexes are separate.

Amphipoda Small, frequently colorless crustacea. Most species are marine but there are a few freshwater and still fewer terrestrial forms. These small animals are characterized by the lack of a carapace and a laterally flattened body. Amphipoda are for the most part scavengers, although there are some suspension feeding or sucking species.

Amundsen, Roald (1872–1928) The Norwegian explorer best remembered for being the first man to reach the South Pole. Amundsen was a naval officer whose first experience in exploration was with the Belgians on an antarctic voyage in 1897.

Amundsen led his own expedition in 1901 in the sloop *Gjøa*, to the northeast coast of Greenland. Two years later, again in the *Gjøa*, he determined the location of the North Magnetic Pole, in the Northwest Territories of Canada. He then continued westward and was the first navigator to traverse the Northwest Passage.

Amundsen attempted to be the first person at the North Pole, but after Peary achieved that goal, Amundsen attempted to be first at the South Pole. He used Frijdtof Nansen's ship, the *Fram,* and sailed into the Bay of Whales in early 1911. Amundsen beat the Scott expedition, which was trying for the same goal, arriving at 90° south latitude on December 14, 1911.

Amundsen spent the rest of his life as an explorer and airplane pilot. He felt he had "found his vocation" in aviation. Flying from Spitsbergen, Norway, in 1925, he attempted to be first over the North Pole in an airplane. That attempt was unsuccessful, but he eventually achieved his goal. In 1928, while engaged in an air-rescue operation for a lost dirigible, he and his crew were lost in the Arctic. See *Peary, Robert; Fram.*

Amur River A river formed in the southeastern part of the Soviet Union by confluence of the Argun and Shilka rivers; it is the river boundary between the easternmost reaches of the Soviet Union and China, then turns north and east and flows into the Tatar Strait. The Chinese name for the river is Heilung Chiang. The Amur forms a large estuary at its mouth in the Tatar Strait, south of the Sea of Okhotsk, delivering large volumes of freshwater into the Strait. The monsoon climate of the river's drainage basin creates spring torrents, which lower the salinity in the Sea of Okhotsk and result in its lack of sediment. The Amur is an important source of freshwater fish and is navigable for much of its 2,800-km (1,700-mile) length. See *Okhotsk, Sea of.*

anadromous fish   Fish that spend most of their adult lives in the ocean, but move up rivers to spawn. See *salmon.*

anchovy   A small (10 to 15 cm or 4 to 6-inch-long) herring-like fish. This filter feeder is representative of the entire clupeform (herring) order. Anchovies are found in enormous schools in the Mediterranean Sea and Pacific Ocean, both in northern waters and off the western coast of South America. Anchovies have great economic importance as food and as a raw material in the manufacture of oil, fertilizer, fishmeal, and animal food. They are also an important link in the food chain, since they support a good deal of bird life.

Anchovies are slender, silvery fish that feed on the great plankton populations that bloom in the spring. The Peruvian anchovy, *Engraulis ringens,* and the Atlantic and Mediterranean anchovy, *E. encrasicholus,* produce prodigious numbers of eggs. The female matures at a year old or less, and may spawn as many as 10,000 eggs. The eggs have very little yolk, so the anchovy hatches quickly and must immediately find food. Mortality among the young is therefore high. See *El Niño, herring.*

Andaman Sea   A marginal tropical sea connected to the Indian Ocean. The Andaman Sea is west of the Malay Peninsula and east of the Malacca Strait. Its average depth is about 900 m (3,000 feet), with its deepest point, 4,180 m (13,800 feet), lying to the east of Nicobar Island. The northern end of the Sea is dominated by the rapidly expanding Ir-

rawaddy Delta. The great volume of water entering the Andaman Sea causes significant changes in its salinity, which falls to 2% in the monsoon months of June to October.

The uplift that produced the Andaman-Nicobar Island probably dates to the Cretaceous Period. Volcanic activity continues on the island and extends south to an area in which numerous seamounts have been found. See *Indian Ocean.*

andesite line   A boundary that marks the Pacific edge of the Eurasian Plate. The line is not continuous, and is delineated by a series of deep trenches, associated with volcanic and seismic occurrences.

angiosperms   Flowering land plants with enclosed or covered seeds. The angiosperms are the dominant terrestrial plant forms and also the most recently evolved land plants. In the context of the marine ecosystem, beach grasses are angiosperms that have evolved from basic terrestrial forms. They can now tolerate a higher than usual salt environment. See *beach, Zostera.*

Angola Basin   A basin located in the South Atlantic Ocean, off the southwest African coast. It lies south of the cone formed by the debris carried into the South Atlantic by the Congo River. See *Atlantic Ocean.*

anglerfish   (lophiformes)   Mainly deep-sea creatures, anglerfish are noted for a dorsal fin adaptation attached to the head which functions as a lure. In anglerfish living in very deep ocean waters, the lure contains light-emitting organs called photophores. In some species of anglerfish the males become parasites on the larger females, either permanently or for the breeding season.

Anguilliformes   See *eel.*

ANGUS   The acronym for Acoustically Navigated Underwater Survey System. It is an unmanned sled on which cameras are mounted. This equipment was designed and operated by the Scripps Laboratory for Oceanographic Research at La Jolla, California. In 1977 it made the first observation of vent communities.

Annelida The phylum of segmented worms. These soft-bodied animals have not left a good fossil record but probably arose in the Paleozoic Era. The annelids are characterized by a fluid-filled tube within a tube body plan (a coelom), bilateral symmetry, a head-like projection at one end, and a closed circulatory system. The latter included a pumping mechanism and respiratory pigments that contained iron resulting in red blood. Many annelids have hemoglobin. Blood moves forward in the large, dorsal vessel and toward the tail in the ventral one. Tiny vessels bring oxygen, absorbed on the epidermis to the interior. The excretory system, present in each coelomic segment (somite) acts as a filter for nitrogenous wastes, water reabsorption unit, and discharge unit. Aquatic species usually discharge ammonia. The nervous system is highly organized with a ganglion per somite and a continuous double nerve cord that terminates in a large anterior ganglion—not quite a brain, but a control center.

This phylum is divided into two large classes: Polychaete and Clitellata (or Oligochaeta and Hirudinea). The polychaetes are marine organisms and for the most part are sexually separate whereas Oligochaetes and the leeches are hermaphrodites. The name Polychaeta indicates many bristles or setae, the chitin structures that protrude from each segment. Polychaetes bear their setae on appendages that extend from the body wall. These are parapodia (almost feet). The parapodia and setae aid in locomotion. They serve to anchor the body while the longitudinal muscle—which elongates the body—moves it along forward; the circular muscle contracts the body and brings it up to the location of the parapodia. This mode of locomotion enables the animal to swim, creep, or otherwise manipulate its environment.

Polychaetes are the most highly diversified of the annelids, both in shape and lifestyle. They occur as sedentary tube worms, primarily filter feeders, or swimmers, or predaceous burrowers, or as ecto-parasites on echinoderms. See *Polychaetea, tube worm.*

Antarctic Circle The 66°30' South Latitude line. See *Antarctic Convergence, Arctic Circle, latitude, polar biome, solstice.*

Antarctic Convergence A band encircling Antarctica between 50° and 60° South Latitude. In this area, relatively cold surface water of low salinity sinks and spreads northward as Antarctic Intermediate Water. The cold Antarctic water can be detected by the measurement of salinity as far north as 35° North Latitude in the Atlantic, but has a much lesser range in other oceans. The Convergence was discovered by the German meteorologist Meinardus during the Gauss expedition of 1901 to 1903, and the Convergence for a time was called the Meinardus' Line.

Antarctic Current The principal circumpolar current, known as the West Wind Drift, flows around the Antarctic continent in an easterly direction, while the coastal current, the East Wind Drift, moves west around the land mass. The West Wind Drift is a deep and fast-moving current, which moves a considerable volume of water thought to be greater than that of the Gulf Stream, through the Drake Passage. The currents around Antarctica are global and affect the temperature, salinity, and dissolved gas content of the water in every ocean of the Earth. See *Antarctic Ocean, ocean currents, seawater.*

Antarctic Ocean An ocean region to the south of South America, Africa, and Australia. The Antarctic serves as the salt, gas, nutrient, and temperature-equilibrating mechanism for the three major oceans. Although the cold water temperature restricts the number of plants and large animals, plankton and krill populations are large and support complex food chains that include birds, (petrels, penguins, skuas) large fish, invertebrates, and cetaceans. The high rate of planktonic growth is supported by the combination of nutrient-rich warm water moving south on the ocean's surface and cold, oxygenated water moving north from the continent's edge. See *Antarctic Current*.

Antarctic Water   Several distinct types of water differing in temperature and salinity. The coldest, least salty water is the Antarctic Surface Water, derived from melting ice, deep-water upwellings, and precipitation on land. Antarctic Surface water reaches its greatest volume in the antarctic summer, December to March. The upwelling of Circumpolar Deep Water, derived from Atlantic, Pacific, and Indian Ocean deep waters, is of higher salinity, is relatively warmer, and does not vary in volume with the seasons. Surface water sinks and continues flowing northward until it encounters the Subantarctic Surface Water, which is warmer and less salty, at the Antarctic Convergence. The mixing of these surface waters produces the Subantarctic Intermediate Water, which continues moving north. Antarctic Bottom Water, formed in the antarctic winter, is cold and highly oxygenated, originating in the salty shelf water and the low-salt Weddell Sea water. This large volume of cold, saline water moves north and also in an easterly direction, away from the Weddell Sea. See *Antarctic Convergence, Weddell Sea*.

anterior   The head or leading end of an animal. The term is also used to denote position. Thus, the head of an arthropod is anterior to its thorax. See *dorsal, posterior, ventral*.

Anthozoa   A class of marine coelenterates (phylum Cnidaria), such as sea anemones and corals. Anthozans are animals with radial symmetry. The dominant form is the polyp, which can be large and quite complex. See *coral, sea anemone*.

anticyclone   An upper tropospheric high-pressure system. Anticyclones are generally subtropical high-pressure zones marking either the boundary between easterly trade winds and prevailing westerlies or that between polar and arctic weather systems. See *meteorology*.

Antilles   An island chain that forms the northern and eastern borders of the Caribbean Sea. The Antilles chain is composed of hundreds of islands, most of them tiny. The Greater Antilles are Cuba, Hispaniola (the Dominican Republic and Haiti), Puerto Rico, and Jamaica. The Lesser Antilles are an eastern arc of very small islets, divided into two groups: the Leeward Islands St. Christopher (St. Kitts), Antigua, Guadeloupe, to the north, and the Windward Islands, Barbados, St. Lucia and Grenada closest to the South American mainland. The islands are rocky, often mountainous outcrops with poor soil and inadequate ground water, particularly on the leeward side. The climate is tropical, with the trade winds bringing wonderful weather, making the entire region a center of tourism. Short rainstorms are most frequent in the Antilles summer. It is the occasional summer Atlantic storm that becomes a hurricane. Hurricanes are more likely to invade the northern portion of the region, and most occur in late summer, especially in September. See *Gulf of Mexico, hurricanes, West Indies, winds, names of individual islands*.

Antilles Current   A branch of the Equatorial Current which moves surface water toward the West Indies in a northwesterly direction. See *Caribbean Sea, Equatorial Current*.

aphelion   The point in the orbit of a planet or comet that is farthest from the Sun. The Earth reaches its aphelion on July 3 in the Northern Hemisphere's summer. The position of the Earth and

Sun affects the ocean's tidal ranges. *See solar tide, tide, tidal range.*

apogee   The point in the orbit of a moon (or satellite) that is farthest from the planet it revolves around. The path of the moon's orbit affects terrestrial tides; the closer the moon to the Earth, the greater the tidal range. *See aphelion, perigee, tide.*

aquaculture   The cultivation of plants and animals in water. Types of aquaculture include growing algae or fish (carp, catfish, buffalo fish, trout) in freshwater pools and raising marine animals such as oysters and mussels, and seaweed on a commercial scale in protected environments. Acquaculture is an ancient practice: mollusks have been cultivated in frames that deter predation at least since Roman times; every medieval village had its fish ponds.

While it is possible to raise some animals in captivity, others, notably most ocean fish, present many problems. For reasons not yet understood, the cultured fry of oceanic fish do better once they are liberated. Some shellfish, particularly lobsters, are not amenable to the techniques of aquaculture. *See mariculture.*

aqualung   An underwater breathing apparatus pioneered by Jacques-Yves Cousteau. The aqualung is a self-contained air delivery system designed to be carried by the diver, thus freeing him or her of the need to be tethered to an air line running to a surface vessel, and providing much greater freedom and range of movement. *See Cousteau, Jacques-Yves; diving; SCUBA; undersea exploration.*

Arabian Basin   A region in the southern part of the Arabian Sea, a part of the Indian Ocean. The Carlsberg Ridge separates the Arabian Basin from the Somali Basin lying to the southwest. The northern boundary of the Arabian Basin is the Gulf of Oman. The deepest point in this basin is 3658 m (12,657 feet).

The Indus River—one of Asia's largest and most significant waterways—feeds the Arabian Basin, and its cone of sediment is the dominant undersea feature. The continental shelf extends west of India for distances ranging from 120 km (75 miles) to over 300 km (190 miles). *See Arabian Sea, Indian Ocean.*

Arabian Plate   A small intrusion between the growing African Plate, the Iran Plate (a subduction zone), and the Eurasian Plate. The northward push of the Arabian Plate impels the Turkish Plate westward.

Arabian Sea   The northwestern arm of the Indian Ocean, lying west of India and along the coast of the Arabian peninsula and the east coast of Africa. The Arabian Sea lies west of the Mid-Ocean Ridge and consists of two deep basins, the Arabian and Somali, which are separated by the Carlsberg Ridge. The Persian Gulf, and Gulf of Aden along with the latter's extension, the Red Sea, are all parts of the Arabian Sea. The undersea topography of this area is still the subject of research and conflicting opinion. It is generally accepted by oceanographers that The Arabian Sea developed in the Mesozoic–Cenozoic era, although some features are as recent as the Pliocene.

Currents in the Arabian are the direct result of the monsoon pattern. From November to March the northeast monsoon brings light winds and little rain, and the best weather in the region. The summer monsoon is from the southwest, bringing with it rain and a change in current patterns. The South Equatorial Current turns north at about 5° South Latitude and moves into the sea. At this point it becomes known as the Somali Current. The Current continues along the coast and bends east, then south along the Indian subcontinental shore. Areas of upwelling along the African and Arabian coasts are well marked.

The salt concentration of the Arabian Sea is high, particularly in winter. During the rainy season there is a drop of more than 0.1% from an average of 3.6% to 3.5% or less. See *Indian Ocean*.

Arafura Sea A body of water that separates North Australia from West Irian (formerly Netherlands New Guinea) and Indonesia. A deep trough of 3,000 m (10,000 feet) runs northeast toward West Irian. The area nearest Australia is a shallow shelf 50 to 80 m (160 to 260 feet) deep. The Aru Islands are on this shallow shelf.

Aral Sea A lake in the Soviet Union east of the Caspian Sea. Both the Aral and the Caspian were once part of the Tethys Sea. Both seas are now river-fed lakes. In earlier times the Aral and Caspian were connected to the Mediterranean Sea. See *Caspian, Mediterranean, Tethys Sea*.

Archaeocyantha A group of extinct, lower- to mid-Cambrian organisms. They are taxonomically difficult to place and may be either in a phylum of their own or related to the hard (stony) corals. They seemed to have ingested food as do the sponges. The Archaeocyantha seems to represent an intermediate stage between sponges and corals. Like both of these phyla, it flourished in warm, tropical, shallow seas. See *Anthozoa, Porifera*.

Archiannelida A small class of the phylum Annelida. These littoral bottom dwellers are an ancient, relict group of ciliated annelids. See *Annelida, cilia, littoral*.

Archimede (research vessel) A French submersible used in the 1974 Mid-Atlantic Ridge survey known as FA-MOUS (French-American Mid-Ocean Undersea Study). The *Archimede* descended to depths between 2,400 and 3,000 m (8,000 to 10,000 feet), but was not as maneuverable as the smaller American submersible *Alvin*. See Alvin, *FAMOUS, submersible*.

archipelago A group of islands. An archipelago may consist of isolated volcanic seamounts or mountain tops whose lower slopes and valleys have been submerged. The Hawaiian Islands are an example of the first type of archipelago; the Caribbean Islands are an example of the second. See *atoll, island arcs, seamount*.

Arctic Circle The 66°30′ North Latitude line. See *Arctic Ocean, latitude, polar biome, solstice*.

Arctic islands Islands at or above the Arctic Circle. The climate of arctic islands is moderated by their proximity to water and two fairly permanent low pressure areas, one near the Aleutian Islands and the other near Iceland. Some arctic islands are glaciated, most receive a good deal of precipitation, and a few are mountainous. The highest peak on any of these islands is Mt. Forel in Greenland. The dominant groundcover is taiga (grassland). Arctic islands vary greatly in size. The largest is Greenland, with 2,175,000 km² (840,000 square miles), followed by Baffin Island, with 500,000 km² (196,000 square miles). The size then drops off sharply to Victoria Island, Ellesmere Island, the Spitsbergen islands, Novaya Zemlya, and the tiny islands of the Franz Josef Land archipelago.

Arctic Ocean The ocean area north of North America and Eurasia. The Arctic Ocean is never totally navigable and in winter is largely covered by ice. The ice cover on some of the large islands may be more than a mile thick (1,300 m).

The bottom of the Arctic Ocean is a highly varied terrain. The North Pole itself is on the Polar Abyssal Plain and about 4,000 m (14,000 feet) deep. Continental shelves extend toward the pole, most particularly off the Siberian coast, where shallows of less than 15m (50 feet) exist. The bottom is divided into the Canadian and the Eurasian basins. The Lomonosov Ridge, which runs from the Laptev Sea just to the west of the Pole to the Lincoln Sea on Greenland's north coast, is in the Canadian Basin. The Alpha Ridge is in the Canadian Basin close to the Lomonosov Ridge. The Mid-Atlantic Ridge continues as the Nansen Ridge. It divides the region into the Eurasian Basin and the Canadian Basin. The Eurasian is, on the average, deeper than the Canadian. Water throughout the arctic is fairly constant in temperature and of low salinity.

The plant and animal life of arctic waters is much more varied than that of the Antarctic, particularly with regard to large animal populations. Plankton, algae, and diatoms all exist and provide food for larger organisms, such as crustacea and mollusks, which are frequently larger than their temperate relatives. In the northernmost waters there is less sunlight penetrating the water due to the ice cover, which limits sea life. See *abyssal floor, mid-ocean ridges.*

Argentine Basin   A deep, cold-water area lying west of the Mid-Atlantic Ridge off the continental shelf of South America and terminating in the shallower Scotia Basin. The average depth of the Argentine Basin is 5,000 m (16,000 feet).

argon   A gas found in seawater at a near-saturation concentration (33.6 ml/1,000 g H_2O at 20°C, 1 atm). The ratio of nitrogen to argon in seawater is fairly stable at 37 to 39. This is used as a base for the determination of the concentration of other gases.

Aristotle (384–322 B.C.)   A classical Greek philosopher and natural scientist; teacher of Alexander the Great. Aristotle was the son of a physician and as such had early exposure to medicine and the materia medica of his day, which consisted entirely of botanicals. He produced his *Natural History and Zoology* from 344–342 BC. The *Historia Animalium* is based on the belief that there are "necessary causes" for structures, and that when these structures are observed in a plant or an animal, their function is apparent.

Aristotle attempted an elaborate classification scheme, known in Latin as the *Scala Naturae*. He ranked animals as more complex and important than plants, and man as the most important and complex animal. He put animals into two large categories, blooded and nonblooded. The blooded animals included man, viviparous and oviparous quadrupeds, cetaceans, fish, and birds. These were subdivided according to the degree of perfection of the young at birth. The bloodless animals were mollusks, crustaceans, and insects. Aristotle realized that he had omitted sponges and snakes from his scheme, but was not sure where they belonged.

He named about 500 kinds of animals, of about 550 to 600 species. Most of these were known in Greece either as native or available in menageries. His scheme contains about 130 fish and excellent descriptions of many invertebrates, crabs, lobsters, and cephalopods. The sea urchin's mouth parts, now known as the "lantern of Aristotle," remain today as he described them.

Examining live animals, Aristotle observed sense perception in scallops, fish and sponges. He saw and commented on the ability of the cuttlefish to adhere to rocks and not be swept away in currents. His statement that sea urchin eggs are largest at the full moon has been confirmed in the 20th century. He opened developing eggs to observe the state of the embryo.

Aristotle was aware of his limitations. He knew that he had not seen everything, nor did he totally understand everything he observed. He noted "suspect entries" and tried to comment on all of the work he did with regard to precision. In addition to his very careful biological work he also observed and commented on the ocean's tides and currents.

The influence of Aristotlean work on European and Arab science endures to this day. Aristotle became the outstanding scientist of his time, and after the decline of science in general, the preservation of his work—however fragmentary—made it the only basis for later workers in science. His methodology was also important: he repeatedly relied on observation—an approach often ignored as philosphy and science became more confused. Not until the idea that observation must be the beginning of science and that philosophy can only follow the facts became firmly established, was Aristotle's contribution to man's knowledge of the world really appreciated.

artesian water   Water from a spring. If artesian water issues forth from the seafloor with sufficient force, it rises to the surface as fresh and sometimes hot water. Springs drowned by sea risings or collapsed limestone caves may produce artesian water. See *blue holes*.

Aschelminthes   A phylum of roundworms. Several rather disparate classes make up this phylum, the taxonomy of which is subject to discussion and revision.

The similarity among the aschelminth classes Nematoda, Rotifera, Gastrotricha, Kinorhyncha, and Priapulida is in their body cavities. Most of these animals are small scavengers. See *individual classes, scavenger, taxonomy*.

ascidian   See *sea squirt*.

aseismic ridge   Oceanic uplands that are not volcanic sites but may have been built up from volcanic products. The Walvis Ridge and the Emperor Seamount are examples of aseismic ridges. The Lomonosov Ridge, once believed to be an aseismic ridge, may be the result of separated continental rock that moved from its original site by seafloor spreading. See *Arctic Ocean, mid-ocean ridges*.

Asteroidea   Starfish or sea stars. The asteroidea are invertebrates with a worldwide distribution, living along coasts and in deep ocean waters. The radially symmetric adult develops from a bilaterally symmetric juvenile. Asteroideans have a flattened central disk surrounded by five arms. The opening that regulates internal water—the madriporite—is on the spiny upper surface. The mouth, anus, gills, and tube feet are on the lower surface. The carnivorous starfish either swallow prey whole and then expel the indigestible parts through their mouth or, more commonly, eject their stomach into or onto their prey and digest it *in situ*. The usual food of starfish is coral or bivalves. They move by retracting and extruding the papillae on their underside, which are part of their internal water canal system, and a distinguishing feature of the Asteroidea. The canal is also the means by which the interior of these animals is aerated.

asthenosphere   The semi-fluid layer in the Earth's mantle. It exists at a depth of 80 to 200 km (50 to 125 miles). Its fluidity makes possible the lateral motion (sidewise slide) of crustal plates. See *crust, mantle*.

Atlantic Ocean   The Earth's largest body of water, accounting for about 20% of the Earth's surface. The Atlantic Ocean is surrounded by more continental land than any other ocean, and receives the most riverine runoff. It is divided into the North and South Atlantic Oceans by the equator. The outstanding geological feature of the Atlantic is the Mid-Ocean Ridge, which runs from the Arctic Ocean to the Scotia Sea. The Ridge breaks at the Romanche Deep (or Fracture Zone) near the equator. There are two transverse ridges, the Walvis Ridge off the South African coast and the Rio Grande Ridge near the Brazilian coast. The large islands of the Atlantic are for the most part of continental origin, while the small ones are volcanic or, in the case of the Bermudas, coral.

The winds and weather in the Atlantic are also divided into northern and southern segments, with the wind-generating currents that cross the ocean moving clockwise. The Labrador Current is the cold current in the north that moves south along the North American coast. In the South Atlantic, warm water moves north and south as it approaches the Americas,

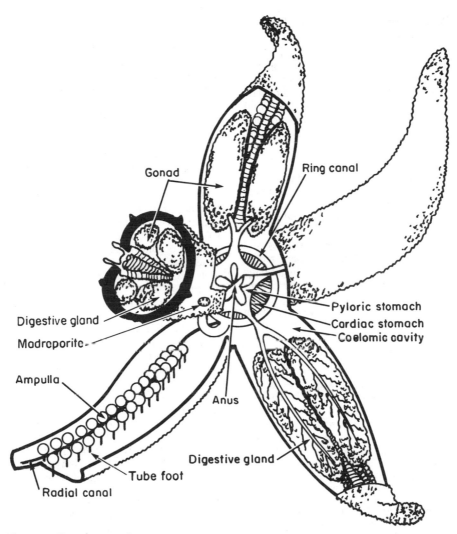

Gonad

Ring canal

Digestive gland

Madreporite

Ampulla

Pyloric stomach

Cardiac stomach

Coelomic cavity

Anus

Digestive gland

Tube foot

Radial canal

Cross section of a sea star

and arches counterclockwise. There is considerable sinking of cold, dense, southern water in winter, and thus considerable stirring. The result is water of fairly constant temperature and salinity, the latter about 3.5%. The deep water of the Atlantic is oxygen and nutrient rich, supplying the raw materials needed by plankton, pelagic (open sea) communities in general, and the predators that feed on them.

Radionuclide studies have shown that the Atlantic Ocean dominates the world's oceans by means of the Antarctic Circumpolar Current, which moves a considerable proportion of the world's ocean water through the Drake Passage. While this thorough mixing makes the global ocean fairly homogeneous, there are temperature and salinity variations that become more intense in shallow or restricted waters. The extent of variation in salinity for the period 1960 to 1981 was only 0.02 parts per million. See *Arctic,*

Coriolis force, currents, Drake Passage, Gulf Stream, pollution.

Atlantic-type margin   The trailing edge of a continental mass which is subsiding under sediment load. The cooling of the lithosphere (the rock of the Earth) is a factor in creating the subsidence and in its extent. See *Pacific-type margin, plate tectonics.*

Atlantis   A legendary island thought to be somewhere in the Atlantic Ocean near Gibraltar. Plato mentioned it in two of his *Dialogues* as a wealthy island with an ideal society, and the legend of Atlantis remained popular in ancient times and the Middle Ages. Some modern archeologists believe that the story might have arisen after the destruction of the eastern Mediterranean island of Thera (Santorini) in a volcanic eruption that occurred about 1450 B.C. That catastrophe destroyed the island and inundated all of the eastern Mediterranean. See *Thera, volcano.*

Atlantis (research ship)   A United States vessel used for research. The first ship of this name was built in 1930, and the second, *Atlantis II,* sailed in 1963 and was used in the exploration of the Red Sea. The Atlantis Deep, named for the *Atlantis II,* is a site of hot water overlying a rich mineral deposit in the Red Sea.

Atlantis Seamount   A group of undersea peaks on the eastern slope of the Mid-Atlantic Ridge, south of the Azores islands. See *Azores, islands, Mid-Atlantic Ridge, seamount.*

atmosphere   The envelope of gases, vapor, and aerosol particles surrounding the Earth. The major components of the atmosphere vary with altitude, with most of the oxygen and almost all particulates being closest to the Earth. The atmosphere is divided into layers or zones. The layer closest to the Earth, the troposphere, is the most dense and is composed of approximately 78% nitrogen, 21% oxygen, 0.9% argon, and 0.02% carbon dioxide, although these percentages vary with local conditions. In addition, the troposphere contains varying amounts of water vapor as a result of oceanic evaporation. The evaporation of surface water supplies the energy that moves air currents and winds, which in turn contribute to the movement of water. The troposphere is the zone of the greatest circulation of air, as warm air rises from the Earth's surface to the cooler, thinner upper layers beyond, and warm tropical air moves from the equatorial regions toward the poles. The troposphere is thus the layer where almost all weather occurs. The interlocking effect of wind and water and air and water currents demonstrates the unity of the ocean-air fluid system.

The average temperature of the atmosphere falls with increasing altitude from the Earth's surface through the troposphere to the tropopause, then rises in the stratosphere. In the upper stratosphere is the layer containing ozone, which filters the ultraviolet radiation from the Sun. Above the stratosphere is the mesosphere and above that the ionosphere, a zone containing layers of charged particles important for the transmission of radio waves. At even greater altitudes is the exosphere, which merges with the interplanetary medium.

atoll   A ring of land surrounding a tropical lagoon, and which is in turn surrounded by a coral reef. An atoll is frequently formed of basaltic rock of volcanic origin, and is capped with limestone covered with coral. The islets that form atolls range in size from those less than a mile (1 km) in diameter to those over 78 miles (120 km) in diameter.

Darwin proposed a theory of atoll formation that is now generally accepted. He postulated that atolls arose from a volcanic base which subsided and then became the nucleus for a calcareous and coralline buildup that ultimately produced the atoll. See *coral and coral reef.*

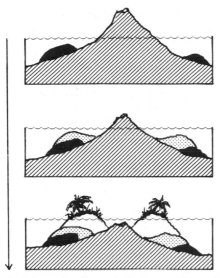

Atoll formation (arrow indicates passage of time)

atom   The smallest complete unit of a chemical element. Atoms are themselves composed of smaller entities, some of which have mass and charge, and others of which have mass but no charge. The major subunits of atoms are protons, electrons, and neutrons. The protons, neutrons and other subatomic particles excluding the electrons make up the nucleus of the atom. They are electrons distributed in a cloud around the nucleus.

A proton has an atomic mass of one dalton, a unit devised to represent the mass of a proton. It has a positive electrical charge.

The electron, which has a negative electrical charge equal in magnitude to the positive charge of the proton, has a mass of $1/1,800$ that of the proton.

Neutrons are uncharged parts of the atomic nucleus that have a mass comparable to that of protons.

The other component parts of atoms are considerably smaller than the three major ones. The entire structure is held together by a series of strong and weak forces.

Most elements are not found as their free atoms; this is true only for a few atmospheric gases: xenon, neon, krypton, and argon, formerly called inert gases. The simplest elements, such as hydrogen (H_2), oxygen (O_2), occur as diatomic molecules. Elemental minerals, such as carbon, (graphite or diamond), gold, and silver are crystalline aggregates of atoms held together by chemical bonds. See *isotope, molecule, radioactivity.*

ATP   (adenosine   triphosphate) The energy storage and transferring compound of most biological systems. ATP transfers energy by the reaction:

ATP + a hydroxyl group →ADP (adenosine diphosphate) + a phosphate ester (found in sugars). ATP is a phosphorylating agent which can transfer a phosphate group to another molecule, releasing energy that is utilized by the living cell in which the transfer occurs. When a phosphate group is added to ADP, ATP is regenerated. The importance of this chemical cycle lies in the role that it plays in metabolism, i.e. the conversion of energy derived from sunlight and fuels (such as carbohydrates) into energy utilized by the cell to do chemical, osmotic and mechanical work.

auk   An arctic marine bird (family Alcidae) varying in length from 18 to 90 cm (7 to 20 inches). The auk is not an adept flyer and roosts on rocks in huge colonies. Auks feed on fish, crustaceans, and plankton. The great auk, which resembled a penguin and was about the size of a goose, became extinct in the 1840s as the result of human activity.

Aurelia   Jellyfish of the order Semaestomeae. *Aurelia* look like fringed umbrellas and are composed largely of water. They are poisonous. See *Cnidaria.*

Australia   The only continent wholly in the Southern Hemisphere. Australia has a land mass of about 7.5 millon km^2 (almost 3 millon square miles). See *Arafura Sea, Great Barrier Reef, Pacific Ocean, Sahul Shelf.*

Australian Plate　A large crustal section underlying Australia, New Zealand, and the eastern portions of the Indian Ocean. The eastern boundary of this plate, the Macquaire Ridge, is a subduction zone that causes the seismic activity of New Zealand. See *Macquarie Ridge, mid-ocean ridges, Pacific Ocean.*

authigenic sediment　A fine grained deposit, such as common salt and phosphorites, found on continental shelves. Authigenic sediment is the result of chemical precipitation.

autotrophs　Organisms that, together with heterotrophs, constitute the two categories of living organisms. Autotrophs use simple inorganic compounds, found in the environment, to synthesize their own food and protoplasm. The energy source that drives these synthetic reactions is usually sunlight, and those autotrophs that use sunlight are called photosynthesizers. Sulfur bacteria and other autotrophs that use the energy stored in the bonds of simple compounds of sulfur and nitrogen to obtain energy for their metabolic reactions are known as chemosynthesizers. See *chlorophyll, heterotrophs, photosynthesis, sulfur bacteria.*

Azores　An archipelago in the Atlantic Ocean lying 1,500 to 1,800 km (900 to 1,200 miles) west of Portugal. The islands are the peaks of undersea volcanoes and are arranged in three widely separated groups. The islands have been known, fought over, and used by Europeans since the 14th century. Columbus stopped at the Azores to wait for favorable winds before his first voyage across the Atlantic. The islands had been known to ancient mariners; Carthaginian coins have been found on some of the Azores. See *Mid-Atlantic Ridge, seamount.*

Azov, Sea of　An extension of the Black Sea north and east of the Crimean peninsula and connected to the Black Sea by the Kersh Strait. Both the Sea of Azov and the Kersh Strait are very shallow, the maximum depth in the Sea of Azov being about 13 m (40 feet).

The Sea of Azov is poor in marine fauna. It is less saline than the Black Sea and freezes in winter, which the Black Sea normally does not. However, a large fish population does live in these waters, and sturgeon are taken in commercial quantities. There are many canals, as well as navigable river channels, leading to the Black and Caspian seas. See *Black Sea.*

B

Bache, Alexander Dallas (1806–1867) An American scientist and superintendent of the U.S. Coast Survey. He helped establish a continuing American interest in marine research. A great-grandson of Benjamin Franklin, Bache was a physicist trained at the U.S. Military Academy at West Point, New York. After serving in the army, he was appointed professor of natural science and chemistry at the University of Pennsylvania. Bache was also in charge of research at the Franklin Institute in Philadelphia.

In 1836 Bache was appointed president of Girard College, and was also a regent of the Smithsonian Institution while connected with the Coast Survey.

backshore The most landward part of a shoreline. It is on the land side of the average spring high water mark and is inundated only during storms and storm-driven high tides. See *beach, coast, tide.*

bacteria Unicellular organisms with no distinct cell nucleus. Bacteria are important in the nitrogen and carbon cycles; they are found in both marine and terrestrial environments. Some are surface-dwelling heterotrophs (i.e., organisms that use organic material from other organisms as a source of nutrients). Among the most recently discovered bacterial marine organisms are the bacteria living in thermal upwellings in a number of places in the Pacific and Atlantic oceans. These organisms are dependent on sulfur compounds, not oxygen, for their metabolic processes. They are the beginning of the food chain in the rift-deep-sea-bed ecological system. See *food chains, heterotrophs, sulfur bacteria, volcano.*

Baffin, William (1584?–1622) An English explorer of the arctic, and the first English explorer to attempt a scientific determination of longitude. In 1615 Baffin piloted the *Discovery*, which was charged with finding the Northwest Passage through North America. The voyage was inconclusive. The following year Baffin sailed again on the *Discovery* to determine whether the Davis Strait would be a promising route. He went to 78° north latitude, mapped Jones, Lancaster, and Smith sounds, and returned to announce that the project had not been fruitful. He suggested that the London merchants fund a voyage starting in the Far East to sail east, but they declined.

Baffin's next employer was the East India Company. Between 1617 and 1621 he was captain of several ships in which he explored the eastern Mediterranean, the Red Sea, and the Persian Gulf, reaching India. He was killed near Hormuz, in the Persian Gulf, where the East India Company was fighting with the Portuguese on behalf of the Shah of Persia.

Baffin Bay The body of water that separates Greenland from Canada. The Davis Strait, to the south, connects Baffin Bay to the Atlantic Ocean. The Bay is about 1,100 km (700 miles) long and varies from 112 to 640 km (70 to 400 miles) in width, with depths ranging from 360 to 2,740 m (1,200 to 9,000 feet). The surrounding islands are bleak, cold, and mountainous. The Bay itself is ice-covered most of the year. The Labrador Current brings icebergs through the Bay in summer.

Baffin Island A landmass north of Hudson Strait, the passage leading to

Hudson Bay. This mountainous, snow-covered island forms the western boundary of Baffin Bay, an arm of the Arctic Ocean.

Bahamas An Atlantic Ocean island group southeast of Florida and northeast of Cuba. Andros, the largest island of the Bahamas, is about 160 km (100 miles) long and about 60 km (40 miles) wide at its widest point. While most of the Bahamas sit in shallow water on a broad bank, the tongue of ocean east of Andros is over 4,000 m deep. Also to the east of Andros are collapsed caves or blue holes. See *blue holes, Caribbean Sea, Gulf Stream.*

Balearic Islands An island group in the Balearic Basin of the Mediterranean. The Balearics lie east of Spain, in relatively shallow water. See *Mediterranean.*

baleen A horny material that hangs in a fine fringe from the roof of the mouth of the Mysticeti or baleen whales. The triangular plates of baleen are used as food filters. See *filter feeders, krill, whale.*

Baltic Sea An arm of the North Atlantic, separated from it by the Skagerrak, Kattegat, and the Danish landmass. The largest island in the sea is Gotland (Sweden). The Baltic is fairly shallow, particularly on its eastern edges. Although the Gotland Deeps are greater than 450 m (1,500 feet) deep, the Gulf of Riga is about 50 m (165 feet) deep and the shallows near the Danish landmass may be as little as 10 m (33 feet) deep at low tide.

The Baltic is fed by large arctic and subarctic rivers, which accounts for the low salt and nutrient content of its surface level waters. The salt content is so low in some places that freshwater fish species can make the transition to the sea and back.

With the retreat of the Scandinavian Ice Sheet, a large lake formed, covering what is now the Baltic and extending beyond its present boundaries. By about

8000 B.C. the Yoldia Sea extended from the Skagerrak, over southern Sweden to Lake Ladoga, east of the Gulf of Finland. Land rebounding from glacial pressure closed the link and the Yoldia became a lake, known as Lake Ancylus, about 7000 B.C. A warm period about 2000 years later reopened the sea connection (Litorina Sea). The most recent warming has caused the sea level to continue to rise. See *Skagerrak.*

Banda Sea A deep-water area in Indonesia southeast of the Sulawesi Sea (Celebes Sea) and north of the Timor Sea. The numerous islands in the Banda Basin are of volcanic origin. The area is noted for deep, cold water which flows westward to the Indian Ocean. See *currents, Indian Ocean, island.*

Banks, Sir Joseph (1743–1820) A British naturalist who began his career as a scientist while still a student at Oxford. He was a member of the prestigious Royal Society by 1776. In that year he traveled to Newfoundland and Labrador, collecting plants and insects. When James Cook made his first voyage around the globe (1768–1771), Banks accompanied him on the *Endeavor* as an unpaid naturalist.

Banks found and brought back exotic biota from Australia. In 1772 he went on an expedition to Iceland and the Hebrides, where he studied the local geology. He was elected president of the Royal Society in 1778 and continued to carry on his collecting of new specimens. He willed his collection to the British Museum. See *Cook, James; Royal Society.*

Barents, Willem (1550?–1597) A Dutch navigator who sailed for Dutch merchants. He sought the Northeast Passage as a quick route to the Indies from Europe. Instead of finding the Passage, he explored Novaya Zemlya, Svalbard (Spitsbergen), and Barents Island. He died during his third voyage.

Barents Island in the Svalbard archipelago, and the Barents Sea, are named for him. See *Barents Sea.*

Barents Sea A part of the Arctic Ocean named for the Dutch Explorer Willem Barents. The Sea lies east of Spitsbergen and is separated from the Atlantic Mid-Oceanic Rise by the Barents Abyss. Scandinavia (North Cape) and the U.S.S.R. (Murmansk) lie to the south, and Novaya Zemlya lies to the east.

The sea bottom of the Barents Sea averages between 100 and 350 m (300 to 1000 feet) deep and then drops off to about 600 m (2,000 feet) where the Barents meets the Norwegian Sea. The Barents is a boundary region: both the remains of the warm Gulf Stream and very cold arctic water enter it. The air above the water is part of the polar front. The vigorous mixing of the two results in a well-aerated nutrient mixture that supports a planktonic bloom in summer, producing a great deal of oxygen. The phytoplankton population supports a large and varied ecosystem comprising the plankton and many invertebrates. See *Arctic Ocean, polar biome*.

barnacle A highly modified crustacean (subclass Cirripedia) long thought to be a mollusk because of its "shell." Barnacles are very successful as a life form: they may have evolved as early as the Silurian Period and comprise four orders with over 800 species. The free-swimming larvae pass through several molts before assuming the adult sessile pedunculate (stalked) form. Some barnacles are enclosed in calcareous plates while others are "naked." Some are parasites on corals or other crustacea, other barnacles are commensals—independent species that live in communities attached to turtles, whales, or coral. The feathery cirri of the barnacles—the remnants of the arthropod leg—beat food into their stomachs. Respiration is accomplished by the passage of gases through a mantle, for there are no real gills. Barnacles are hermaphrodites.

Barnacles live in almost all marine environments, and their negative impact on commercial enterprises is considerable. They are responsible for the extensive fouling of ships, piers, and offshore installations. See *Cirripedia, Crustacea*.

barracuda A carnivorous fish usually found in tropical waters. There are about twenty species in the family Sphyraenidae. They are usually 1.2 to 1.8 m (4–6 feet) long, although some are much larger, notably the Great Barracuda of Caribbean and West Pacific waters. The family as a whole are aggressive feeders, with large mouths, many teeth and a heavy underslung jaw. The Great Barracuda has been reported to attack swimmers.

Barracuda are taken by commercial fishermen and also sought after by sport fishermen, who are attracted by the barracuda's fierce battling ability when hooked.

barrier island A narrow, sandy landform separating open ocean from a lagoon or embayment. The beach usually has a natural canal on its landward side. Barrier islands are separated from one another by shallow inlets. The entire barrier configuration is subject to rapid change as a result of currents and storms. The Virginia barrier islands, for example, have retreated landward in the last 100 years, and some no longer exist as islands. See *coast, long shore currents*.

barrier reef A partially submerged coral outcrop on the seaward side of a lagoon. The barrier reef follows the contour of the land, of which it may be an extension. Water in the lagoon is shallow, permitting coral growth, but the descent is steep on the ocean side. Barrier reefs result from coral growth or a change in the level of the seawater or subsidence of the beach area.

Barrow Strait An arctic passage used by the oil tanker S.S. *Manhattan* on its September 1969 passage through the North Canadian islands from the Atlantic to the Pacific Ocean. The Strait is north of the Somerset and Prince of Wales Islands, and leads to Point Barrow, the most northerly point of Alaska and the start of the Alaska pipeline on the shores of the Beaufort Sea.

basalt   A dark gray to black igneous rock of volcanic origin. Basalt is a rapidly cooled volcanic material and so exhibits a microcrystalline structure. Basalt consists largely of about 50% silicon dioxide (SiO_2), roughly 16% aluminum oxide (Al_2O_3), and about 9% calcium oxide (CaO). Other inclusions are well-known. Darker basalts contain pyroxene, olivine, or both. Basalt is found as basement rock on land and on the sea floor spreading from the Mid-Ocean Ridge. See *mid-ocean ridges, volcano.*

base   A chemical substance that can attract a proton hydrogen ion, H^+ and remove it from an acid. Bases dissolved in water produce a solution that is alkaline. See *acid, acidity and alkalinity.*

basking shark   This fish *(Cetorhinus maximus)*, of the family Cetorhinidae, is a giant, slow-moving animal found principally in the Atlantic Ocean. Only the whale shark is larger. Basking sharks are usually about 8 m (25 feet) or longer, but there was one reported sighting of an individual about 14m (46 feet) long. Basking sharks are sluggish plankton eaters who also prey on small fish, but are relatively harmless to humans. Their color ranges from gray to brown to black, with lighter undersides. They are hunted for their hides and oil. The name comes from their habit of drifting slowly in the sun just below the surface of the water. See *shark.*

bass   A valuable food fish of the perch family, found in marine and freshwater environments. Common species are the striped bass, *Roccus lineatus;* black bass, *Micropterus salmoides;* and sea bass, *Centropristis atrarius.*

Bass, George (1771–1803)   An English sailor and surgeon. Bass joined the Royal Navy, where his ability in navigation led to his assignment to the *HMS Reliance.* Along with Matthew Flinders and William Martin, Bass explored the Botany Bay region of the Australian coast, and recommended that a settlement be established there.

Bass then continued to explore the coast south of Botany Bay for several additional years. He collected specimens of plants and animals, and explored the region south of what is now Sydney. In 1798 he found the strait separating Australia from Tasmania, which is named for him. In 1799 Bass was made a member of the Linnean Society in recognition of his contributions to botany and zoology. In 1803, while en route to South America from Australia to collect specimens, his ship disappeared. See *Botany Bay; Flinders, Matthew.*

Bass Strait   A generally shallow passage separating southeastern Australia from Tasmania, it is named for George Bass, who discovered it in 1798. Water movement is seasonal in the Strait—being eastward in winter and westward in summer. The salinity is lowest in summer, when the snow melt from Tasmania reaches the sea. See *Bass, George.*

bathyal zone   The region of ocean from the edge of the continental shelf to that depth where the water is at a constant temperature of 4°C, usually at about 4,000 m (13,300 feet). The bathyal zone is referred to as the continental slope.

Except for currents or upwellings, the salinity is about 3.4 to 3.6% and the water generally contains less oxygen and nutrients than does shallower water. The bathyal zone has far fewer species and numbers of individuals than does shallower water. See *continental slope, hadal zone.*

bathymetry   The charting of the ocean bottom by taking soundings. See *ocean floor, sounding.*

bathypelagic zone   The portion of the ocean water column 1,000 to 4000 m (3,300 to 13,500 feet) deep and not in direct contact with the ocean bottom.

The bathypelagic region has no sunlight. Plankton are abundant and cnidarians, arthropods and fish move freely. See *Arthropoda, Cnidaria, fish, plankton.*

bathyscaphe A free-moving, three-man, deep-sea research submersible designed by Auguste Piccard in the late 1940s. His first effort was called *FRNS I* (Fonds pour la Récherche Nationale Scientifique). Iron pellets controlled by an electromagnet regulated the vertical motion of the bathyscaphe; a small electrically powered propeller moved it laterally. The first bathyscaphe had limited success. It was improved and the second attempt, the *FRNS II*, dove in 1954 to the 4,050 m (13,365 feet) mark.

Piccard and his Italian coworkers incorporated some of the features of the bathyscaphe into a new submersible, the *Trieste*, which was acquired by the U.S. Navy. In 1958 it dove in the Guam Trench. See *Piccard, Auguste, submersibles.*

bay A wide area of the ocean; part of the sea extending landwards. For specific bays, see the geographic name.

beach A stretch of sand or rocks along the shore, extending from the highest high-tide point to the lowest low-tide point. A beach is formed by erosion caused by wind and waves, and by deposits of sediment brought to the site by ocean waves and rivers. The structure of a beach is subject to considerable variation, depending on the tide, season, storms, and long-term geologic processes.

Sandy beaches are usually wide and have gentle slopes. They are found on the leeward sides of atolls or on continental shelves. They may be bar or pier-shaped. Pocket or crescent beaches have sand trapped between cliffs that border the sea.

On a worldwide scale, sandy beaches have been lost due to a complicated series of geological events aggravated by human activities, such as damming and restructuring river channels, and the construction of sea-walls which, if incorrectly located or not maintained, can increase destruction. See *breakwater, coast, erosion, littoral, longshore currents, sand.*

beach grass Members of the genus *Ammophilia,* beach grasses are coarse, reedy plants found on temperate sandy coasts in Europe, North America, and Africa. The stalks are about 1m (3.3 feet) tall and grow in clusters. Beach grasses have extensive underground stem and root systems and send up new shoots meters away from the original plant. This network of roots and stems acts to stabilize the sand and prevent its erosion by wind. See *beach.*

***Beagle,* H.M.S.** A survey ship sent by the British Navy under the command of Captain Robert Fitzroy to chart the South American coast and then circumnavigate the globe. The ship carried naturalist Charles Darwin on board, and his observations during the four-year long voyage, particularly of the fauna on the Galapagos Islands, led him to develop his theory of the evolution of species. See *Darwin, Charles; evolution; Wallace, Alfred Russell.*

Beata Ridge Part of the Caribbean seafloor lying south-southwest of Beata Cape on the island of Hispaniola. The Ridge is between the Colombian and Venezuelan basins, and terminates in the Aruba gap. The Beata Rise is east of the Ridge in the Venezuelan Basin. See *Caribbean Sea.*

Beaufort, Francis (1774–1857) A British naval commander and later admiral who was interested in natural science. In 1805 Beaufort designed a wind scale which, although much modified, is still used. The wind scale was used to determine the effect of wind on a man o'war in full sail. Beaufort was elected a member of the Royal Society in 1814.

Beaufort scale A standardized, descriptive scale of wind velocity outlined by Admiral Sir Francis Beaufort of the

Force	Name	Km/Hr	Mi/Hr	Description
0	Calm	0–2	0–1	Flat sea, slack sails
1	Light airs	2–6	1–3	Wavelets
2	Light breeze	7–11	4–7	Small waves, sails filling
3	Gentle breeze	12–18	8–12	Distinct waves, filled sails
4	Moderate breeze	19–30	13–18	Wave caps, branches bending on land
5	Fresh breeze	31–39	19–24	Some spray
6	Strong breeze	40–50	25–31	Cresting waves, holding an umbrella is difficult
7	Stiff breeze	51–62	32–38	High sea, walking is difficult
8	Fresh gale	63–75	39–46	Difficult to stand unaided
9	Strong gale	76–88	47–54	Rolling sea
10	Storm	89–100	55–63	Violent seas, high waves
11	Hurricane-like storm	101–117	64–72	Blown-foam-covered sea, high sea, impeded visibility
12	Cyclonic storm	118+	74+	Hurricane or typhoon

Royal Navy in 1858. See *meteorology, wind.*

Beaufort Sea A branch of the Arctic Ocean lying north of Alaska, with Banks Island on its eastern boundary and the Chukchi Sea to the west. There is a great deal of floating ice in the Beaufort Sea, making its surface the coldest water layer in it. The intermediate layer of the water is warmer and moves north through the Bering Strait. The Alaskan coast of the sea has very little shelf, and the bottom drops abruptly into the Beaufort Deep. The coast is dotted with small islands. See *Bering Strait, Chukchi Sea.*

Beebe, Charles William (1877–1962) An American ornithologist, explorer and writer, active in a number of scientific fields. Beebe's contribution to marine science involved deep-sea exploration. He descended to a depth of 918 m (3,028 feet) in the Puerto Rico Trench in a submersible vessel of his own design in 1934. The vessel was essentially a hollow steel ball lowered by its mothership. This bathysphere was able to withstand the pressures of the depths but was not maneuverable. See *submersible.*

Belcher, Edward (1799–1877) Surveyor of the arctic coasts of Canada for the British Admiralty, Belcher was born in Halifax, Nova Scotia, and devoted his entire career to coastal surveying. In 1825 he mapped the Pacific coast of Canada and the Bering Strait. He then worked along the northern and western coasts of Africa from 1830 to 1833. Between 1836 and 1846 he mapped the Pacific coasts of North and South America, the southern Pacific Ocean, and the coasts of China, Borneo, and the Philippines.

Belcher's surveying career came to an abrupt end as a result of his voyage to find John Franklin, the explorer lost in an arctic expedition in 1848. He was given command of the expedition in 1852, but by May of 1854 he had abandoned his ships, which were frozen in the ice of the arctic, and was removed from command. Belcher wrote his version of what happened in *The Last of the Arctic Voyages* (1855). He was knighted in 1867 and promoted to admiral in 1872. See *Franklin, John; Northwest Passage.*

belemnite   Also known as a belemnoid, this extinct cephalopod was characterized by a large internal shell similar to that of the modern cuttlefish. The belemnites appeared in the Mississippian period and achieved their greatest diversity in the Jurassic and Cretaceous. They disappeared from the fossil record during the Eocene Epoch. See *Cephalopoda.*

Belle Isle, Strait of   A strait that connects the Gulf of St. Lawrence to the Atlantic Ocean. Despite its extensive ice cover, due to its chilling by the Labrador Current, the strait is an important commercial waterway leading to the deepwater ports of the Great Lakes. See *St. Lawrence River.*

Bellingshausen, Fabian Gottlieb von (1778–1852)   A Russian explorer of the antarctic. Bellingshausen was the leader of the expedition to the antarctic commissioned by Czar Alexander I in 1819. His two sloops, the *Mirny* and *Vostok,* circumnavigated the antarctic continent for the first time. The expedition touched the two South Sandwich islands, which Bellingshausen named for Czars Peter I and Alexander I. These islands were once thought to be part of the antarctic mainland.

On his return to St. Petersburg (Leningrad), Bellingshausen was made an admiral and later governor of the city of Kronstadt, near Leningrad. The Bellingshausen Sea in the Antarctic Ocean is named for him. See *Antarctic Ocean.*

Belon, Pierre (1517–1564)   A French biologist, and with Gesner and Rondelet one of the three "Encyclopedic Naturalists." Belon was first an apothecary and then studied with the great botanist of his day, Cordus. Belon's protector, the Bishop of Le Mans, introduced him to the royal court society of France, and he eventually developed a medical practice.

Belon traveled widely to the Middle East and also to England, where he did research at Oxford. He wrote on coniferous trees and, most importantly, on marine subjects. His *Natural History of Strange Marine Fish* (1551) classified fish and included cetaceans, which he recognized as air-breathers that produced milk, yet he still classified these animals as fish.

Belon also wrote *On Aquatic Life,* in 1553, and the *History and Nature of Birds,* in 1555. He is known as "the father of comparative anatomy" because of this carefully researched work. He was murdered while walking in the Bois de Boulogne in Paris. See *Cetacea; Gesner, Konrad, Rondelet, Guillaume.*

bends   See *diving, nitrogen.*

Bengal, Bay of   The northern portion of the Indian Ocean, separating India and Burma. Its area is about 2 million km^2 (800,000 square miles). The major characteristic of the Bay is the change in the direction of the current as a function of the monsoon: the water moves clockwise in spring and summer and then reverses its direction in autumn. The salinity of the water in the Bay of Bengal is relatively low, since it is the recipient of torrents of river water from both India and Burma. The rivers also carry tons of sediment into the Bay. See *delta, Ganges, Indian Ocean, Irrawaddy River, monsoon.*

Benguela Current   Wind-driven, cold bottom water of low salinity flowing northward along the west coast of South Africa to about 15° south latitude. The upwelling of this cold water makes the Benguela Current rich in nutrients, resulting in an environment favorable to plankton and other marine organisms. See *nutrients, plankton.*

benthic ecosystems   Ecosystems of the ocean floor, whose flora and fauna vary widely from region to region, depending on available light, temperature, salinity, and pressure. The benthic zone with the greatest diversity of plant and animal life extends from the high-tide limit of abundant growth of sessile plants. Subcommunities of plants and animals exist in small, isolated areas such as rock pools or coral reefs. The deeps lie at the edges of continental shelves with depths ranging from 400 to about 1,000 m (1,200 to 3,000 feet). They constitute one zone of cold and dark water. Deeper still are the abyssal depths, which are characterized by relatively stable physical conditions of salinity, temperature, and light, and by relatively low nutrient levels. See *continental shelf, coral reef.*

benthic organisms Organisms that live on, near, or in the bottom sediments of the sea. The organisms of the benthos are classified by their environment—the epifauna live on the bottom; the infauna burrow into it—and by their size: Microbenthic organisms—bacteria, unicellular algae and protists—are less than 0.1 mm (0.004 inches) in diameter. The meiobenthos—copepods, amphipods, small worms, etc. range from 0.1 to 1 mm. The macrobenthos are greater than 1 mm (0.04 inches, and include the polychaetes, mollusks, snails, sea urchins, and other echinoderms. Megabenthos are large animals: the decapods and large fish. See *algae, individual multicellular animals, Protista.*

Bering, Vitus (1681–1741) A Dane who sailed for Czar Peter the Great of Russia. Although a Russian named Dezhnev had sailed through the Bering Strait in 1648, his record was overlooked and Bering was commissioned to determine whether or not a land bridge between Siberia and North America existed. Bering sailed in 1724 and found the strait named for him. After Peter's death, Bering continued his explorations for Peter's successor, and mapped Arctic Siberia.

On June 4, 1741 Bering sailed from Kamchatka on a voyage called the Great Northern Expedition. In August 1741 his ship was wrecked in the Gulf of Alaska and Bering died of scurvy and exposure on December 19, and most of his crew also perished. Some survivors straggled back bringing information about the Alaskan islands and coasts, which led to the Russian interest in Alaska as a source of minerals and furs. See *Dezhnev, Semyon Ivanov.*

Bering Sea and Strait The strait that separates Siberia and Alaska, named for Vitus Bering, who explored the Siberian and Alaskan coasts and the Aleutian islands of the North Pacific Ocean. The southeastern region of the Bering Sea near Siberia is the deepest part of the sea; while the northwestern area is a broad, shallow continental shelf. The Bering Strait is narrow and shallow (85 km and 40 m, respectively or 53 miles and 130 feet), which makes water exchange with the Arctic Ocean difficult. The relatively recent (Pleistocene) submergence of the Siberia-Alaska land bridge divides the two portions of a single continental mass.

The Bering Sea is characterized by tidal variability in its northerly portions, with cool or cold and cloudy weather most of the year. The warmest water temperatures occur from August to October. The large quantity of snowmelt contributes to the low-salinity of the Bering Sea water. In the basin, water flow is anticyclonic and comes mainly from the Asian edge. Upwellings of deep water and the nutrients it brings account for the Bering Sea's diversified and abundant fauna. See *currents.*

Bermuda Rise The best example of a gently sloping, non-seismic elevation in the western Atlantic that is neither continental in origin nor connected with the mid-ocean-ridge system. The islands of Bermuda are on a volcanic foundation that is part of this rise; however, the islands themselves are coral. See *mid-ocean ridges.*

Bernoulli, Daniel (1700–1782) A Swiss mathematician, physicist and physician, Bernoulli became an international authority in physics. In his book, *Hydrodynamica,* he established the relationship between pressure, velocity, and density in fluids (if velocity is increased, pressure decreases). This work established the principles of movement in fluids, such as air or water currents.

Bernoulli was a member of the Russian Academy of Sciences in St. Petersburg, and a lecturer there until 1732. In the years 1725–1749 he worked on a number of research topics which were awarded prizes by the Paris Academy of Science. These included such diverse scientific problems as magnetism, ocean currents, tides, and the movement of ships.

beta rays Emissions of electrons that are part of the disintegration process exhibited by radioactive nuclei. See *atoms, radioactivity.*

Biafra, Bight of Part of the Gulf of Guinea, extending from the Niger River delta to Cape Lopez (Gabon) on the west African coast. Historically, the Bight of Biafra was the site of much of the west African slave trade, which centered around the Calabar coast (Nigeria). See *bight.*

bight A bay with a headland at each end. A bight is usually fairly shallow. Examples are the Bight of Benin (Africa) and the Great Australian Bight.

Bikini A classic atoll in the Marshall Islands in the Pacific Ocean group. Bikini was the site of United States atomic testing from 1946 to 1958. See *atoll.*

Bimini A group of islands in the Straits of Florida and part of the Bahamas group. The Lerner Marine Laboratory is in Alice Town on North Bimini. The major industry is tourism. The first tourist on record was Ponce de Leon. He believed this to be the site of his "fountain of youth." See *Bahamas, Caribbean Sea.*

biogenic reef A ridge or rise above the sea floor made up of organisms or debris composed of the hard parts of organisms, such as bits of shell, teeth, or bone. See *atoll, coral, reef.*

biogenic sediment A deposit of once-living organisms or parts of them on the ocean floor. This can be anything from bits of flesh to outgrown shells, or skeletons to plant debris. Biogenic sediments occur in all sizes, and have existed in oceans from the beginning of geologic time, producing a nearly continuous record. See *sediment.*

bioluminescence The production of nonthermal light by living organisms, including species of bacteria, marine invertebrates, and fish. Bioluminescence results from a conversion of chemical energy to light energy. The light may act as a lure; for example, light-producing areas (photophores) on the lower and lateral surfaces of bathypelagic (deep-dwelling) fish, such as anglerfish, attract smaller fish as prey. The bioluminescent glow of dinoflagellates results from wave action, which brings them to the surface. Different species have daily or seasonal light cycles, causing luminescent displays in some seawater areas, such as Phosphorescent Bay in Puerto Rico. Cnidarians (jellyfish), arthropods (shrimp), annelids, and mollusks (squid) all have light-producing genera.

biomass The amount of living matter, or total sum of the plant and animal organisms of a particular area at a given time. The biomass is measured as milligrams of carbon per unit of volume, which in turn is used as a measure of the productivity of an oceanic system and the growth of phytoplankton, which depends on temperature, salinity, nutrients, carbon dioxide, oxygen, and light. The primary productivity of the ocean varies from 0 milligrams of carbon per square meter per day in the midwinter Arctic Ocean to 3,000 to 4,000 milligrams of carbon per square meter per day for the midsummer Walvis Bay region of South Africa.

bird   A warm-blooded vertebrate with modified forelimbs and a feather-covered body.

Most birds fly, although some have evolved into flightlessness, such as the ratites or running birds (the ostrich, emu, rhea, and kiwi), and the penguins, which are non-flying birds that swim. Birds are land based almost everywhere on Earth. Individual orders have adapted different body forms, coverings, and behavior to better suit a particular environment. Various orders of the class Aves live in coastal environments. These include the: spheniscifoms (penguins); procellariforms (albatrosses and petrels); pelecaniforms (pelicans, boobies, cormorants); ciconiiforms (herons, ibises, storks); anseriforms (ducks and geese); grisiforms (cranes and rails); and charadriforms (gulls, terns, auks).

Although there is considerable variation, the common characteristics of birds include a very short, bony tail; very good hearing and sight; a toothless bill; and a calcareous (calcium-containing) eggshell. The skeleton of birds is a light structure with a flexible neck and a large sternum (breastbone), to which the heavy flight or pectoral muscles are attached. The legs and toes of perching birds are also well-developed. The circulatory system is a closed one; there is a four-chambered heart and large, nucleated red blood cells. The syrinx, or sound-creating organ of birds, is an enlarged tracheal (throat) structure where the trachea separates into the bronchi. The lungs are small but have connected air sacs that hold air on reserve. Thoracic movement inflates and deflates the lungs. The digestive system includes a crop, or storage vessel, and a gizzard that macerates food by muscle pressure, since birds have no teeth to chew with. (Flesh-eating birds will pick up stones to aid in this process.) The small intestine leads from the gizzard.

Birds seem to have evolved in the Jurassic Period. The earliest bird fossils show a close resemblance to lizards.

However, birds are seldom found as fossils, possibly due to the lightness of the skeleton and the lack of teeth. Nevertheless, charting the development of the class is difficult. Archeopteryx, a chicken-sized beast with a long bony tail and teeth, along with feathers, has been called the first bird. See *flight, marsh, migration, names of specific birds.*

Biscay, Abyssal Plain and Bay of   The Biscay Abyssal Plain is located in the Biscay Rise, a ridge extending west from the northwestern corner of Spain to the Mid-Atlantic Ridge. The Bay of Biscay is the part of the Atlantic Ocean adjoining France, south of the Breton peninsula. See *Atlantic Ocean.*

Bismarck Achipelago and Plate   Islands and a crustal plate in the West Pacific Ocean northeast of New Guinea, in an area of considerable seismic activity and many small islands. See *Island, Pacific Ocean.*

Bivalve   See *Pelecypoda.*

Black Sea   The easternmost portion of the Mediterranean. The Black Sea has a large eastern bay, the Sea of Azov. Both Seas had more extensive connection with the Mediterranean in earlier geologic time.

The Black Sea is surrounded by Bulgaria and Romania on the west, the Soviet Union on the north and east, and Turkey on its southern and eastern coasts. The area is prone to unpredictable and frequent squally storms, but its temperatures are moderate. The Bulgarian and Crimean (U.S.S.R.) coasts are resort areas abutting broad shelves of shallow water. The Danube is the major river emptying into the Black Sea. The mountain ranges that come to the shoreline are the Balkans south of the Danube delta, the Crimean mountains on the peninsula that indents the sea from the north, and the Caucasus and Pontus mountains on the western and southwestern edges of the sea. The latter two are significant weather-altering ranges of known volcanic activity. These mountains divide the Black Sea from the Caspian Sea, which in Tertiary times were connected.

Circulation in the Black Sea is poor and the interchange with the Mediterranean is limited to the very narrow Bosphorus. As a result, there is less biomass than would be expected due to very low oxygen levels, especially at the center of the sea, where there is almost no oxygen. However, considerable commercial fishing is done, particularly for sturgeon. See *Mediterranean, Tethys Sea*.

Blake Plateau   An undersea shelf extending along the east coast of Florida from the Bahamas to Cape Hatteras, with depths of 200 to 1,000 m (660 to 3,300 feet). The Plateau drops off sharply at the Blake Escarpment. The Blake Plateau is part of the ancient continental shelf which subsided in the Cretaceous Period. See *continental shelf, plateau*.

blowfish   A group of largely tropical marine fish, whose common characteristic is a protective mechanism that allows them to gulp air or water and greatly increase in size. Some species also produce a toxic material or have spiny projections. Blowfish or puffers vary from minnow-sized to meter-long individuals, and feed mainly on small crustacea and mollusca. See *fugu*.

blue crab   An edible crustacean of the order Decapoda. The name is most often used for the *Callinectes sapidus* or *C. hastatus*—crabs well-known on the eastern North American shore. The blue crab's shell is wider than it is long, usually 15 to 20 cm (6 to 6.5 inches) wide and about 7 to 8 cm (3 inches) long. It is a drab green-blue above and has a grayish-white underside. A series of eight short barbs protrude on each side of the carapace.

Crabs are scavengers and live in estuaries, bays, inlets, and muddy shorelines. The females provide minimal care of the young, carrying them around for several weeks after spawning.

Like all other organisms with an ex-oskeleton, the crab outgrows its shell. It molts and tends to hide until the new one grows and hardens. Between the shedding of one shell and the growth of a larger one, the crab is relatively unprotected, surrounded only by a thick, membranous covering. At this stage it is called a soft-shell crab. See *Arthropoda, crab, Decapoda*.

blue-green algae   Cyanophyceae, the prokaryotic (simple, non-nucleated) algae, which occur as part of the marine phytoplankton population. Most blue-green algae are not marine organisms. Those that are, however, are most abundant in tropical waters. Their remains are found in some of the geologically oldest reefs. See *algae*.

blue holes   Features of the Caribbean's Grand Bahama Bank, blue holes are visible from the surface, where they stand out as dark blue spots against sandy or vegetation-covered bottoms. The hole is an opening into a submerged limestone cavern or sinkhole formed by water moving through limestone and dissolving it away. This occurred during the last ice age, when the water level of the oceans was much lower than it is now. As the ice melted, the water level of the oceans rose and the once-dry caves were flooded. The current that moves water in and out of the holes is powerful and lags behind the tides. See *archipelago, Bahamas, Caribbean Sea, Pleistocene Epoch*.

blue shark   Also known as the great blue shark, this large animal, 3 to 4 m (10 to 13 feet) long is known in all of the Earth's oceans, but has its principal habitat in equatorial to warm temperate waters. The blue shark is a great scavenger and a potential danger to man in the sea. See *shark*.

blue whale   The largest animal on Earth, also called the sulfur-bottomed whale because it may carry along a yellow diatom population. The blue whale is a krill-feeder common to all oceans, and like other krill eaters, is a baleen type

(*Balaenoptera musculus*) whale. Since this huge creature (adults are about 30 to 35 m (95 to 100 feet long) was extensively hunted until very recently, even its placement on the endangered species list may not save it from extinction. See *baleen, diatoms, krill, whales.*

BOD   Biochemical oxygen demand, or the amount of oxygen necessary for the functioning of bacteria engaged in decomposing or breaking down organic matter. Bacteria use oxygen to break down complex compounds, (usually proteins). Thus, the more organic material there is in a given volume of water, the higher the BOD.

Since a great deal of oxygen is necessary to support a population of bacterial decomposers in a region of high industrial effluent or domestic waste (sewage), there is often insufficient oxygen in the water to also support normal flora and fauna. A high BOD, therefore, is an indication of pollution. See *pollution.*

bonito   A relative of the tuna and mackerel and like the latter, a member of the family Scombridae. The bonito is a carnivorous fish, found worldwide. It is a sleek fish about 70 cm (30 inches) long, with a sharply forked tail, greenish-gray upper surface, and silvery underside. There are dark lines radiating from the backbone, forming a series of "V"'s when the fish is observed from above. The bonito is hunted by both commercial and sport fishermen. Those species of commercial interest are *S. sarda,* found in the Atlantic and the Mediterranean, *S. orientalis,* of the the Indian and Pacific Oceans; and *S. chilensis* of the western Pacific. The oceanic bonito is called the skipjack tuna. See *tuna.*

booby   A sea bird of the order Pelicaniformes, which like the pelican lives in tropical to temperate waters. The birds are 60 to 80 cm (20 to 30 inches) long. They have long bills and angular, thin wings. They are efficient flyers, skimming above the water hunting squid and fish, and diving precipitously into the water to catch their prey.

Common boobies are the red-footed (*Sula sula*) and blue-faced (*S. dactylatra*) boobies. The latter are frequently found on the coast of western South America and in the Galapagos islands.

The boobies are noted for their elaborate courtship rituals. They nest in large colonies on rocky shores, and the females lay two eggs in messy nests made of stones and debris. The booby is known as a "dumb" bird because it is relatively tame and frequently robbed of its prey by frigate birds. See *frigate birds, pelican.*

boom   A spar (support) holding the foot or bottom of a fore-and-aft sail—the sail is one that is hung either in front of or behind a mast. See *lateen, sail.*

borers   Several unrelated marine organisms that bore holes in the structures of other organisms or into wood and other materials. The shipworm (Teredinidae), a mollusk, and the gribble (Limnoriidae), a crustacean, both bore into wood, and in the case of the gribble into kelps. Boring sponges, the Clionidae, bore into calcareous (calcium-containing) materials. These organisms, however, will not feed on the organism they bore into: they are filter feeders that hollow out mollusk shells, coral, or limestone to make burrows for themselves. These sponges will weaken limestone breakwaters, and the activity of shipworm and gribbles has had a major economic effect. See *kelp, Teredinidae (teredo).*

Botany Bay   An enclosed arc about 1.5 km (1 mile) wide lying between the La Pérouse and Kurnell peninsulas in the Australian state of New South Wales. The Bay is where James Cook first landed in Australia in 1770.

The original name given to the Bay by Cook was Stingray Harbor. Because of the great variety of unknown and exotic flora and fauna found there by the Cook expedition's naturalist, Sir Joseph Banks, the name was changed.

The first penal colony in Australia was

to be established at Botany Bay in 1778, but unhealthy conditions forced its removal to Port Jackson (now Sydney) to the north. The bay is fed by two rivers, the Georges and the Cook. See *Banks, Sir Joseph.*

bottom   See *ocean floor, shelf, trench.*

bottom currents   Water movements at the ocean bottoms that are continuous and have a well-defined direction. Bottom currents are present in almost all waters. Early oceanographers attempted to separate the effects of tide, wind, and surface water movement from that of true subsurface currents. The latter are shown to exist in present seas by ripple patterns in bottom sediments. Fossil ripple patterns imply that the effect is not recent. Other indications of the existence of a bottom current are rocky bottoms swept clear of all but the coarsest rocks. Bottom currents are confined to particular areas by sills—elevations such as that which divides the Eastern Atlantic from the Western. The Antarctic Bottom Current is a relatively fast, large current. It moves at about 0.8 km/hour (0.4/mph). Local currents at major passages can achieve much higher velocities. The Mediterranean water leaving through the Straits of Gibraltar moves at 3.5 km/hour (2 mph).

bottom water   Deep water that moves toward the equator from a particular polar region. Bottom water is generally high in oxygen content and has a low salinity characteristic which distinguishes it from the ambient ocean water above it as it moves through the ocean. Like any other water mass, bottom water eventually merges into the overall body of ocean water and so loses its identity.

Bougainville, Louis Antoine de (1729–1811)   Explorer of the South Pacific and leader of the first French fleet to circumnavigate the globe, in 1766–9. Bougainville described his travels in *Voyage autour du monde* (Voyage around the world), a popular book that helped disseminate Rousseau's vision of the

idealized noble, "natural" man. The voyage was underwritten by the French Crown, and its purpose was exploration and the search for new lands. The ship's company included naturalists. The expedition headed west, through the Straits of Magellan and then northwest to Tahiti, Samoa, and the New Hebrides islands. While the Great Barrier Reef was sighted, the party turned north there away from Australia and on toward New Britain. Bougainville spent some time in the Moluccas and then went to Batavia (Jakarta) and on back to France.

boundary current   See *east boundary current.*

Bowditch, Nathaniel (1773–1838)   An American mathematician and navigator. Bowditch was instantly recognized and appreciated by the sailing community for his publicaiton in 1799 of an improved edition of J. H. Moore's *The Practical Navigator.* He was also known for his translation and extensive commentary on Laplace's *Celestial Mechanics* (1828), an astronomical work.

He had had considerable experience as a sailor and captain before the publication of his books.

Brachiopoda   A phylum of bivalves in which the two valves or shells (dorsal and ventral) are asymmetric, with the ventral shell the larger of the two. The brachiopod shell is also known as a lampshell, since it supposedly looks like a Roman oil lamp. There are now about 300 known species of brachiopod. However, brachiopods appeared in the Cambrian and have had a continuous existence ever since. Over 30,000 species have been identified. Their range then and now is worldwide, and they represent a significant proportion of the antarctic fauna. The present species range in size from 2 to 15 cm (0.75 to 6 inches), while fossils of 35 cm (14 inches) are known. The extant animals have tongue-shaped shells of white, cream, rose pink, brown, or gray.

Brachiopods divide into the classes Ar-

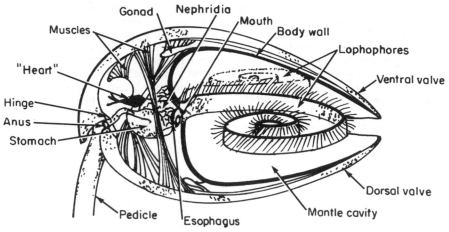

Lampshell

ticulata, those with shells that are hinged by interlocking "teeth," and Inarticulata, in which the shells are held together by muscles only.

Lingula, a representative species from Hawaii, lives in shallow water (about 40 m, 130 feet) deep, buried in a sand burrow. A footlike extension anchors it in the burrow and also serves in locomotion and digging. The distinguishing characteristic is a lophophore, a filter for extracting food from seawater. The digestive system consists of a digestive gland, stomach, and intestine, but no anus. The mantle that builds the calcareous shell of the animal has setae or footlike structures on its fringe. *Lingula* is in the class Inarticulata.

Since their fossil record is so complete, brachiopods are used as index fossils. See *Mollusca.*

brackish water   Water that contains too much salt to be drinkable (potable) but not enough to be seawater. Its average salt content ranges between about 0.5 and 1.7%.

Brahe, Tycho (1546–1601)   A Danish astronomer who built a huge wooden quadrant in Augsburg, Germany, with which he observed the motion of Jupiter and Saturn, and realized the inadequacies of the then-best astronomical tables. He was given an island on which to build an observatory, where he pioneered the use of giant astronomical equipment and the notion of reliability and precision in data collection.

His great work, *Mechanisms of the New Astronomy,* was dedicated to the Holy Roman Emperor Rudolph II. Brahe spent his last years in Prague as a guest of the Emperor.

Branchiura   A subclass of the Crustacea in the phylum Arthropoda, branchiurans are small (less than 2 cm or 0.8 inches), disk-shaped parasites. The rear legs of branchiurans are swimming legs. Their prey is bony fish. They attach to the scales of a fish using modified appendages that have become suckers. The animals may detach themselves from one host and move to another. See *Crustacea, parasites.*

Brazil Current   A current resulting from the South Equatorial Current hitting the South American coast. This warm, salty surface water moves along the Brazilian coast until it is joined by the cold, north-moving Falkland current. The conjoined stream then moves eastward as the South Atlantic Current. See *Atlantic Ocean, Guinea Current.*

breakwater Usually a mound of broken rock parallel to the shore, built up on the ocean bottom and extending upward to average wave height. The purpose of a breakwater is to create an area of calm water behind it for safe anchorage, or to protect coasts or offshore towers. The placement and design of a breakwater is critical, lest it create more problems than it solves. Badly designed or misplaced breakwaters can result in dangerous currents in their gaps. These however, are essential for ship passage. In other situations, normal wave action has scoured away the coastline beneath a breakwater, causing it to crumble and thus destroying the area it was meant to protect.

If the breakwater does its job and does indeed protect the coast, the creation of a calm channel between the beach and the breakwater results in the deposition of sand between the breakwater and the beach. Thus, the coast grows out to the breakwater. The deposit must be dredged if the channel is to stay open, since the sand is not removed by wave action. See *coast.*

Brendan, Saint (485?–577? or 583?) Also called Brendan the Navigator, an Irish monk supposed to be the first European to have traversed the Atlantic. While this is probably untrue, Brendan's voyages were real. In his *Navigatio Sancti Brendan Abbatis* he describes islands which could have been the Canaries, the Azores, Iceland, the Faroes, and the Shetland Islands. Brendan's boats were curraghs, wicker frames covered with caulked hide similar to the fishing boats known and used on the west coast of Ireland for centuries. See *Norsemen.*

brigantine Or brig. A sailing vessel with two masts carrying square sails.

bristle worms Marine annelids. See *Polychaeta.*

brittle star Also called serpent star, one of about 2,000 species of echino-

derms of the class Ophiuroidea. Brittle stars have a small central disk and five thin radial arms, which are fairly easily broken but can regenerate. If the arms of the brittle star are branched, the tangled organism is called a basket star. Brittle stars are most often scavengers or plankton feeders living on the sea bottom at almost every depth. The most common species is the slate blue, luminescent, *Amphioholis squamata.* See *Echinodermata.*

bromine Chemical element number 35, a red-brown liquid sufficiently reactive to ensure that it is never found in the elemental state. Bromine, in the form of bromide ion, is present in seawater at a concentration of about 65 parts per million.

The cheapest commercial source of bromine is seawater. The commercial extraction process consists of several steps: the seawater is concentrated, then air streams are directed over the water to vaporize the bromine. Sulfur dioxide is introduced to react with the bromine vapor and then water is added to produce hydrogen bromide (HBr) and sulfuric acid (H_2SO_4). The hydrogen bromide, on treatment with chlorine, reacts to produce hydrogen chloride and free liquid bromine.

brown algae Mainly marine algae of the class Phaeophycae. Brown algae can live in brackish water and are found in salt marshes or estuaries, although some genera, for example *Fucus*, can exist out of water entirely for longer periods and therefore have colonized intertidal zones. Brown algae grow luxuriantly in colder waters and on rocky coasts, albeit in shallow water. Both annual and perennial types are known. The distinguishing feature of these plants is the xanthophyll pigment known as fucoxanthin, which imparts to them their characteristic brown color. Brown algae do not store starches; their carbohydrate reserves are mannitol and laminarin, a glucose-mannitol polysaccharide. Alginic acid, another polysaccharide, is present in the cell walls of

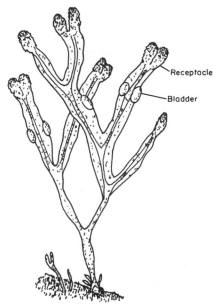

Receptacle

Bladder

Fucus (rockweed)

brown algae. Derivatives of this, known as alginates, are used industrially as emulsifiers, thickeners, and stabilizers in pharmaceuticals, foods, paint and pigments, and paper products.

The brown algae plants seem to exhibit two characteristics of more highly evolved organisms. They have alternating sexual and asexual generations, and differentiated cells and structures, including "roots" and "stems." Three representative types are filamentous algae *(Ectocarpus)*, the kelps *(Laminaria)*, which are large-leaved plants, and the rockweeds *(Fucus* and *Sargassum)*, which are branched.

Bryozoa   Aquatic, mainly marine invertebrates, also called ectoprocts (animals with an external anus). Bryozoans are microscopic, sessile colonial organisms. The colony they form is the result of budding, but begins with a free swimming larva that lands on a suitable piling or rock or animal shell. Ectoprocts are found most commonly in the littoral (shoreline) zones of the sea at all lati-

tudes, but do occur down to depths of over 6,000 m (20,000 feet). Since they have calcareous shells, they preserve well and many fossil forms are known. The original designation "Bryozoa" was applied to a marine animal with two openings to its digestive tract, as opposed to Anthozoa, with one. There was much controversy among zoologists for more than a century over just what the name Bryozoa, as opposed to Polyzoa, meant. In the older literature, then, there are references to Bryozoa that now refer to the phylum Ectoprocta. See *Ectoprocta.*

buffer   A term in chemistry describing a solution made up of a dissolved acid and its conjugate base, or a base and its conjugate acid, that is used to maintain a certain pH. Maintaining a particular pH is essential in biological systems, where small deviations in the pH can be lethal to a particular organism.

A common buffer is one containing HCO_3^-, the bicarbonate ion, and CO_3^{2-}, the carbonate ion.

Depending on the concentration of the components, small additions of strong acid (or base) will not materially change the pH of a buffer.

See *acidity, pH.*

Buffon, Georges-Louis Le Clerc, comte de (1707–1788)   A Frenchman who catalogued life forms. After 1725, George-Louis LeClerc was also known as LeClerc de Buffon. A member of the minor nobility, he studied medicine, botany, and mathematics as a young man.

Buffon published his first scientific work in 1735, a translation of Stephen Hales' *Vegetable Staticks,* a work on plant physiology. In 1740 he translated Newton's *Fluxions* into French.

On his appointment to the post of Curator of the Royal Gardens, Buffon augmented the inventory of the king's collection. This inventory developed into the *Histoire Naturelle, Génerale et Particulière,* Buffon's most famous work. The best known section of this book, *Epoques de la nature,* written in 1778, explained the geologic history of the Earth as a series of stages. Buffon envisioned unknown and extinct species in his works on animals and plants.

Buffon expanded his duties to include a small menagerie, and carried on some experiments on the nature of fire. He was made a member of the French Academy of Sciences in 1753. See *catastrophism, evolution.*

buoy   A moored or anchored object marking a navigational channel or ob-struction. Some buoys collect information about aspects of the environment, such as the large and sophisticated data-collecting buoys FLIP (Floating Instrument Platform) and NOMAD (Navy Oceanographic Meteorological Automatic Device). See *lighthouse, FLIP.*

byssal threads   Threads produced by the byssal gland in the foot of the common bivalves, which serve to attach the animal to a stationary object. The "hairs" or "beard" of mussels are byssal threads.

C

Cabot, John (1450?–1499) Born Giovanni Caboto, possibly in Genoa, navigator and explorer, attempted to find a westward route to the Orient. He was made a citizen of Venice in 1476. He sailed the eastern Mediterranean for the Venetians and became a skilled navigator.

In 1484 he moved to London, taking his family with him. Like Columbus, Cabot was attempting to amass money and influential sponsors for an expedition to the Orient by a westward route. He received his grant from commercial interests in Bristol, England in March 1496, with the understanding that any trade ensuing would be the monopoly of the merchants of Bristol.

Cabot's first attempt to sail, in 1496, was aborted because of bad weather and insufficient food; his Mediterranean sailing experience was not sufficient training for the rigors of the North Atlantic. Trying again the next year, he sighted land in the west, probably southern Labrador. Cabot, however, thought he had found Asia. His report of the abundance of fish was most welcome in England, and the area of the Georges Bank and Grand Banks that he had discovered was opened to European fishermen.

Cabot left England again in 1498 on an expedition that may have reached the American coast, but which was lost and its ships assumed to have been sunk.

Cabot, Sebastian (1476–1557) An explorer, probably born in Venice, who came to England as a boy with his father, the explorer John Cabot. Sebastian may have accompanied his father on the expedition of 1497 in search of the westward passage to the Orient. He was a mapmaker for Henry VIII of England, and accepted a commission in the Spanish navy, which he held concurrently with the post of pilot major in the navy of Charles V, the Holy Roman Emperor.

While on an expedition to the Orient, Cabot diverted his ships to the eastern South American coast, where he hoped to find treasure. The diversion resulted in a court martial which banished him from the Holy Roman Empire, possibly because his attempt was unsuccessful.

Cabot returned to England in 1548 and joined the Merchant Adventurers, a commercial group interested in finding the Northwest Passage. The Adventurers had sponsored several expeditions to find the fabled Passage, all of them unsuccessful. Instead they turned their attention eastward and opened up trade with the Baltic regions. Trade with Russia was one of their successes. See *Cabot, John; Northwest Passage.*

Cabot Strait A strait that connects the Atlantic Ocean with the St. Lawrence River in eastern Canada. The strait, which runs between Newfoundland and Cape Breton Island, is almost 100 km (60 miles) long. See *St. Lawrence River.*

Cabral, Pedro Alvarez (1460 or 1467?–1530) A Portuguese navigator who sailed for Portugal when Vasco da Gama refused to make a second voyage to the Far East. Cabral sailed west in 1500 with a fleet of thirteen ships. They sailed southwest and made landfall at a place Cabral called Porto Seguro (now Baia Cabrailia) in Brazil. The fleet then left for India, but that portion of the voyage proved a disaster. A storm drove the flotilla east and a number of ships were lost. The remaining ships required considerable repair when they reached

Mozambique. The fleet reached Calicut (Calcutta) in September 1500. It had had only limited trading success because of Arab competition. The seven remaining ships returned to Portugal separately from June to July 1501. Five were loaded with Far Eastern goods, but the other two were empty.

While the goal of the voyage—finding a reasonable jumping-off place from which to make the run around the Cape of Good Hope—was not realized, Cabral is credited with being the first European in Brazil. See da Gama, Vasco.

Calanoida An order of free-swimming planktonic copepods, equipped with very elaborate feather-shaped setae (hairlike projections) that act as flotation devices. See Copepoda.

Calcarea A class of the phylum Porifera, consisting of sponges that have spicules (support structures) composed of calcium carbonate. See Porifera, spicule, sponge.

calcium The fifth most abundant element on Earth. Its compounds form the skeletons of animals, including marine animals. Calcium phosphate forms the bones and teeth of vertebrates, while the shells of bivalves and gastropods are formed from calcium carbonate. See carbon cycle, calcium carbonate.

calcium carbonate The calcium salt of carbonic acid. Calcium carbonate ($CaCO_3$) is the structural constituent of many marine organisms. Crystalline calcium carbonate occurs as the minerals calcite and aragonite. Coral, algae, and the planktonic foraminifera secrete calcium carbonate. The carbonate ion [CO_3^{2-}] is in equilibrium with the CO_2 and calcium ion [Ca^{2+}] that are dissolved in ocean water. The ability of water to dissolve calcium carbonate (and therefore the shells of clams, oysters, or coral) increases with lower temperatures and increasing depth (pressure). Thus, in deep ocean the sediment is carbonate-free, although carbonate-containing organisms

are present on and near the surface. Their calcareous remains dissolve on sinking below about the 4,000-m (13,000-foot) level. This varies with specific locale and the deep circulation encountered. The depth at which $CaCO_3$ dissolves is known as the carbonate compensation depth.

California Current A wind-driven Pacific Ocean current that moves in a southeasterly direction until it meets the North Equatorial Current. Surface water in the California Current moves northwest near the coast in winter and southeast in summer. Subsurface water flows north all year round. Upwelling contributes to seasonal changes in the temperature, nutrient content, oxygen, and salinity of the water of the Current. See Equatorial Current.

California, Gulf of A branch of the Pacific Ocean, the gulf separates the Baja California peninsula from the Mexican mainland. The gulf was first named the Vermilion Sea by its circumnavigator, Francesco de Ulloa in 1539. The reddish color of the gulf is the result of planktonic bloom. The gulf is about 1,200 km (700 miles) long and 150 km (100 miles) wide at its widest point. The channel is divided by a constriction into an upper, fairly shallow basin with maximum depths of about 200 m (660 feet) and a more interesting lower basin. The bottom sediments are a combination of Colorado River delta and wind-blown terrigeneous materials. The gulf itself is an arm of the Pacific, not a drowned river valley. It is the junction of the East Pacific Rise and the North American continent, and probably formed as a result of the slide to the northwest of all of California west of the San Andreas fault.

There are several islands in the upper basin; Angel de la Guarda and Tiburon are the largest. Navigation in the gulf is hazardous because of sudden storms and a strong tidal bore. The extremes of temperature of the surrounding region and its aridity are mirrored by conditions in the gulf, where the temperature fluctuates considerably. In winter, cool, nutrient-rich Pacific water enters and wells up. This results in the growth of plankton, which in turn supports a large fish and mollusk population. See *East Pacific Rise, San Andreas Fault*.

calorie   A unit of heat. It is defined as the quantity of heat required to raise the temperature of 1 gram of water by 1° centigrade at atmospheric pressure. Calories in a nutritional context are kilocalories (1,000 calories).

Cambrian Period   One of the oldest periods dating to about 570 million years ago in the Earth's geologic record, it was preceded by the Pre-Cambrian and followed by the Ordovician. It is in the earliest portion of the Paleozoic Era. The Cambrian was named by Adam Sedgwick a British geologist, for Cymru ("Wales," in Welsh), where such formations are found.

The distinctive zoological development of the Cambrian was the appearance of shelled marine invertebrates. The dominant animals were the trilobites. Brachiopods, stromatolites, and fragments of jawless fish (the last in the very latest Cambrian) were also known.

Geologically, the Atlantic Ocean had begun to form between Laurentia and Gondwanaland, the two supercontinents. Africa, South America, Asia, India, and Antarctica, made up Gondwanaland, while Laurentia comprised the major part of North America, Greenland, Spitsbergen, parts of Scotland, Ireland, and Norway. See *brachiopod; Gondwanaland; Laurentia; Sedgwick, Adam; stromatolite, trilobite*.

çamouflage   See   *chromatophore, color*.

Canadian Basin   Part of the Arctic Ocean nearest to Canada, the Basin is fairly shallow near the shore, which in many areas is the result of the submergence of steep valleys. Near Patrick Island it is a drowned (subsided) headland,

as is the Baffin Bay area. See *Baffin Island, subsidence*.

Canaries Current   A branch of the North Atlantic Current that is deflected off Northern Europe and tempers the climate of the Canary Islands, which would otherwise be tropical. On its way south, the cold water of the Canaries Current meets warm wind and water from the Iberian peninsula, producing the fogs common on the western edge of the Pyrenees. See *currents*.

Canary Islands   A group of islands belonging to Spain in the Atlantic Ocean, off the coast of North Africa at about 28° north latitude. The island closest to the continent is about 100 km (65 miles) from the coast. The Canary Islands were known in classical times: Ptolemy used Hierro, the most westerly island, as his line of prime meridian (0° longitude). Arabs later occupied the islands, and they were repeatedly "rediscovered" by Genoese, Portuguese, and French sailors before the Spanish conquest in 1496. Explorers used the island landfall as a waystation for ships before attempting the run around Africa or to America.

Geologically, the western islands in the Canaries group are volcanic mountain peaks, and the eastern ones are on an undersea plateau. The climate is very mild, with almost constant temperature maintained year-round. There is very little rainfall.

Cape Verde Islands   A group of islands in the Atlantic Ocean off Cape Verde on the Senegalese coast of Africa. The islands form two chains, the Barlavento (Windward) and Sotavento (Leeward), lying on either side of 10° north latitude. These islands were first noted by Europeans in 1460, and Portuguese colonists arrived in 1462.

The climate is hot and dry year-round. The Cape Verde Islands are volcanic peaks, and some are spectacularly eroded.

capillary wave   A ripple. Ripples are wind generated and of short wavelength, and can occur on any body of water. A pond is large enough to have ripples on its surface if the wind velocity is sufficient to overcome surface tension. The surface tension of the water restores a smooth surface. Waves of greater wavelength are called gravity waves. See *waves*.

Carangidae The carangid fish, the family of pompanos and jacks. Carangids are fished commercially and by sport fishermen. The family consists of about 150 species, most often found in tropical seas. Only a few species live in temperate waters, and a few are freshwater fish.

carapace A relatively large, hard shell covering the back of an animal. Several unrelated animals have carapaces, including turtles, most crabs, all horseshoe crabs, and sea spiders.

caravel A 15th century Spanish or Portuguese trading vessel, relatively wide for its length, and with a rounded stern

and a sharply pointed bow. Caravels had two masts, a main and a mizzen, the latter behind the former. The sails were either triangular (lateen) on both masts or a triangular sail on the mizzen and a large square one as the mainsail. The poop, or rear castle, extended over the stern. Columbus' *Nina* and *Pinta* were caravels.

The caravel was an important engineering advance in the construction of ocean-going vessels since it made sailing into the wind possible. Thus the Portuguese traders moving along the African coast could go as far as they wished into the southern waters and still get back home by tacking into the prevailing westerlies.

carbohydrate A compound of carbon, oxygen, and hydrogen. The empirical formula for this large group is $C_nH_{2n}O_n$, where n is any number greater than three. The carbohydrate compounds include the glycogens, starch,

cellulose

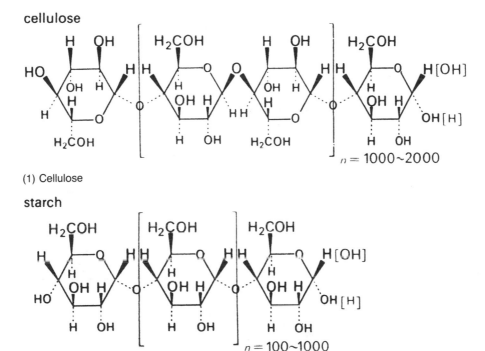

(1) Cellulose

starch

(2) Starch

amylose, and cellulose. Carbohydrates may be polymers made up of monomers that are saccharides. Sugars are saccharides. See *glucose, polymer, sugar.*

carbon   The element essential to all living structures. Carbon is present on land as the free element, where it exists in one of two forms—amorphous (coal) or crystalline (graphite and diamond)—or in an array of compounds.

The compounds of carbon range from the simple gases carbon monoxide and carbon dioxide, present in air and water, to carbonates in minerals and to the complex proteins and carbohydrates of living organisms.

Carbon is a unique element in that it can form an almost infinite number of chemical compounds by the bonding of carbon atoms to other carbon atoms. All sorts of structures, rings and branched or linear chains are thus possible. Carbon also bonds to other elements, such as hydrogen, oxygen, sulfur and nitrogen, to form hundreds of classes of compounds; some are of biological importance. The specific chemical compounds that are made by an organism are determined by the genetic makeup of the organism that makes and uses them. See *carbon cycle, carbon monoxide, carbon dioxide, hydrocarbon, nitrogen cycle, carbohydrates, proteins.*

carbon cycle   A linked series of reactions, triggered by sunlight, by which carbon, obtained from the atmosphere as carbon dioxide, is absorbed by green plants, used to create organic compounds, and then returned to the atmosphere as carbon dioxide. Light energy, through the action of the pigment chlorophyll, initiates the splitting of the car-

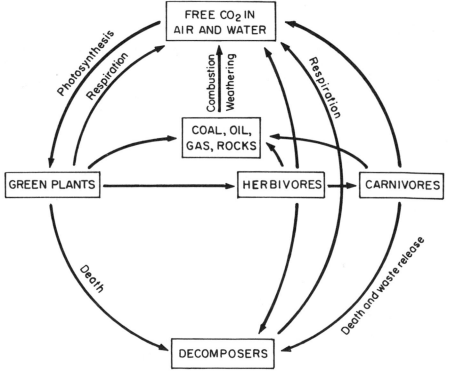

The carbon cycle

bon dioxide molecule. Reaction of the resulting carbon atom with water ultimately produces sugars with oxygen as a byproduct. This process, known as photosynthesis, is what converts energy into the chemical compounds required for life.

Glycolysis is the means by which plants and animals obtain energy from sugars. The sugars react with oxygen under the influence of enzymes, releasing energy in the process. The major energy-releasing step is respiration, and it is through this process that the ultimate end product, carbon dioxide, is returned to the atmosphere. See *nitrogen cycle, photosynthesis, respiration, sugar.*

carbon dioxide (CO$_2$)   A colorless, odorless gas present in the atmosphere at a concentration of less than 1%. Without this simple compound, today's life forms could not exist.

When dissolved in water, carbon dioxide forms unstable carbonic acid.

$$CO_2 + H_2O \rightleftharpoons H_2CO_3 \text{ (carbonic acid)}$$

which is in equilibrium with the bicarbonate ion and hydronium ion.

$$H_2CO_3 + H_2O \rightleftharpoons HCO_3^- + H_3O^+$$
(bicarbonate)   (hydronium ion = water
    ion   + hydrogen ion)

The bicarbonate ion is in turn in equilibrium with hydronium ion and carbonate ion.

$$HCO_3^- + H_2O \rightleftharpoons H_3O^+ + CO_3^{2-}$$
        (hydronium)   (carbonate ion)
            ion

Both the carbonate and bicarbonate ions are soluble, much more so than is CO_2. These ions become incorporated into the calcareous shells and skeletons of animals including coral reefs and plankonic tests, in the form of calcium carbonate (CaCO$_3$). The ions are also incorporated into the blood-buffer systems of numerous species of animals.

Carbon dioxide is produced in the respiratory processes of sea plants and sea animals. Any CO_2 that doesn't dissolve in the surrounding water is lost to the air.

Conversely, CO_2 in the air is taken up and dissolved by sea water to the saturation point. The overall turnover time for CO_2 in air to CO_2 in seawater and back to CO_2 in air may be many years.

The role of CO_2 in regulating climate has been discussed since the late 19th century. There is considerable disagreement over measuring the capability of the ocean as an absorbing agent and the long-term effects of both human activities and volcanic eruption on the CO_2 content of air and water. See *calcium carbonate, coral, greenhouse effect.*

carbon monoxide (CO)   A colorless, odorless gas which is normally present in air is produced by living organisms where insufficient oxygen is present. It is also produced by the incomplete combustion of carbon compounds. Carbon monoxide is highly toxic because it combines with hemoglobin in the blood, destroying the oxygen-carrying capacity of hemoglobin that is necessary to support life.

In the presence of sufficient oxygen, carbon monoxide is readily oxidized to form carbon dioxide:

$$2C + O_2 \longrightarrow 2\,CO$$
$$2\,CO + O_2 \longrightarrow 2\,CO_2$$

Carboniferous Period   The fifth period of the Paleozoic Era. It began about 250 million years ago and lasted roughly 8 million years. Its name is derived from the vast coal beds that were put down in this time. In North America, the Early and Late Carboniferous Period are called the Mississippian and Pennsylvanian epochs.

During the Carboniferous Period, as seas advanced and then retreated, the Atlantic Ocean continued to open, separating North America and Europe, or as they still were then, the protocontinents of Laurentia and Gondwanaland. The seas were populated by a variety of invertebrates including sea urchins, crinoids, brachiopods, corals, bryozoans, clams, and trilobites. The last, however, were disappearing in the Carboniferous Period. The dominant vertebrates in the seas were fish, which were very diverse, both in number of individuals and in number of species.

Caribbean Current A westward-moving warm-water flow that is the result of the joining of the North Equatorial and Guyana currents. The Caribbean Current moves past the Yucatan Peninsula of Mexico and past Cuba, then continues on to form the Florida Current and ultimately the Gulf Stream. See *Florida Current, Gulf Stream.*

Caribbean Sea A part of the Atlantic Ocean bounded by South and Central America on the south and west and by the Antilles island chain on the north and east. The Caribbean is a 2.6 million km² (1-million-square-mile) basin.

The underwater terrain is marked by alternating basins and ridges. The greatest average depths are in the Cayman Basin (between the Cayman and Jamaica Ridges) and in the eastern region of the Venezuelan Basin (2,300 and 1,750 m, respectively, or 7,000 and 5,500 feet). Bartlett Deep in the Cayman Trench is about 8,000 m (25,000 feet) below sea level.

The climate is tropical in most of the region. Rainfall depends upon wind direction, and since the prevailing westerlies flow almost all the time, rain is more likely to fall on the windward side of the islands. On mountainous islands, this may produce almost desert-like characteristics on the leeward side.

Water enters the Caribbean from the Atlantic Ocean over the sills marked by the Anegada Passage between the Virgin Islands and the Lesser Antilles, or the passage between Cuba and Haiti. The surface water of the Caribbean is very warm and the bottom water is also warmer than that of the Atlantic, because of the sills, which are high enough to exclude the very cold bottom water of the Atlantic. Tide levels near the Mexican coast are generally higher than they are elsewhere in the Atlantic, since water is driven westward by both tide and wind, and

tends to "pile up" on the Mexican and Central American coasts, a phenomenon observed early in the 19th century. The warm, shallow water of the Caribbean is an ideal environment for the luxuriant growth of both plant and animal life. See *archipelago, coral, shark, sill, turtle.*

Carlsberg Ridge Part of the mid-ocean ridge system, this segment of that undersea mountain chain is in the Indian Ocean. It lies roughly northeast to southwest, with its northern end pointed toward the Red Sea. See *Indian Ocean, mid-ocean ridges.*

carotenes A group of pigments related to the pigment known as visual purple that is found in the mammalian retina, and to vitamin A. These pigments are characteristic of the yellow algae and of diatoms. Their color ranges from pale yellow to red. See *algae, diatom.*

Carpentaria, Gulf of A large bay in northern Australia, between Arnhem Land and the Cape York Peninsula. The Gulf was explored by the Dutch in the early 17th century, and was named for one of the explorers, Pieter Carpenter (or Carpentier).

The rectilinear outline of the Gulf suggests that it is a graben—a fault line depression—which probably dates from the Pleistocene Epoch. The Gulf is fairly shallow, with average depths between 50 and 70 m (150 to 210 feet). The islands of the western edge, Groote Eylandt and the Pellew Islands, are bottom hills.

The climate of the land surrounding the Gulf is hot and dry, with a significant wet season that is monsoonal. The runoff in August is large and drops the salinity of the Gulf sharply, which may explain the lack of coral formation in an area that is otherwise suited to coral growth. See *fault, graben.*

carrack A large European trading vessel in regular use from the 14th to the 17th century. Carracks were rigged similarly to the caravel, but were larger and wider. The carrack was three-masted and square-rigged on its fore and mainmasts, with lateen (triangular) sails on the mizzen (the mast behind the mainmast). This style of rigging was a nautical compromise between the square-rigged ships of northern Europe and the triangular sails of the Mediterranean. Carracks had large castles fore and aft. All subsequent large sailing vessels before the age of steam were based on this design. The carrack evolved into the galleon. See *galleon, caravel.*

Cartier, Jacques (1491–1557) The French master pilot who discovered the St. Lawrence River of Canada. Cartier came from St.-Malo in Brittany, a port with a long association with French shipping. He made three voyages west beginning in 1534, ostensibly to find the Northwest Passage. He may have traveled west earlier, however, with a fishing fleet bound for the Grand Banks.

Cartier's first voyage in 1534 was financed by the the French King, François I, in the hopes of finding gold and other treasure. Instead Cartier encountered ice in the Straits of Belle Isle, and moved south. He rounded the coast of Newfoundland and eventually discovered the mouth of the St. Lawrence. His second voyage began in 1536 and continued his exploration of the Gulf of St. Lawrence. The three-ship squadron sailed upriver as far as the present site of Montreal.

Cartier's third voyage in 1541 was a major effort in which he was to rendezvous on the coast with Jean François de La Roque, sieur de Roberval, a military commander. Together they were to defeat the native nation of Saguenay. The two commanders missed each other but finally met in 1542. However, the nation they had been sent to subdue did not exist. They returned to France with fur and wood.

Cartier's achievement was the establishment of a French presence in the New World. The then-dauphin of France, later Henri II, was uninterested in exploration, but a precedent had been set which was

later exploited by Champlain, the colonizer of New France.

cartilaginous fish Fish with a skeleton consisting of cartilage; the sharks, rays, and chimaeras. See *Chondrichthyes, Elasmobranchi, sharks.*

Caspian Sea The largest lake in the world. It is almost a rectangle about 300 km (165 miles) wide and 1,200 km (750 miles) long, and lies about 30 m (85 feet) below sea level. The short southern side is in Iran; the rest of the Sea is in the Soviet Union. Two rivers feed the Caspian, the Ural and the Volga, the latter supplying the bulk of the water inflow. The Caspian was known in ancient times as Caspium Mare or Hycanium Mare.

Since it is a body that loses more water by evaporation than it receives in river inflow, the Caspian is saltier than the average freshwater lake. The major fishing industry centers on the famous sturgeon and its roe, caviar. The largest port is Baku on the western shore. Like the Aral Sea, the Caspian is a remnant of the Tethys Sea. See *Mediterranean, sturgeon, Tethys Sea.*

catadromous fish Fish that live in freshwater and migrate to salt water to breed. See *eel.*

catastrophism An 18th- and 19th-century concept of the history of the Earth, used to explain geologic features and the existence of fossil forms which either resembled existing biota or which were totally different from any plant or animal known.

Catastrophists believed that all organisms were created at once, at the Creation, and that some were demolished in cataclysmic events. The remnants remained in evidence as the biota currently known. Noah's flood was the most recent of these cataclysmic events.

This theory of Earth's history contrasted with the gradualist idea that land formations took millions of years to come into being. The gradualists were far more likely to adopt evolutionary theories. See *Cuvier, Georges; evolution; Hutton, James; Lamark, Jean-Baptiste Pierre Antoine de Monet, chevalier de; Lyell, Charles.*

Cavendish, Thomas (1560–1592) [also known as Candish] An English explorer and adventurer, who made the embarrassment of the Spanish the basis of his career. Cavandishes' first expedition was an attempt to colonize Virginia. On his second voyage in 1586, he sailed to the Cape Verde Islands and then to the coast of South America. His party went through the Straits of Magellan and up the Pacific Coast of South America. In November of 1587 they intercepted the Spanish Manila galleon. Knowing that he would be hunted by the Spaniards all along the American coast, Cavendish headed west and, after touching land in the Philippines, arrived back in Plymouth, England in September 1588, having been at sea for 25 months.

Cavendish was well rewarded, but spent most of his prize money recklessly and used the remainder to finance a third voyage (1592). This time his stated objective was the exploration of a trade route to China and Japan. However, bad luck, no wind, and numbing cold plagued the voyage. The ships entered the Straits of Magellan in April 1592, but were forced back. Cavendish died on the return trip to England.

Ceara Abyssal Plain A great deep in the Atlantic Ocean due east of the mouth of the Amazon River and east of the Mid-Atlantic Ridge. See *mid-ocean ridges.*

Celebes Sea See *Sulawesi Sea.*

Cell The basic structural unit of most living organisms. All cells contain the genetic material, known as deoxyribonucleic acid (DNA), which holds the assembly code for all of the protein molecules that make up the cell's contents and structural features. In the cells of bacteria and blue-green algae, this genetic mate-rial lies free in the cytoplasm (the content of the cell other than the genetic material), whereas in plant and animal cells the DNA is enclosed by a membrane to form a nucleus.

[The cytoplasm of the cell, surrounding the nucleus, consists mainly of water, in which are distributed ions (such as sodium, potassium and chloride), sugars, amino acids, the energy-supplying molecule known as asenosine triphosphate (ATP), and proteins, lipids and nucleic acids.]

Cells without a nucleus are called pro-karyotic cells and those with a nucleus are called eukaryotic cells.

Cells were discovered by English physicist Robert Hooke in 1665, but it was not until 1839, when Matthias Schleiden and Theodor Schwann, two Germans, the first one a botanist, the latter a zoologist, presented their cell theory, that it was realized that all living organisms are cellular.

cellulose A linear polymer composed of glucose units. Cellulose is the major constituent of plant cell walls. See *glucose, polymer.*

Celsius, Anders (1701–1744) A Swedish astronomer who designed the temperature scale named for him in 1742, using the boiling point and freezing point of water as the markers on his scale, and dividing the difference by 100. On the Celsius scale (prior to 1948 also called the Centigrade scale), the boiling point of water is 100° and its freezing point is 0°. The Celsius thermometer is now standard laboratory equipment everywhere.

Celsius also published works on the aurora, and worked on means to measure the brightness of stars. See *temperature.*

Cenozoic Era The third era in the Earth's history. During this era the shape of the landmasses we now recognize was determined. The Cenozoic is divided into the Tertiary and the Quaternary Periods, which are further subdivided into epochs.

The major new developments in the

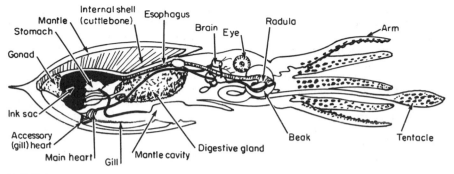

Cuttlefish

Cenozoic took place on land: the rise in importance of mammals and of flowering plants.

Cephalopoda The most highly evolved class of the phylum Mollusca. The most common species of cephalopods today are the octopuses, cuttlefish, and squids. In Paleozoic and Mesozoic times, nautiloids, ammonoids, and belemnoids were widely prevalent. With the exception of the nautiloids, cephalopods have no shell and move by means of a muscular foot that is divided into several appendages. Cephalopods have eyes (and excellent eyesight); a mouth with the characteristic molluscan rasping organ (the radula), and a horny beak. The circulatory system is closed and efficient.

The nervous system is complex but not cephalized (not concentrated in a head). An octopus can be taught and learns tasks and responds instantly to stimuli by flight or color change.

Squid and octopuses are voracious feeders who prey on crustacea or fish, using their sucker-covered arms to catch and hold the prey. In turn, both squid and octopuses are caught and used as a food source by humans and other large mammals. The cephalopods are widely distributed in the ocean, inhabiting water from the equator to the poles and from abyssal depths to tidal pools. They range in size from pigmy squid to giants over 15 m (50 feet) long. See *cuttlefish, Mollusca, nautilus, octopus, squid.*

cephalothorax The combined head and midbody segment of an organism's exoskeleton. Examples of organisms with exoskeletons (skeletons outside their bodies) are spiders and crabs. See *crab, horseshoe crab, spider.*

ceriatite A primitive cnidarian with modern descendants. The present ceriatites—the order is Ceriatharia, of the subclass Zoantharia—are simple, solitary polyps. *Ceriathus* is a representative genus. See *Cnidaria, polyp.*

Cetacea The order of aquatic mammals that includes the whales and porpoises or dolphins. In general, this order is characterized by a broad tail, tiny ears, and hairlessness. Cetaceans live in practically every marine environment. See *dolphin, mammals; porpoise, whales.*

Chaetognatha The arrow worms, present in relatively shallow water as one of the groups of organisms comprising plankton. Although some are 10 or more centimeters (over 4 inches) long, most individuals are between 1 and 3 cm (0.4 to 1.2 inches) long and are shaped like darts. These free-swimming, hermaphroditic predatory organisms have numerous chitinous teeth. Externally they have swim fins and a distinct tail. Although only a few species are known, they are biologically interesting because of their ecological importance and their digestive system. The chaetognatha are very primitive examples of coelomites—organisms that have a tube-within-a-tube body form.

This phylum in turn provides food for larger organisms and is therefore a vital link in the food chains of the continental shelves. See *chitin, hermaphroditism.*

Chain (research vessel) A converted tugboat that in 1966 charted the depths of the Red Sea. The expedition observed very hot and very salty areas in the Sea. See *hot springs, Red Sea.*

Challenger, H.M.S. A ship that sailed in 1872 on the first totally scientific circumnavigation of the world. George Nares was in command, C. Wyville Thomson and John Murray were the leading scientists on board. The ship spent almost four years at sea collecting specimens of plant and animal life and taking temperature, salinity, and other measurements of the water from surface to ocean bottom. New organisms were identified from sediments, and typified according to specific locale.

The voyage's findings confirmed earlier work on winds and currents, and the *Challenger* studied the circulation of water, charting the movement of very cold Antarctic water and mapping underwater mountain barriers. The *Challenger* also discovered the Mid-Atlantic Ridge. Although sections of this undersea mountain range had been reported earlier, the *Challenger's* voyage established the single, continuous nature of the Ridge.

Because of the large number of experiments done, specially designed instruments used, and far-reaching results achieved, the voyage of the *Challenger* has been called the beginning of the science of oceanography. See *mid-ocean ridges; Murray, John; Nares, George; ocean floor; oceanography—history; sediment; Thomson, C. Wyville.*

Chancellor, Richard (? –1556) A British navigator and the pilot on Sir Hugh Willoughby's expedition to find the Northeast Passage to the Indies. Chancellor left England in 1553 aboard the *Edward Bonaventure,* and was later separated from Willoughby by a storm while attempting to sail around the North Cape of Norway. Chancellor waited for the rest of the group and then continued to the White Sea. He landed and went overland to Moscow, where he met Czar Ivan IV (the Terrible). The result of this visit was the establishment of the Muscovy Company, a trading venture between England and Russia.

On a second voyage, in 1554, Chancellor retrieved Willoughby's body and possessions from Archangel. On the return from an embassy to Russia in 1556, the pilot-ambassador was shipwrecked off Aberdeen and drowned. See *Northeast Passage; Willoughby, Sir Hugh.*

chart A map for the use of navigators. The first known seaman's chart was drawn by Marinus of Tyre in the 1st century. This was based on the work of Eratosthenes, who had placed the island of Rhodes at 0° latitude and longitude. From this time to Ptolemy's, two centuries later, there were only minor changes in seamen's charts, and then almost none for centuries. The great developments in charts finally occurred after the Renaissance, with the renewal of interest in exploration. Before the end of the 13th century, sailors relied on the magnetic compass (in use in the Mediterranean since early in the century) and pilots' books as guides to coastlines. By the end of the 13th century portolon or portulan maps appeared in the Italian trading cities of Venice and Genoa. These showed coastlines and distances between points. The *portolano* depicted anchorages, shoals, bars, and other navigational features.

The increasing sophistication of mariners' maps continued and spread from country to country. Aragon's maps were superseded in quality and clarity by those of Catalonia, then Portugal. The best portolans date from the 16th century: by 1530 Pedro Nunes, Portugal's Chief Hydrographer, realized that meridians of longitude converge at the poles.

The mapmakers of the northern European countries accepted the work of Ptolemy as the basis for their charts. Thus, the mapmaker Ortelius of Antwerp published a chart in 1570 which showed the Canary Islands at 0° longitude, as had Ptolemy. Northern sailors were much more dependent on anecdotal information and the use of the lead line than Southerners until the time of Lucas Wagenaer, a native of the Low Countries who combined the ideas of Ortelius and the portolan maps of Spain and Portugal. He published his results in an atlas of sea charts, and attempted to standardize the notation in sailing directions.

In the 17th century, both the English and French surveyed their waters and produced superior series of charts.

By the mid-18th century the British Admiralty was sponsoring surveys of British coasts and nearby island groups, producing an atlas called *The Atlantic Neptune* in 1777. This atlas was the source for sailors in North American waters for over half a century. By 1801 the Admiralty produced the first charts based on its own surveys. The Office of the British

Hydrographer of the Navy is now an outstanding producer of navigational charts with a worldwide scope. See *compass, map, navigation.*

chela   A large pincer claw that is the specialized first leg of arthropods. The lobster claw and the fiddle arm of the fiddler crab are examples. See *Arthropoda, lobster.*

chemical defenses   Chemical compounds produced by organisms to protect themselves against predators. A number of these compounds are also of pharmaceutical interest.

Intensive exploration since the 1960s has produced over 100 chemical defense compounds of pharmaceutical interest. Repellants, natural and synthetic, particularly shark repellants, are also being investigated.

Some of the compounds isolated, the organisms that produce them, and their intended use are shown in the table below.

Compound	Producing Organism	Use
Manoalide	*Luffariella variabilis* (sponge)	Analgesic, anti-inflammatory agent
Stypoldione	*Stypopodium zonale* (brown alga)	Cell-growth inhibitor
Bryostatin 8	*Amanthia convoluta* (bryozoa) *Bugula neritians*	Antineoplastic agent (mouse leukemia)
Claviridenones	*Clavularia viridis* (soft coral)	Anti-tumor prostanoids
Aplysistatin	*Aplysia angasi* (sea hare, mollusk)	Cell-growth inhibitor, antineoplastic
Stichostatin I	*Stichopus chloronotus* (sea cucumber)	Cell-growth inhibitor antineoplastic agent
Dolastatin 3	*Dolabella auricularia* (shell-less mollusk)	"
Strongylostatin 1	*Strongylocentrotus drobachiensis* (sea urchin)	"
Sphyrnastatins	*Sphyrna lewini* (hammerhead shark)	"

(*continued*)

Compound	Producing Organism	Use
Bromophenols	*Aplysina songelii* and *A. lacunosa* (sponges)	Antimicrobial agents
Muricin	*Muricea fruticosa* (California gorgonian)	Cell-growth inhibitor
Homarine	*Leptogorgia virgulata* and *Listetacea* (Atlantic gorgonians)	Cell-growth inhibitors
Terpenes	*Ostracion lentiginosis* (boxfish) *Ladophyrsis quelar* (smooth truckfish) *Pardachirus marmoratus* (Red Sea or "Moses" Sole) *P. pavoninus* (Japanese sole)	Ichthyotoxic materials lethal to fish; shark repellants

Various terpenes have also been isolated from soft corals and nudibranchs: sponges from carotenoids; *Nereis* toxin, from a sea worm, is a toxic mixture of sulfur compounds; several marine plants and animals produce haloterpenes; some bryozoans yield toxic alkaloids. The chemical identification of products of organisms from the sea is a new and challenging branch of marine biology and pharmacology. See *Bryozoa, cone shell, coral, Nereidae.*

Chesapeake Bay   The largest embayment of the Atlantic Ocean on the North American coast, the Bay varies from about 5 to 50 km (3 to 30 miles) in width and separates the Delmarva Peninsula from the rest of Maryland. The greatest deeps are about 100 m (330 feet).

A number of rivers, all of them significant in American history, empty into the Bay. They include the James, York, Rappahannock, Potomac, and Susquehanna. The very irregular shore extends for more than 5,000 km (3,000 miles). Its relatively mild weather and many inlets have made the Chesapeake Bay area one of the most important sources of shellfish in the world. Crabs and oysters are taken commercially, as are fish—notably shad and herring. The waterfowl population that feeds on the fish and shellfish is attractive to hunters and birdwatchers alike.

This region has a good deal of wild country in spite of the fact that it is very close to the large urban centers of Baltimore and Annapolis, Maryland, and Washington, DC. Nothwithstanding the efforts of both the commercial fisheries and conservation groups, there is pollution of Chesapeake Bay from industry and from human habitation and encroachment on the area for housing and recreational use. The plants and animals of the region need protection. See *coast, estuary.*

Chile Rise   Also known as the Chile Ridge, this is a branch of the East Pacific Rise, running northwest from the Taitao Peninsula of Chile toward Easter Island. Its existence was predicted on the basis of seismic activity before it was charted in 1958. See *evolution of oceans, mid-ocean ridges, Nazca Plate, Pacific Ocean.*

chimaera   Also know as the rat fish of the class Chondrichthyes, order Holocephalus, this is a deep-bottom-dwelling fish living on mollusks. The chimaera has a fixed upper jaw and a movable lower one, and most specimens are about 1 m (3 feet) or more long, much of which length is a long, whiplike tail. Chimaeras may have evolved from early sharks. See *Chondrichthyes, benthic ecosystem.*

chinook   A local wind that moves in a southeasterly direction in the Rocky

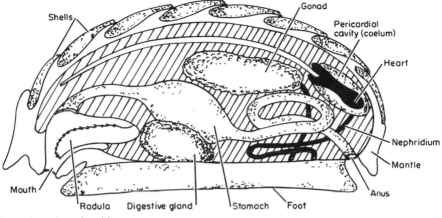

Lateral section of a chiton

Mountains of North America, carrying a dry air mass which in spring is very warm and creates a great temperature change in a short time span. See *weather, wind.*

chitin   A polymer structurally related to the sugar glucose, chitin is a key component of the exoskeletons of arthropods and the shells of crabs, lobsters, and insects. Chitin is also present in the hard structures of some coelenterates, hard corals, and sea anemones.

chiton   Marine mollusk of the order Polyplacophora. Mostly chitons are found in shallow water or intertidal pools. Chitons are very primitive animals, highly adapted to adhering to rock. They are relatively flat, ovoid creatures, 2 to 7 cm (1 to 3 inches) long, although specimens of the tropical species may be more than 30 cm (1 foot) long. Most individuals are dull gray-green or brown. The shell is a construction of overlapping, broad plates, making it possible for chitons to adhere tightly to rocks, where they may stay for months at a time. The mantle or girdle and the foot create an excellent suction device. Chitons, like snails, are herbivores. Their nervous system is a primitive network with some regression, as the free-swimming larva loses its eyes upon maturing. See *Mollusca.*

Chlorella   See *green algae.*

chlorine   A chemical element of the halogen family (the other members of the family are fluorine, bromine, and iodine). In its elemental state chlorine is a greenish-yellow gas (Cl_2). In nature, however, it always occurs in combined form as chloride salts. The chloride ion is present in seawater in concentrations of about 19 grams per kilogram of water (about 2%). The chloride ion is the most abundant dissolved negative ion in seawater. About 54% of the total weight of all dissolved inorganic substances in oceanic water is chloride ion. See *seawater, chlorinity.*

chlorinity   A measure of the concentration of chloride ion in seawater. It is usually expressed as the mass (weight) of chloride (e.g., as $NaCl$ or KCl) in 1 kilogram of seawater. The average chlorinity of seawater is 19 gm/kg, or 2%. See *chlorine.*

chlorophyll   The name given to a series of pigments that produce the green color of plants. These compounds absorb blue-violet and red light, and therefore reflect green light. They play a basic role in photosynthesis—the use of the energy of sunlight in making carbohydrates from carbon dioxide (CO_2) and water (H_2O)—

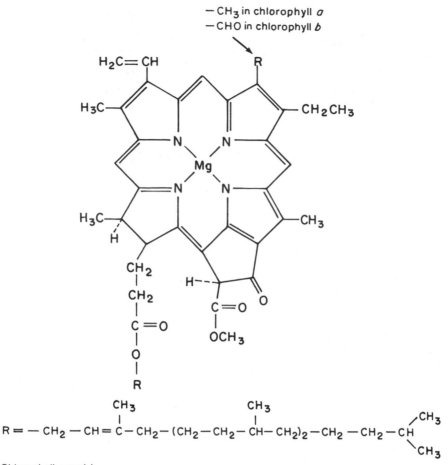

Chlorophyll *a* and *b*

by absorbing light energy and converting it into chemical energy. The chlorophylls are oxidized by sunlight and in turn reduce some other compound in the leaf of a plant. All chlorophyll molecules are large rings (porphyrins), made up of four smaller rings that contained nitrogen and carbon. The porphyrin ring of chlorophyll has a magnesium atom at its center. There exist different chlorophylls, designated a, b, c, and d. The most common in terrestrial plants is Chlorophyll a, where the $R = CH_3$ (see illustration). In Chlorophyll b, $R = CHO$. These two types of chlorophyll are present in most plants and algae in the ratio 3:1. Type c chlorophyll, in which $R = H$, is present in most marine algae. In type d chlorophyll a CHO replaces the $-CH=CH_2$ group.

The porphyrin ring occurs in a number of biologically significant compounds besides chlorophylls. Heme, the core of the hemoglobin molecular blood segment, is a porphyrin with an iron atom at its center. The molecules forming arthropod the blood of arthropods are porphyrins with copper at their center.

Chondrichthyes   A primitive class of fish, named for its cartilaginous skeleton. Chondrichthyes first appeared in the Devonian Period and are well-represented by fossil remains. The sharks, rays and chimaeras are chondrichthyes.

While these fish have no true bones, they do have enamel-covered teeth and toothlike processes on their scales, called placoid scales. All chondrichthyes are predators; they have real jaws and paired gills, each in a separate cleft, arranged in groups of five pairs for sharks, five to seven pairs for rays, and three pairs for chimaeras. The group has the usual two-chambered heart of fishes. See *chimaera, ray, shark.*

chromatophore   A pigment cell in either an animal's skin or a plant. In plants, the chromoplasts, or color granules within a cell, are most often chloroplasts, containing chlorophyll. In animals they are most likely to contain melanin—a dark brown or black pigment. Other possible pigment colors are reds, blues, yellows, and iridescent colors. In animals, the nervous connections to chromatophores are vital. They allow for fairly rapid (minutes to several hours) color alterations in amphibia, fish, and most strikingly, cephalopods. See *Cephalopoda, fish, octopus, squid.*

Chrysophyta   A phylum of algae. These plants contain carotene, which colors them yellow because the carotene masks the chlorophyll (green) that is also present in the algal cell. The Chrysophyta are photosynthesizing organisms. See *algae.*

Chukchi Sea   A shallow area north of the Bering Strait, its easternmost end reaching Point Barrow, Alaska. Wrangel Island is the western boundary and the 75° north latitude line is the rough boundary of the Chukchi Sea with the Arctic Ocean to the north. Because of its shallow bottom, relatively warm currents entering from the south, and the low salinity typical of arctic waters, the Chukchi is rich in plant and animal life despite the year-round presence of ice. The large arctic mammals are all represented in it. Typical fish are large cod and char. In summer the bird population consists of several species of duck, goose, and gull.

The area of the Chukchi is of interest to naturalists because of the abundance of wildlife and the fairly recent geological history of the sea as part of the Asia-to-America land bridge. See *Bering Strait, polar biome.*

cilia   Short, hairlike extensions of certain cells. Cilia are capable of vibratory motion, and are used by the cell for feeding itself, for locomotion or both. They are frequently based on centriole fibers (tendrils of protoplasm). Nonmotile, ciliated cells in the mammalian respiratory and reproductive tracts propel mucus and dust out of the former and help propel the egg cell along the fallopian tube in the latter. All motile cells have cilia or the long, hairlike appendages called flagella. The difference between flagella and cilia is one of size and number, their functions being similar. The distribution of cilia on the body of a cell is part of the genetic "package" of the cell. Structurally, cilia contain microtubules—thin, cylindrical, unbranched tubes that consist of similarly arranged bundles of fibers. The cilia of a cell operate in concert: a single beat is like that of the oars of a galley: stroke—recovery—stroke—recovery. The beat of the cilia can move the cell forward or backwards or rotate it.

Cilia contain glycoprotein molecules—proteins with carbohydrate chains attached to them—that are used in testing for a particular type of cell, and especially for such single-celled organisms as bacteria and protozoa.

Ciliophora   Protozoans that have cilia; also known as ciliates. These organisms are distributed widely in both fresh- and salt-water habitats. The familiar *Paramecium* is the most well-known genus of this order. Ciliates are usually free-swimming, although some live either fixed in place or are colonial. The exterior or pellicle of a ciliate is a complex structure of organelles with the cilia extending through it.

Raptorial (carnivorous) ciliates feed on other protozoans. Sessile ciliates, which live attached to a single location, beat into their buccal cavities (mouths) anything that comes along, including bacteria and living or dead organic matter. Ciliates reproduce asexually by fission most of the time. They do have a sexual phase in which male and female gametes are produced and fertilized eggs result. See *cilia, Protista, Protozoa.*

circulation   The movement of masses of water on the surface or in the depths of the oceans. Circulation on ocean surfaces can be described as permanent westerly flow of water near the equator, toward the poles along the western edges of oceans, easterly in mid-latitudes, and toward the equator along the eastern edges of oceans. These large gyres exist in the Atlantic, Pacific, and Indian oceans. There is a subpolar gyre in the Northern Hemisphere, whose closest counterpart is the Antarctic Circumpolar Current. Circulation in the Indian Ocean is seasonal, depending on the monsoon wind direction.

Deep circulation in the oceans is much slower and very much less dependent on winds or seasons than are surface phenomena. Not only is deep water moved laterally over great distances, it rises and sinks to produce considerable interchange between waters of differing regions. See *bottom water, currents, gyre, individual currents, west-flowing currents.*

Cirripedia   An order of marine crustaceans that use thoracic legs to trap waterborne food. These animals are free-swimming as larvae, but are parasitic as adults, and become permanently attached until they no longer look like crustaceans. Barnacles are an example. See *barnacle.*

Cladocera   Water fleas, a mainly freshwater order of the Crustacea. *Daphnia* is the most widely known of the freshwater species.

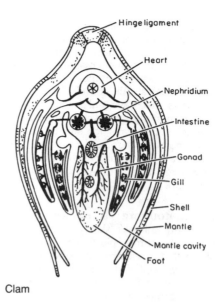

Clam

clam   A member of the class Pelecypoda, phylum Mollusca, comprising common marine and freshwater bivalves. The greatest number of species of clams are marine. Clams can range in size at maturity from barely visible to the giant *Tridacna*, which can weigh upwards of 250 kg (550 pounds). Clams usually have two shells of equal size and shape. Like other mollusks, the clams use their siphons to bring in and then expel water, from which they remove food and oxygen. Clams are generally found on or just below water on muddy or sandy bottoms, and use their powerful, muscular "foot" to dig themselves in, to a depth of 50 to 75 m (2 to 3 feet). This ability to "dig in" and the two adductor muscles that close the shell tightly, are the clam's defense against predation by birds and echinoderms. See *Mollusca.*

clastic   Made up of rock fragments; unconsolidated sediment or beach deposits. See *sediment.*

clay   Usually aluminum silicates. The particles of clay are larger than silt and smaller than colloidal particles of mud. See *sediment.*

climate   Weather conditions prevalent in an area over a period of time. Weather conditions include temperature, rainfall, sunshine, wind, humidity, and cloudiness.

The interaction of atmosphere and water is enormous, and each is affected by the other. The effect of ocean water is based on its heat capacity. As contrasted to land, the ocean heats up more slowly and cools more slowly. The atmosphere on an average day absorbs heat that is re-radiated from the Earth, raising the atmospheric temperature by about 2°C. This is equivalent to the heat capacity of the uppermost 3 m (11 feet) of ocean. Water also radiates heat to the atmosphere, and as a result, it cools considerably in local areas. This cooler, and therefore denser, water, sinks and from then on is heated only very slowly from above. As a result, the cold bottom water stays on the bottom for very long periods.

Since the overall effect of the oceans is to act as a heat reservoir, any land mass near a large body of water is protected from great swings of temperature. Comparing the high and low temperatures of three isolateral (same latitude) Canadian cities shows this.

There is evidence that sunspot activity has an effect on climate that, in turn, will affect the mean temperature of deep ocean waters, and eventually the composition of the foraminiferal populations. Orbital variation is another cause of climate change, and in the past has influenced arctic weather. This has produced increased volumes of North Atlantic Deep Water, which eventually surfaced in the South Atlantic and were traced in the microfossil record. See *atmosphere, circulation, Coriolis, Ekman spiral, weather, wind.*

clipper   A sailing ship of the late 18th and 19th centuries built to satisfy the demands for speed between American coastal cities. The successful design was then adopted by shipyards worldwide.

There were clipper barks, brigs, and schooners. The usual clipper ship built for cargo was a large vessel with three or more masts rigged with square sails and lateens on the bowsprit.

The great age of the clipper was 1845 to 1860, when these ships carried freight worldwide. Record travel times, such as from San Francisco to New York via Cape Horn in 89 days and from Hong Kong to New York in 74 days, made the clippers the "monarchs of the seas." While clippers continued to be built and used for cargo well into the 20th century, they were displaced as fast transportation by steamships, which in addition to the advantage of greater speed, required a smaller and less skilled crew.

cloud   An accumulation of water droplets or ice crystals in the atmosphere. Clouds were assigned names according to their shapes in the first decade of the 19th century by Luke Howard, an English scientist. "Cirrus" means curl, "cumulus" means heap, and "stratus" means layer.

Most clouds are in the troposphere; few ascend higher than this. Cumulus clouds continually rise in warm regions. This movement, and the rain that falls from them, are the chief causes of air warming as cool air moves toward the equator from the poles.

Precipitation occurs when the droplets of water or ice crystals in a cloud reach a critical size, usually about 0.2 mm (0.01 inches). See *atmosphere, ocean-atmosphere, water, water cycle.*

City	Victoria	(Pacific Coast) Winnipeg (Inland)	St. Johns (Atlantic Coast)
Average maximum temperature in July (°C)	20	22.5	20.5
Average minimum temperature in January (°C)	2	−22.3	−7.5

Clymenias A genus of the Maldani-
dae, the family of "bamboo worms."
Clymenias are polychaete (segmented)
animals with a worldwide habitat. They
are found at all depths, and can form
great beds of attached organisms on continental shelves and slopes. They are of
very ancient lineage, and were a common feature of Paleozoic seas 300 million
years ago. See *Polychaeta*.

Cnidaria or Coelenterata A phylum whose members have tentacles, nematocysts, and radial symmetry. They
are, for the most part, marine organisms
that include sea anemones, corals, and
jellyfish. Some of the characteristics of

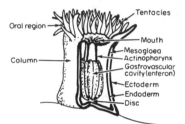

(1) A colonial polyp (soft coral)

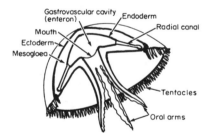

(2) A solitary polyp (sea anemone)

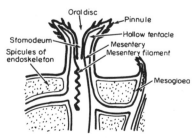

(3) A medusa (jellyfish)

this phylum are a central body cavity in
which digestion occurs, both sexual and
asexual reproduction, an epithelial layer
in which the "stinging cells" or cnido-
blasts are embedded, and broad distribution in fairly shallow, warm seas. Cnidarians have existed since the pre-
Cambrian. Although they consist mostly
of soft tissue, enough material has survived or exists in some sort of calcareous
or silaceous shell to have preserved a
fossil record of them. See *chitin, cnido-
blasts, medusa, nematocyst, polyp, individual animals*.

cnidoblast A stinging cell of a cnidarian. It contains the nematocyst. See *Cnidaria, nematocyst*.

coast Seashore or land near the shore;
the border between land and ocean. The
coast is an area of change. Materials
brought seaward by rivers and then rear-
ranged by waves and currents form
coastlines. In the context of plate tectonics, coasts are either the leading edges or
trailing edges of plates. The western United
States coast, where mountainous cliffs
come down to the water, is an example
of a leading edge. The Atlantic coast
south of Sandy Hook, New Jersey, with
its broad, flat beaches is an example of
a trailing edge. The trailing-edge coasts
in general have more varied forms both
onshore and offshore.

A current flowing parallel to the beach,
the littoral drift, carries beach sand with
it. The reshaping of Fire Island, New
York, is an example of such drift, as is
that of Cape Cod, Massachusetts. Fire
Island has, in about a century, grown
considerably longer through extension at
its western end. Coastal engineers have
done considerable experimenting with
materials to hold sand in place. These
trials include artificial seaweed, old
Christmas trees, floating breakwaters and,
in Texas, artificial sand dunes. The overall effect is still uncertain.

The organisms found in a coastal community are dependent on the type of
local physical habitat and the temperature. On rocky coasts the type of rock

best suited for maintaining an ecosystem is a hard one. To such rocks the plants (algae) and animals can attach firmly. Crumbly shores of shale or sandstones are unlikely foundations for ecosystems. Rocky pools on coasts either near the poles or near the equator have small populations, the former because they are scraped away by the movement of ice in winter and spring, the latter because of drastic temperature changes between tides. The middle latitudes support an array of animals, some on exposed rocks, such as limpets, and others in very shallow water, including barnacles, snails, mussels, echinoderms such as starfish, and urchins. They are capable of withstanding tidal surges or live just below them. Crabs, snails, some segmented worms, and mollusks are capable of living temporarily on land either on or in the damp sand of a beach. The temperature of tropical beaches makes this transition difficult, particularly in the daytime. Most intertidal animals are therefore nocturnal. See *barrier islands, beach, littoral.*

Coast and Geodetic Survey, U.S.
An outgrowth of the U.S. Navy's Department of Maps and Charts, now part of NOAA, The National Oceanic and Atmospheric Administration of the National Ocean Survey. See *NOAA.*

cobalt   A hard, silver-white magnetic metal present in seawater in much lower concentration than in continental rock. It is assayed at about 1 ppm (part per million) in rock and 0.5 ppm in seawater. Cobalt is an essential micronutrient for plants. It is present in deep-sea sediment and in the ferro-manganese micronodules present on the sea floor. Its origin is uncertain, but it may have precipitated from the seawater or been brought up from subcrustal layers by volcanic activity.

coccolith   A crystal of calcite (calcium carbonate, $CaCO_3$) which is deposited as a protective plate by the Coccolithophoridae, a group of unicellular, flagellated, tiny brown algae. The coccoliths

are simply bipyramidal or hexagonal crystals formed by a single group of organisms, the Homococcoliths. Heterococcoliths, a category that includes several different genera, form complex crystals that resemble cut glass.

Most of the Coccolithophoridae are photosynthesizers, and are usually found in tropical and warm-temperate seas. Both haploid generations, with half of the animals' usual full number of chromosomes, and diploid generations, with the full complement of chromosomes may be found in the same heterococcolith. Many simple Coccolithophoridae have a viable gamete (haploid) generation, which usually does not resemble the diploid form.

The deposition of fossil coccoliths occurred worldwide. These plants are found in limestone deposits of marine origin such as the Dover Cliffs of Britain, and in undersea sediments. Their continuous fossil record reaches back to the Jurassic Period, and is used as a means of dating sediment.

cockle   A clam of almost universal marine distribution. Cockles are found in waters ranging from the tropics to the polar regions; from intertidal zones to ocean depths of 1500 m (about 1 mile).

The European cockle, *Cerastoderma edula,* is eaten by humans. Many other species are so muscluar that they are too tough to be chewed. The average cockle has a shell 2 to 15 cm (0.4 to 6 inches) long, usually gray or gray-white, and ribbed. There are several very beautiful species of cockle collected by hobbyists. See *bivalve, clam, Mollusca.*

Cocos Islands   A group of islands in the Indian Ocean, south of Malaysia and east of Christmas Island. They are the peaks of mountains of the Cocos Rise.

Cocos Plate   One of the smaller plates forming the Earth's surface. It is northeast of the Galapagos Islands and is moving northeast, pushing under the plate bearing Central America.

cod   A gadiform fish. The major North Atlantic species is *Gadus morhua,* an extremely important fish commercially. This species is usually found in cold water on continental shelves, mainly in the Northern Hemisphere. It is best known around the Grand Banks and St. George's Bank, both areas off the Canadian coast that have been international fishing grounds for centuries.

The cod is a bottom dweller, feeding on other fish and invertebrates. It is usually dark green or brown, and has a lighter underside. It has a characteristic chin barb that looks like a sharp tooth. The cod usually landed by a commercial boat are 5 to 8 kg (12 to 20 pounds) in weight, but specimens of well over 50 kg (110 pounds) exist. Traditionally, cod were taken in small boats on individual lines, but the advent of fish factory vessels has changed commercial fishing and possibly the cod population on a long-term—if not permanent—basis.

The economic importance of the cod is underscored by the fact that many of the 16th century voyages of exploration aimed toward North America were at least in part attempts by European nations to establish fishing rights. Dry salt cod is still a formidable ingredient of the Mediterranean diet, particularly in Portugal.

coelacanth   A primitive teleost fish. About 25 genera of fossil coelacanths are known. They first appeared in the Devonian Period; the most recent specimens date from the Cretaceous. The group was assumed to be extinct. The name coelacanth is derived from the hollow spine in the dorsal fin.

A living specimen was first caught in 1938 and named *Latimeria chalumnae.* It was about 2 m (6.5 feet) long and weighed 40 kg (90 pounds). Several more have since been caught, and others photographed in their normal habitats. They are a nocturnal carnivores and caught very infrequently, but nearly always off the South African coast. The body is purple-brown with pink and blue splotches.

The significance of the coelacanth, besides its great age, is its place as a precursor of land vertebrates. The fossil specimens were often small, and are thought to have had fat-filled, lunglike structures that may have been flotation devices, as well as fins that were protolegs. The original habitat of the coelacanth may have been in fresh water. See *fish.*

Coelenterate   See *Cnidaria.*

coelom   The central body cavity of an animal. It contains the digestive, excretory, reproductive, and circulatory organs.

cold-blooded animals   A misnomer, since the blood of these animals is actually slightly above ambient temperature. However, their body temperature is not constant, depending instead on the environment. See *homeotherms, poikilotherms, temperature in animals.*

colonial animals   Animals that live in groups of attached and closely related individuals. They range from autonomous individuals to highly specialized interactive groups. The corals are representative of the former, the Portuguese man of war of the latter. See *coral, Portuguese man of war.*

color (of water) (of organisms)   The color of water is a function of its depth, its content, and the condition of the sky. Sediment or dissolved animal and plant pigments or both will alter the color of water.

The color of marine organisms is a function of sunlight, and of their habitat, mineral content, and distribution. The most brightly colored corals, fish, and invertebrates are found in warm and relatively shallow water. Fish and many invertebrates, notably the octopus, can rapidly alter their color as a defense mechanism. This involves an extensive nervous system, instant response to stimuli, and the ability of specialized cells to produce different colors on the animals' body surface. See *chromatophore.*

comb jellies   Animals in the phylum Ctenophora. There are about 90 species of these widely distributed planktonic organisms. They feed on smaller plankton such as arrow worms, and are in turn fed upon by fish, such as herring.

Comb jellies are spherical or teardrop-shaped, and motile. The outstanding characteristic of the comb jellies are rows of hairlike projections, called cilia, that are fused together, and which beat synchronously to propel the animal along. These fused cilia are arranged in eight longitudinal plates which, outlined on the gelatinous, almost transparent body of a comb jelly give the animal the appearance of a minute gooseberry. The internal organs are visible through the body wall. The comb jelly has two retractable tentacles which are lined with colloblasts—specialized sticky cells that entrap prey. See *Ctenophora.*

commensal relationship   A relationship between two living organisms of different species in which one partner is unaffected by the relationship while the other partner benefits. See *parasitism, symbiosis.*

compensation depth   That point in a column of water at which the oxygen generated by photosynthesis equals the oxygen demand of the flora. Below this point plants cannot live. In nearshore areas where, because of high turbidity, light penetrates the water to a lesser degree than it does in the open ocean, the compensation depth is closer to the surface than it is in the open sea. See *BOD, photosynthesis.*

conch   A tropical marine snail characterized by a heavy, coiled shell and a large, flaring, wide lip. The common Caribbean conch *(Strombus gigas)* has a pink, smooth, pearly interior and whitish-gray, grainy surface. Conchs are taken for food and their shells are sometimes collected.

Average specimens are about 30 cm (1 foot) long. Conchs of the Earth's more temperate zones range in size from 7 to 25 cm (3 to 10 inches). The largest conch is the Triton's trumpet *(Charonia)* of the western Pacific. It may exceed 45 cm (18 inches) in length and 15 to 20 cm in diameter. This shell is a prized collector's item. See *Gastropoda.*

condensation   The change of state by which a gas becomes a liquid. This physical change is always accompanied by a loss of energy. When the gas becoming liquid is water vapor, it changes into liquid water. If condensation occurs around a nucleus of dust or other very fine particulate matter in the atmosphere, fog or clouds form. If condensation occurs very high in the atmosphere, the condensate is ice rather than water. See *cloud, fog, water.*

cone shell   A marine snail named for its colorful, heavy shell. The shell is anywhere from 1 to 20 cm (0.5 to 8 inches) in length, and has been found in a variety of patterns. It may be banded, spotted or of a single color with dark lines. The colors range from gray to dark brown; the lip is usually white.

Cone shells are carnivores. There are about 450 species of these organisms, which are most often found in shallow, tropical waters. They feed on worms, fish, and other mollusks, which they kill by injection of a neurotoxic venom. All cone shell venoms are dangerous, and those of some species are fatal to humans.

Some cone shells are quite rare, and the shells of some Indo-Pacific species are quite costly. Because of this there is a trade in the shells despite the danger in their collection. See *Gastropoda.*

conger eel   One of fourteen eel species found worldwide. They are black or gray on the dorsal surface and white or gray-white on the ventral. The usual size of the adult that is caught is about 1.5 m (4.5 feet), but some grow to 2.5 m (7.5 feet). The average specimen weighs 2.5

kg (5.5 pounds). Conger eels are caught commercially along continental shores. As food fish, they are more popular in Europe than in America. The species *Conger oceania* breeds in the Sargasso Sea. See *eel*.

conodont   A tiny fossil composed of calcium phosphate. These remains range in age of origin from the Cambrian to the Cretaceous periods. Their pointed shapes led to the name "conical teeth," but what they were and from what organism they came was long unclear. Geologists and paleontologists have arranged them in groups that are associated with specific geological formations. For this reason, conodonts are index fossils.

Very recently an entire conodont organism was identified as a minute, ancient relative of the hagfish.

continent   A large landmass; that part of the Earth that stands above the water of the oceans. About 30% of the Earth is covered by the continents. They are Eurasia (Europe and Asia), North America, South America, Africa, Australia, and Antarctica. Most of the landmass of the Earth is in the northern part of the globe, only Australia and Antarctica are wholly below the equator.

Evidence from geological and seismic studies has indicated that at one point in geological history all the landmass of the Earth was concentrated in one supercontinent (Pangaea), and has since drifted to its present positions. This drift still occurs. The granitic continents float on a denser base of the Earth's crust, made largely of basalt.

The material of the continental crust is very old. Each continent has an area called a shield that is the oldest rock, 2 to 4 billion years old. The rock of the ocean floor is youngest. Mountain-making is the result of the upthrusting of relatively young rock at the edges of the floating platforms that carry the continents. As plates of the Earth's crust collide with one another, buckling and folding occurs. The Himalayan Mountains, for example, are the result of the collision of the Indian landmass with the Eurasian one. See *basalt, continental drift, crust, evolution of oceans, granite, magma, ocean floor, plate tectonics.*

continental drift   The name given by Alfred Wegener (1880–1930) to his theory that there was originally only one continent (called Pangaea) which broke up, and that the resulting, newly formed continents then moved away from one another to form the present continents. Wegener also believed that mid-ocean sediments were the oldest rocks on Earth, compacted from continental sediment. The continents themselves float on a semifluid mantle or viscous basement rock. While Wegener's idea concerning the age of the oceanic bottom is wrong, the idea of continents moving away from each other and away from the ancient single continent Pangaea is now supported by considerable evidence. See *evolution of oceans, oceans; ocean floor; Pangaea; plate tectonics; Wegener, Alfred Lother, sediments.*

continental margin   That part of a continent that extends from or borders on a steep cliff that falls off toward the ocean floor. The margin is considered to be that portion from the lowest low-tide line to the cliff's edge. The cliff and its component parts—the rise, slope, terrace and shelf—are extensions of the nearby continent and receive materials from it. The dominant feature of the continental margin is terrigenous sediment covering a plain that may be broken by canyons. See *continental shelf, sediment.*

continental rise   A name proposed in 1959 for the cliff leading from the continental slope in a seaward direction to the plain of the ocean floor. The Atlantic gradient (east of North America), for example, is steeper near the continent and becomes a more gentle slope as it descends to the ocean floor. Near South Carolina, this submerged coast resembles a broad staircase. In other areas the slope is a more precipitous one. See *Blake Plateau, ocean floor.*

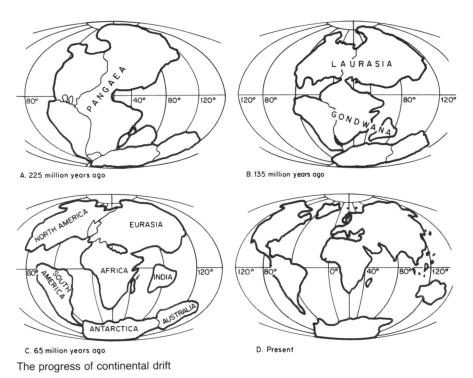

A. 225 million years ago

B. 135 million years ago

C. 65 million years ago

D. Present

The progress of continental drift

continental shelf and slope The shallow sea area closest to a continent. This area is about 70 km (45 miles) wide on the average, and deepens as it extends seaward to an average depth of about 130 m (400 feet). Although most shelves are geologically rather featureless, they are very important biologically and ecologically. The edge of the shelf, or shelf break, is the sharply descending cliff or continental slope. Geologically it is less well-explored than the shelf, and is frequently deeply cut by canyons. The slope also seems to be the area of greatest faunal change, with new species arising near the shore and spreading out across the shelf. The new species are abruptly stopped from speading by the drop down to the ocean floor. Continental slopes generally extend from the shelf break to the continental rise.

Shelves in areas that were once glaciated, such as the Grand Banks and Georges Bank, off the coast of New-

foundland, are highly irregular. Continental slopes in tropical seas may be covered with corals and their attendant ecosystems, as in the Bahamas and the Great Barrier Reef. See *coral, mid-ocean ridges, plateau.*

Cook, James (1728–1779) A British navigator, one of the great 18th century explorers. Cook made three voyages to the Pacific Ocean, exploring the Australian and Tasmanian coasts and circumnavigating the South Pole to the limit of the floating ice, as far as 71° 10′ south latitude. During his first voyage (1768–1771) Cook charted New Zealand, Australia, and the Indian Ocean. On the second voyage he again visited New Zealand and circumnavigated the pole. On the third voyage he stopped at Tasmania, New Zealand, and the Society Islands.

Cook was charged with collecting and preserving specimens of all unknown plants and animals found on his voyages, and his ships carried naturalists, among whom the best known was Joseph Banks, who were on board to collect specimens and sample the water. Cook was killed on his third voyage by natives on the Sandwich Islands—today the Hawaiian Islands—which he discovered. See *Banks, Joseph; explorers and explorations; longitude.*

Cook Inlet   An extension of the Gulf of Alaska, the inlet separates the Kenai Peninsula from the mainland. Captain James Cook explored the area in 1778 while attempting to find an Arctic Passage. Anchorage is the significant city at the head of the 240-km (150-mile)-long inlet. The area is known for its sudden and often violent storms.

Cook Islands   A group of islands in the South Pacific about 2,600 km (1,600 miles) northeast of New Zealand. The islands were discovered by Captain James Cook in his voyage of 1773–1777. The total landmass of the islands is less than 250 km (93 square miles). They consist of two separate volcanic groups. Rarotonga is the largest island, and the capital, Avarua, is also on this island. See *Cook, James; Pacific Ocean*

Cook Strait   A body of water that separates the two main islands of New Zealand. It ranges from 25 to 145 km (16 to 90 miles) wide. The Strait was found by Captain James Cook, who sailed through it in 1769. See *Cook, James.*

Copepoda   The most numerous crustaceans, with more than 6,000 species found in fresh and marine water. Some copepods are parasitic, living on fish; others are free-living. Those that are free-living are only millimeters long (0.5 to about 10 mm, or less than 0.5 inches). They have varied body shapes but are recognizable as arthropods, although there is no distinct head-plus-thorax or -cephalothorax body plan. Most copepods have distinct and elaborate antennae. Larval copepods undergo several growth stages called instars. Their appearance changes drastically from one instar (intermediate juvenile form) to the next.

Calanoid copepods, which are representative of the free-living suborders, occur in both fresh and salt water.

Copepods are an important link in many food chains; they feed on smaller microorganisms and are in turn food for larger animals. Some are herbivores that eat algae, others are carnivores that eat other zooplankton. They are filter feeders that screen large quantities of water for food, although many species capture individual food particles. They expend great energy in propelling themselves through water and orient themselves by the gravitational field of the Earth, as do decapods such as shrimp. The other suborders of copepods present in the greatest numbers of both individuals and species are the Cyclopoids, planktonic organisms, and Harpacticoids—long and slim bottom dwellers. The suborders differ in the shape of the thorax and in their flexibility. See *Arthropoda, ecosystem, filter feeders, food chains.*

Copernicus, Nicolaus (1473–1543)   A Polish astronomer who displaced the Earth from the center of man's conceptual universe and made it orbit a stationary sun. Niklas Koppernigk was the original name of this churchman who specialized in canon law but was also a skillful astronomer who originally set about to refine some of the observations in Ptolemy's astronomical work, the *Almagest.* In his search for perfect motion he rejected the Ptolemaic geocentric cosmos, which nevertheless explained astronomical phenomena and predicted events such as eclipses. Copernicus' rejection of this model was based on the ideal of perfect motion, because he believed— and this was central to his work—that uneven motion was not possible for heavenly bodies.

Copernicus knew of the ancient Greek philosophical view of the cosmos that placed the Sun at the center of the universe. Using that concept as an exercise in philosophy, he proposed it in his work, *De Revolutionibus Orbium Coelestium*

(The Revolution of Celestial Spheres), which did not appear in print until the year of his death. This book profoundly rocked astronomy. Although its practical effect on navigation and on the sciences that became oceanography were in the short run minimal, in the long view the "new astronomy" changed everything. See *Galilei, Galileo; Ptolemy.*

copper   A metallic element, and a micronutrient necessary for many enzyme reactions in living systems. Copper is the central metal atom in the blood of some crustaceans, where it is the analog of the iron atom in the hemoglobin of mammalian blood or the magnesium atom in chlorophyll.

Free copper in excess is toxic to many marine organisms, including those in which it is part of enzyme systems. The giant kelp is an example of a plant that requires copper in some quantities and is poisoned by an excess of this element. See *chlorophyll, Crustacea, kelp.*

coral and coral reefs   Corals are cnidarians of the class Anthozoa; they are polyps, predominantly with six-fold or eight-fold symmetry, and are either solitary or colonial.

The coral line coelenterates construct calcium-containing shells of characteristic shapes. Colonies of coral and their associated foraminifera and symbiotic algae live best in tropical and semitropical environments. The "coral reef belt" is within the boundaries of 30° North Latitude and 30° South Latitude. Some slow-growing, non-reef-building coral organisms are found in colder waters. Maximum growth of corals occurs at depths less than 50 m (160 feet), and is optimum at a level of about 7 to 20 m (20 to 60 feet) down.

The colonial organisms utilize the skeletons of dead coral, undersea boulders, the continental shelf, and submerged peaks to build coral reefs.

There exist three main geomorphological categories of reef: fringing reefs, barrier reefs, and atolls. Fringing reefs are formed close to shore on rocky coastlines. Barrier reefs are separated from land by lagoons or channels produced as a result of subsidence. Atolls are found around subsiding volcanoes.

The location of barrier and fringing reefs on the eastern edges of continents is the result of both current flow and terrestrial spin. Warm surface currents move close to the eastern side of landforms that lie away from the equator. This maximizes the growth of algae. Deep polar upwellings move toward the equator along the western edges of continents.

The symbiotic relationships between coral and photosynthesizing algae and coralline algae, which also form hard structures, are part of a complex ecosystem. In addition to these organisms there are others that live on a coral reef, using it as support, shelter, or hunting ground. Its collection of fish, echinoderms, mollusks, algae, and arthropods constitutes a large ecosystem unique to each reef.

Fish present in and around coral reefs feed on sea grasses and planktonic organisms. In turn they excrete nitrogen-

Reef Types

	Fringing	Barrier	Atoll
locale	Continental shelf	Ocean shallows	Oceanic rise
shape	Long bar, parallel to continent	Multiple bars separated by deep channels	Circular, surrounding a lagoon
growth pattern	Outward from the continent to the edge of the shelf	Rapid in shallows, shifts with opening of new channels by wave action	Enlargement and eventual rise above sea level to form islands
example	Florida Keys	Great Barrier (Australia)	Bikini

Representative Corals

Class, Anthozoa	Scientific Name (representative)	Common Name
Order: Coenothecalia	Heliopora	Blue coral
Alcyonacea	Alcyoncean	Soft coral
Gorgonacea	Gorgonian, Corralium	Sea fan
Rugosa	Petraia	Compound coral
Scleractinia	Fungia, Astrangia	Stony or true coral
Tabulata	Favositea	Honeycomb coral

and phosphorus-containing materials that are recycled and used by the coral and other organisms living in and on the reef.

Reefs have characteristic populations that are part of the unique nature of each reef. There is usually a single species that controls the population of "its" reef. These determining species are called keystone species. They are predators and prevent the possible intrusion of another species. Damselfish, for example, eat algae, exposing coral, but in turn aggressively maintain territory and keep other predaceous populations, such as parrotfish, in check. This insures their use of the algae population of a given reef. The resulting oversupply of algae feeds other organisms.

The coral organisms are most vulnerable to wave action and storm surges, which destroy the platforms that support them. There are also organisms that prey on live coral, such as the crown-of-thorns starfish and coral-eating parrot fish.

The reproduction of corals is usually asexual or gamete-producing in the typical mode of the phylum. There are, however, highly specialized variations. Acropora is an organism that produces gametes which are externally fertilized at brief periods during the year. Spawning is triggered seasonally by the sea temperature, and by specific dark periods. In other species the majority of the gametes are produced in a one or two day period. Fertilization is not random, since many species produce gametes at the same time. Eggs are both attractants for compatible sperm and toxic to the sperm of other species. The reproductive advantage of this mass spawning is that fertil-

ization by gametes from another colony of the same species is possible.

The soft corals, like the gorgonians, are in the subclass Octocorralia; they are eight-sided. The true or stony corals are in the subclass Hexacorralia; they have a six-sided structure. The gorgonians are most prevalent in the Atlantic and Pacific Oceans; the soft-bodied corals are found most often in the Indian and East Pacific oceans. See individual corals, island, reef, Zooxanthellae.

Coral Sea   An area of the Pacific Ocean bounded by Australia's Great Barrier Reef to the west, Papua to the north, and the New Hebrides and New Caledonian island groups to the East. The coral of the Great Barrier Reef gives the sea its name. There are two other major coral reefs in this sea. It is also characterized by trenches and rises that indicate a relationship to Queensland, the northeastern part of Australia. A considerable portion of the Coral Sea plateau may be thought of as related to Tasmanian formations which once were terrestrial.

The Coral Sea was the site of a definitive naval battle in May 1942, which marked the limit of Japanese naval movement and prevented an invasion of Australia. See coral, Great Barrier Reef, Pacific Ocean, plate tectonics, trench.

coralline algae   Red algae encased in transparent calcareous (calcium containing) shells. Coralline algae are colonial, and like coral polyps build up hard structures that become reefs. In 1984 a coralline alga, a photosynthesizing plant, was found at a depth of 280 m (884 feet) in the Bahamas, a depth at which photosynthesis had previously been assumed impossible. See *coral, photosynthesis, polyp.*

Coriolis force   A concept used to describe the motion of an object in a rotating system. In geophysics the term is used to indicate lateral displacement in what would otherwise be a north-south migration of air or water masses. For example, air or water moving from north to south over the surface of the Earth would be seen as moving in a straight line by an observer on the Earth. However, to an observer in space, the path of the moving air or water would seem to be curved because the Earth rotates. Looking in the direction of motion of the air (wind) or water, they would be seen to veer off to the east, or to the right, in the Northern Hemisphere, and to the west, or left, in the Southern Hemisphere.

In addition, the fluid mass of the air or water is accelerating. The cause of this acceleration is the rotation of the Earth. As the entire Earth moves about its axis eastward, the fluids on it are pushed along at an increasing rate. This effect is called Coriolis acceleration. See *wind.*

cormorant   An aquatic, duck-like bird, 50 to 100 cm (20 to 40 inches) long belonging to the same order as pelicans. Cormorants have long necks and slender bills, and are usually black. The 26 species of cormorants both swim and fly, with the exception of the flightless Galapagos variety. Cormorants have a worldwide distribution, and usually feed on fish, with some crustaceans and mollusks for variety. See *duck, pelican.*

Cortes,   Hernan   (1485–1547)
Spanish explorer-adventurer, conqueror of Mexico. At the beginning of his career he went to Cuba to make his fortune. The governor of Cuba sent Cortes to the Yucatan Peninsula of Mexico in 1519, but then reversed this order almost immediately. However, Cortes proceeded with the conquest of Mexico City and its

surrounding areas and later claimed not to have received notice of the change.

Cortes made several forays into other parts of Central America, going overland into what is now Honduras. He crossed Mexico and had two primitive ships built on the Pacific side. He used these to explore the region around Baja California. He was the first European to see California (1536). The voyage, looking for spectacular treasure, the fabled lands of the Amazons, or the Strait of Anian (Northwest Passage), was unsuccessful. The politics of New Spain made it unsafe for him to continue his exploration. He returned to Spain in 1540.

cotidal points   Those points along a coast that exhibit high tide at the same time. See *tide.*

Cousteau,   Jacques-Yves   (1910–)
A naval officer, undersea diver, and marine explorer who, together with Emile Gagnon, developed SCUBA (self-contained underwater breathing apparatus) diving in 1943. After World War II, Cousteau and others created the Undersea Research Group at Toulon, France, which became a center for research in undersea study and techniques.

Through his books and films, Cousteau has promoted interest in life in the sea and the great need for careful use of the marine environment. His works have repeatedly stressed the need for intelligent use of the seas, control of pollution and overfishing, the use of mariculture, and the preservation of coastal areas. He is the director of the oceanographic institution founded by Albert I of Monaco. See *diving, mariculture, SCUBA.*

cowrie   A marine snail of the family Cypraeidae. There are about 300 species of these beautiful animals. They are usually found in warm seas; the greatest number of species are from the eastern Pacific. Cowries have been traded extensively for many years; their fairly heavy, glossy shells have been used as decoration and coinage, including African decorations that have used Asiatic cowries. See *Gastropoda*.

crab A crustacean with a broad, flat body, a tiny abdomen usually tucked under the cephalothorax, and five pairs of walking legs. The crabs range in size from 1.5 mm (0.1 inch) to 3.5 m (12 feet) in diameter. True crabs, of which there are about 4,500 species, are distributed worldwide and have a range from mudflats to abyssal depths of 3,500 m or 12,000 feet. Some freshwater and terrestrial forms also exist, although most species are marine. They feed on a range of organisms from plankton to live prey or detritus.

True crabs have up to 19 pairs of appendages, such as antennae, antennules, mouth parts, gill ventilators (mandibles and maxillipeds), walking legs (pereopods and chelipeds), and swimmerets (pleopods) which are copulatory organs on the male and egg-carrying devices on the female. In the true crab only, the skeletal plate above the mouth is fused to the chitinous shell, or carapace.

The classification of crabs is subject to differing opinions. One scheme is as follows:

Phylum: Arthropoda
Class: Crustacea
Subclass: Malacostraca
Order: Decapoda
Suborder: Repantia
    Macrura—crayfish and lobsters
    Anomura—hermit crabs
    Brachyura—true crabs

The true crabs are then subdivided into the following groups;

1. Gymnopleura, primitive crabs of which there are about 30 species. They have long bodies and undeveloped chelipeds.
2. Dromiacea, also primitive but numbering about 250 species. On these crabs the fifth pair of walking legs is on the back. These legs hold a piece of camouflage, such as a leaf, clump of algae, or sponge.
3. Oxystomata, burrowing crabs consti-

tuting about 500 species. Some of these crabs also hold onto pieces of camouflage.
4. Brachygnatha, comprising about 80% of all crabs, and numbering about 3,600 species. These crabs have a square mouthframe and have lost the first pair of swimmerets. This group is further divided into the (a) Oxyrhyncha, or Spider crabs and other quick-moving crabs with hard shells, which have a triangular body shape, narrow head, and a body covered with spines and some camouflage; and the (b) Brachyrhyncha, the best known group of crabs, which range from pea crabs that live in the mantles of oysters and clams, feeding on whatever the oyster dropped, to the commercially important edible crabs, stone crab, blue crab, and Dungeness crab. The tropical, terrestrial Gecarcinidae return to the sea only to breed, the Jamaican tree crab and the Japanese spider crab are part of this large group.

See *Arthropoda, cephalothorax, chela, Decapoda, individual names.*

crabeater seal *Lobodon carinophagus,* is a misnamed antarctic animal. This seal feeds on krill, not crabs, and has teeth that act collectively to filter these tiny animals from the seawater. See *krill, seal.*

Cretaceous Period A period in the history of life on Earth, the Cretaceous began about 136 million years ago and lasted for about 70 million years. It is the most recent period of the Mesozoic Era.

Geologically and geographically, the Cretaceous was a very eventful segment in the Earth's history. The southern continent, Gondwanaland, was still essentially intact, but the Indian subcontinent had detached and started drifting northwards. By the mid-Cretaceous, South America had moved away from Africa. North America began to move west, away from Europe, opening the Atlantic, while Africa moved north toward Europe, closing the Tethys Sea and beginning the folding that resulted in the Alps.

The folding of the North American continent that eventually produced the Rocky Mountains also began in the Cretaceous. By the end of the Cretaceous, New Zealand had separated from Australia and Australia itself was moving away from Antarctica.

The climate during most of the Cretaceous was warm. Shallow seas extended over large areas of all continents. The dominant sea animals were the ammonites. Belemnites, brachiopods, echinoderms, and mollusks were widely dispersed. Seagoing reptiles such as the ichthyosaur and the plesiosaur were alive. On land, the dinosaurs reached their zenith during this period, and by its end were extinct. The end of the Cretaceous was marked by the extinction of large numbers of species. See *extinction*.

Crinoidea   A class of echinoderms especially abundant in the western Pacific Ocean. Once thought to be extinct, this class of sea creatures includes the stalked, sessile (stationary) sea lily and the mobile feather stars. In fact, the sea lily comprises about 80% of the 600 living crinoid species. The Mesozoic Era was the highpoint in crinoid development. The greatest number of species and individuals lived then.

Crinoids are covered by unique exterior calcareous plates held together by collagen ligaments. Broken body parts regenerate. These animals are filter feeders generally left alone by predators. See *Echinodermata*.

Cromwell Current   An equatorial, subsurface flow that appears in three major oceans. The Cromwell Current is the equatorial eastward-flowing undercurrent embedded in the westward moving surface current. It is about 300 m (900 feet) wide and about 20 m (70 feet) deep, and moves along at a rate of 2 to 3 knots. See *currents*.

Crossopterygii   An ancient subclass of the bony fishes. These animals are related to lungfish and are sometimes called lobefish. The coelacanth is the only survivor. The Crossopterygii are important because they were the probable progenitors of amphibia. The dorsal fins have fleshy pedicels from which the fin rays emerge. These may have been the precursors of limbs. See *coelacanth*.

crown of thorns   A large starfish, *(Acanthaster planci)* averaging 40 to 45 cm (16 to 18 inches) in diameter, with 12 to 20 arms and covered with red, thick, strong spines. This starfish feeds on living coral. A large increase in the crown of thorns population in the 1960s in Australia resulted in the destruction of large areas of coral on the Great Barrier Reef. One possible reason for the population boom may have been the removal of many triton snails by shell collectors. More normal reef conditions have brought the starfish population into better control. The crown of thorns overpopulation had also affected other areas in the Pacific where the coral is now recovering. See *coral, Echinodermata, starfish*.

crust   The topmost layer of the Earth. The ocean floor is basaltic; the continental crust, which is much thicker than that of the ocean, is largely granitic. See *discontinuity layer, Moho, ocean floor*.

Crinoid forms

Crustacea A world-wide class of arthropods that includes crabs, shrimps, lobsters, and barnacles. There are eight subclasses of the Crustacea and over 25,000 species. They are present in every type of marine biome and are the aquatic analogs of insects. Both fill their respective ecological niches with incredible numbers of species and numbers of individuals. Most crustaceans are aquatic and breathe through brachioles leading from gills. They have five pairs of cephalic appendages: antennae or antennules (or both), mandibles, and maxillae or maxillules or both. The body is divided into a head, thorax and abdomen, and is covered by a chitinous exoskeleton. The last body segment (somite) is the telson or tail, which lacks any appendages. The crustaceans have the typical arthropod circulatory system, with an open heart.

The excretory system consists of a primitive network leading to a pair of glands under the antennules. There is one simple median eye and two compound lateral eyes. Generally, the antennae are more effective in testing the environment than are the eyes. Most crustaceans are dieceous (having distinct male and female forms). Fertilization of the eggs is internal. The female is the usual brooder, carrying the eggs about on her abdominal appendages, which affords some protection for the eggs until they hatch into a larval form. The larvae are called nauplia (singular, nauplius). Successive molts and growth spurts produce the adult. The rate of growth varies with different species, and depends on the availability of food and the water temperature.

While the larvae of most crustaceans are pelagic filter feeders, the adults may develop into an astonishing variety of diggers, borers, or raptors while some remain filter feeders. The adult forms include, in addition to the usual motile animals, parasites (copepods) and sessile (stationary) forms (barnacles). Most crustaceans are grazing and scavenging animals, although the larger, more advanced decapods (lobsters, crabs, shrimp) and stomatopods are more likely to be predators.

The crustacea and other arthropods seem to share an ancestor with annelids. They were certainly present and well-differentiated in the Cambrian Period. Shells from this and other periods are well-preserved, and the resulting fossil record is quite complete. See *Annelida, Anthropoda, crab, chitin, lobster, shrimp, specific crustaceans.*

Ctenophora A phylum of marine plankton abundant on the surface and also found in deep water. About 90 species have been described. Ctenophora are macroplanktonic, voracious, carnivorous eaters, using their radial arms, which have coleblasts (adhesive cells), to trap food.

Ctenophora have biradial symmetry and a medusa-like three-layer body, with the exterior and interior composed of epithelial cells. The intermediate layer, the mesoglea, is gelatinous.

Ctenophores are divided into two classes and five orders, based on their body shapes. They can be rather large: the girdle of Venus, a ribbon-like organism of transparent blue and green, is over 1 m (3 feet) long. See *medusa, plankton.*

Cubozoa A class of the Cnidaria. These animals are in intermediate form that falls between the Hydrozoa and the Scyphozoa. The eggs hatch into four-tentacled polyps that grow fairly rapidly into the adult form. When adult, they are a four-tentacled meduse with a "cube-shaped" body. The Cubozoa are among the most toxic of marine organisms. See *Anthozoa, Cnidaria, Scyphozoa, toxins.*

currents Parts of a fluid body (air or water) moving continuously in a certain direction. In oceans, currents are responsible for the circulation of vast quantities of waters. They are caused by several factors. The primary one is wind, which creates a friction effect by pushing the mass of water. Another factor is temperature, because cold water, like cold air, flows toward the equator and also sinks, since (like cold air) it is denser than when warm.

Local currents are produced by the activity of volcanoes, a sporadic phenomenon; by river systems, whose effect is permanent; and by tidal currents. The last are periodic. In general, oceanic circulation is horizontal. The movement of water is anticyclonic (clockwise) in the Northern Hemisphere. Water moves eastward at higher latitudes (i.e., the Gulf Stream) and returns (moves west) closer to the equator. The east-to-west movement in the Southern Hemisphere is cyclonic (counterclockwise). This gyre is typical of the Atlantic and Pacific oceans. It reverses itself semiannually in the Indian Ocean due to the monsoon winds. The antarctic region has a single current: the Antarctic Circumpolar Current. On the whole, there is less water transported by gyre in the Southern Hemisphere compared to the Northern one.

The speed of currents in open ocean is relatively slow, less than 1 knot (0.5 m/second) as opposed to the Gulf Stream at 2 to 4 knots or 1 to 2 m/second. The Pacific counterpart of the Gulf Stream is the Kuroshio. This current moves northeast out of the South China Sea to warm the Japanese islands and then moves east to the Aleutians. It, too, moves rapidly in comparison to the speed of the average ocean gyre.

In addition to surface currents, subsurface equatorial currents such as the Cromwell are known. Deep water circulation is slower than surface currents. It is the result of polar water sinking, a phenomenon observed in the Norwegian and Weddell Seas. The cold deep-water mass then moves towards the equator.

While this is the overall ocean pattern, it is very much influenced by conditions in particular locales. The conditions of tide, temperature of the abutting land mass, and the size of the land formation all have profound influences on deepwater currents. The arrangement of continents in the northern hemisphere makes it easier to see the circulatory patterns in the ocean—the gyres. See *Antarctic Current, Ekman Spiral, Equatorial Current, gyre, wind, individual currents.*

cutter A small ship built for speed. Historically the cutter was a deep, narrow sailing vessel carrying a single mast rigged fore-and-aft, and had a long bowsprit. Today the U.S. Coast Guard uses dieselpowered steel-hulled cutters that are 25 m (83 feet) or less long.

cuttlefish A cephalopod of the order Sepioidea, related to the octopus and squid. Its origins are in the Miocene. Cuttlefish comprise about 100 species of warm-water bottom dwellers. The cuttlebone is really a shell that has become internalized. It acts as a flexible support rod. Like all other mollusks, cuttlefish produce a calcareous shell by secretions from their mantle.

The various species of cuttlefish range in size from 2.5 to 9 cm (1 to 36 inches) long. They have eight arms, two longer tentacles, and a pair of lateral fringing fins. They eat small fish, crustaceans, and their own young, and are in turn fed upon by large fish and mammals. See *Mollusca, octopus.*

Cuvier, Georges-Leopold Dagobert, baron de (1769–1832) The French founder of the science of paleontology. Cuvier's interest in natural science was evident early in his life; he studied comparative anatomy at Stuttgart, graduating in 1788. For the next seven years he taught privately and studied the Mollusca. This work led to a post at the Museum of Natural History in Paris. He remained at that post for the rest of his long career.

Cuvier's early published works were popular successes. The *Tableaux Elementaire de l'Histoire Naturelle des Animaux* (1797) was followed in 1800–1805 by *Leçons d'Anatomie Comparée.* In this work Cuvier correlated the structures of a body and showed how they were related and interdependent, concluding that the body plan of an animal was determined by the particular ecological niche it inhabited.

Cuvier created four categories of animals: vertebrates, articulates, mollusks, and radiates. Each, according to him, had distinctive features that were common within the group, and different from the distinctive features of the other groups. He devised a table of organization, placing all known animals, living and extinct, on it. This work was a major improvement over existing attempts at taxonomic charting.

In explaining the origins of animals, Cuvier was a catastrophist, holding that species were all created in the form in which he knew them. Thus, fossils were remains of species that had been wiped out in some catastrophic event in the past.

While Cuvier's theory would not have been able to compete with evolutionary theory as an explanation of interrelationships between organisms, his work on comparative anatomy is still basic to that science. His attention to minute features of the structure of an unknown bone could lead him to describe the organism it came from and the habitat it lived in.

In addition to his scientific work and publications, Cuvier was instrumental in establishing universities in the French provinces. He also achieved a general standardizing and improving of the level of public instruction. See *Buffon, Georges-Louis Le Clerc, comte de; evolution.*

Cyclades   A group of more than fifty islands in the Greek archipelago. They are of volcanic origin and have quite different aspects. The probable cause of the tsunami, or tidal wave, that shook the ancient world—and inundated most of it—was a volcanic eruption on Thera (Santorin), an island in this group. Other islands are Siros, Delos, Naxos, and Paros (where Parian marble comes from).

The beginnings of long-distance sailing in the Mediterranean probably developed here in the 3rd century B.C. See *Aegean, Atlantis, islands, Thera, tsunami.*

Cyclostomata   A class of jawless chordates, the lampreys and hagfish. The other agnathans or jawless fish are all extinct. See *Agnatha, conodont, hagfish, lamprey.*

cypris   The larval form of the barnacle. It is freely moving. See *barnacle.*

D

Dampier, William (1652 1715) An English buccaneer, navigator, and surveyor, Dampier sailed to Newfoundland and then to Java on commercial vessels before serving the English crown in the Dutch Wars. He joined the pirates of Campeche (Mexico), and with them sailed to the South Pacific in 1683. It was an adventurous trip which took him to China, the New Hebrides, the Australian mainland, and New Guinea. Dampier was shipwrecked and skillfully navigated a canoe to Sumatra, finally returning to England in 1692. His account of this voyage was published in 1697.

After the publication of his memoir of the voyage, *A New Voyage Round the World,* the Admiralty sent Dampier out as captain of the *Roebuck,* with the charge of exploring New Holland (Australia). The unsuccessful voyage included a mutiny and the loss of the ship. Dampier, who seemed to have a charmed life, limped back to England and embarked on yet another unsuccessful voyage in 1703. The real objective of this expedition was piracy. It was on this trip that Alexander Selkirk, the real-life Robinson Crusoe, was put on an uninhabited island. Dampier's last voyage in 1711 was a financial success, and also brought Selkirk back to England. See *Juan Fernandez Islands.*

Daphnia A freshwater crustacean (Cladocera) also called a water flea.

Darwin, Charles Robert (1809–1882) An English naturalist who first formulated the theory of evolution by natural selection. He also produced major works on barnacles, orchids, and climbing plants. Darwin came from a distinguished family: his grandfather, Erasmus Darwin, was a physician and bota-

nist; his father was a physician; and his mother was a daughter of Josiah Wedgwood, the potter, chemist, and ceramicist. After abandoning a medical education, Darwin intended to be a clergyman when J.S. Henslow, a prominent natural scientist, suggested him for the post of accompanying naturalist on the H.M.S. *Beagle,* a position for which he was accepted by the captain, Robert Fitzroy. Darwin's voyage on the *Beagle* turned out to be one of the most fruitful "field trips" ever undertaken in terms of data acquired and the far-reaching conclusions that could be drawn.

Darwin's later work was based largely on observations made during the long voyage on the *Beagle,* 1831–1836. He also proposed a theory of coral reef and island formation which is in large part accepted today. He believed that the coral atoll began as a fringing reef surrounding a high island. The reef accreted and the island eventually subsided. While the sequence may not follow Darwinian mechanics, both the processes of accretion and subsidence do occur. Darwin did not know of tectonic plates and ridge-building mechanisms.

The Darwinian explanation of the differentiation of species, the evolution of current forms from earlier ones, and the process of change was based on his studies of fossils in South America, isolated bird communities on the Galapagos Islands, and careful observation of organisms in tide pools. His work, *On the Origin of Species,* finally appeared in 1859, at the urging of friends. See *H.M.S. Beagle, coral, Eniwetok, evolution, fossil, islands, Wallace.*

dating The placement of a specimen in time. The dating of marine specimens relies on the differences in the ratio of the concentration of radioactive to stable isotopes of certain elements in surface samples and core samples. This, in turn, is based on several assumptions about the constancy of radionuclide decay and its inclusion into the living systems of ocean waters. One assumes that a radioactive nuclear sample will decay (break down into smaller nuclei and emit alpha, beta, and gamma rays) at a constant rate. This is a "clock." One also assumes that if a certain element is incorporated into a living plant or animal, that too happens at a constant rate. The latter is subject to much more change. Plant and/or animal populations move. They also mutate. Both of these factors can alter the uptake of a particular element considerably.

The comparison of various ocean waters and the loads of radioactive species of different elements in these waters is a clue to the movement of water from one oceanic area to another, and to the rate of water turnover in any particular locale. See *radioactivity.*

Davidson Current   A deep countercurrent (200 m or 650 feet deep) in the temperate Pacific. The current runs north, parallel to the California coast. It appears on the ocean surface north of 35° North Latitude in winter. See *bottom waters, California Current, currents.*

Davis, John (1543–1605)   An English pilot, master mariner, and arctic explorer. Davis' voyages contributed considerably to the knowledge of the arctic.

His first voyage to the far north began in June 1585, with the stated purpose of finding the Northwest Passage. Instead, he charted many features of the arctic, including the strait that is named for him. He returned to England, and made two similar voyages in the two succeeding years.

Davis took part in the fight against the Spanish Armada in 1588. His next exploratory voyage was with Cavendish in 1591. After Cavendish turned back to Europe, Davis continued on into polar (southern) waters, and was also driven back by weather. On his return trip to England, he discovered the Falkland Islands.

After his return, Davis found employment in the Far East, where he sailed for both Dutch or English commercial venturers. He was killed by pirates near Sumatra.

Davis was a superb pilot and a skilled navigator. His permanent contribution to seamanship included two books on navigation and development of the backstaff. The French called this navigational aid the English quadrant, the English called it the Davis quadrant. With it one could sight the Sun and measure elevation above the horizon, and therefore calculate latitude. See *Davis quadrant, navigation, sextant.*

Davis quadrant   Also called the backstaff. It is an instrument developed in the 16th century by John Davis as a means for measuring the altitude of the Sun. There were two versions: one measured altitudes of angles less than 45°, the other angles greater than 45°. The latter is the simpler instrument. It consists of a horizontal vane and a graduated staff held horizontally. With the viewer's back to the Sun, the shadow of the horizon vane was aligned with the graduations on the staff.

The 90° quadrant, which was a later development, consists of two half-transoms, one straight, the other an arc. The straight part is held vertical and perpendicular to the staff, and slides along the staff. The arc is fixed to the lower end of the staff; it, too, functions as a horizon vane. As in the simpler instrument, the shadow of the horizon vane is compared to the graduations on the vertical staff. See *astrolabe; Davis, John; latitude; sextant.*

Davis Strait   Part of the fabled Northwest Passage; the strait is the boundary between Greenland and Baffin Island. The strait is navigable for about one month after the summer solstice. See *Davis, John; explorers and explorations; Northwest Passage.*

Dead Sea A landlocked lake on boundary between Israel and Jordan, part of the African Rift Valley, once a part of the Red Sea but cut off from the Red Sea by steep, barren hills. As a result of its high rate of evaporation, the water in this lake is the densest on Earth. Minerals are extracted commercially on the shore of the Dead Sea. See *African plate, evaporates, minerals, Rift Valley.*

Dease Strait A strait in the Canadian Northwest Territories, lying northwest of Hudson's Bay and forming part of the "southern route" of the passage from the Atlantic to the Pacific Ocean. While the passage is possible, it is not practical or commercially viable, since the route is icebound for most of the year. See *Northwest Passage.*

Decapoda The best known order of the Crustacea, comprising the crustaceans most often consumed by humans: the lobsters, crabs, and shrimp. The decapoda have a fused cephalothorax covered by a chitinous carapace, and they have five pairs of legs. The arrangement of the appendages into swimming and walkings legs varies with the individual species. See *crab, lobster, shrimp.*

decomposers Some multicellular animals, fungi, and bacteria that utilize the tissues of dead animals and plants for their own metabolic needs. They eventually break down complex proteins into simple materials that are essential for plant growth. See *ecosystem.*

decompression A term that refers to the controlled or uncontrolled release from pressure that a diver experiences during and after rising to the surface from ocean depths. If an organism is raised from a depth, the dissolved compressed gases in its tissues will rapidly come to pressure equilibrium with the atmosphere and form bubbles as the gases expand. Thus, if a diver rises rapidly, the gas dissolved in his body will accommodate to the ambient pressure, and bubbles will form in his blood or whatever organ contains the

gas. Symptoms of decompression disease, known as "the bends," arise hours after decompression and may be fatal.

By 1907, the British physiologist J.S. Haldane had noted that slow rising in stages will eliminate the bubble formation and thus eliminate decompression sickness.

Most organisms would appear to be subject to decompression distress. The hadal (abyssal) bacteria will die at surface pressure. Whales and dolphins both exhibit deep-diving and surfacing behavior, but for reasons that are not well understood, do not develop decompression sickness. It is theorized that they may have no "nuclei" around which gas bubbles might form, as well as a circulatory system that operates so smoothly that no eddies form in the blood which might give rise to gas bubbles.

An alternative mode of decompression, one that would enable a diver to rise rapidly if this should become necessary, involves the use of a hyperbaric (greater than normal pressure) chamber. If a diver, immediately after ascending, is put into a pressurized environment in which the pressure is gradually dropped to the normal 1 atm (atmosphere), there is no decompression distress. See *diving.*

deep-scattering layers Ocean layers at depths greater than 200 m (660 feet) that produce a sonic effect detectable by echo-sounding equipment. There may be one or more deep scattering layers at any given location, and these layers move up in daylight or bright moonlight. They then descend at night or on heavily overcast days. The movement corresponds to that of schools of small fish, cnidaria (jellyfish), and possibly squid and larger crustaceans. The sonic difference is attributed to the echo produced by gas bubbles within the animals. See *plankton, sounding, vertical migration.*

deep sea A sea zone at depths below approximately 3,000 m (6600 feet), or the 4°C isotherm. The sea zone at depths below 6,000 meters (19,800 feet) is hadal.

The depth of these deep sea zones is set by international agreement. However, there is usually no plant life at depths where there is no light. The bacteria, protists, and animals in deep seas are adapted to both pressure and darkness. They represent all major invertebrate phyla and many fish. The deeper an organism's environment, the greater the likelihood that it is a detritus feeder, living on pieces of dead plants and animals, or a deposit feeder—a mudeater. Some species that are carnivores near the surface may have deposit-feeding hadal relatives; for example, mud-eating sea stars live below 4,000 m. See *abyssal environment, ocean floor, seamount.*

defense mechanisms Means by which organisms defend themselves against predators. Some organisms use camouflage or move away rapidly, some "clam up," and others, particularly slow or non-motile organisms, use chemical defenses. For example, some brown algae use chemical defenses to make them unpalatable to predators, several species of sole produce shark repellants, and some sessile (stationary) organisms produce toxic substances. See *chemical defenses.*

Delaware Bay A bay of the North Atlantic at 39° North Latitude, separating the state of New Jersey from the eastern shore of the Delmarva peninsula (Delaware/Maryland/Virginia). The Delaware River is the major source of terrestrial water to Delaware Bay. The Bay, in spite of its location as the water-catchment basin for water draining past the heavily industrialized areas of Marcus Hook, Pennsylvania, and Wilmington, Delaware, has a wetland and swamp region surrounding it. These wild areas serve as breeding grounds for many species of plant and estuarine animal life. See *Chesapeake Bay, coast, estuary.*

delta A fan shaped plane at the mouth of a river, created by sediment deposited by the river. In general, the foundation of a delta depends on the rate of sedimentation and the currents of the sea at the point at which the river discharges its deposit. Thus, a strong current can prevent the formation of a delta or cause the erosion of an existing one. The movement of sediment in a delta and its rearrangement by local currents causes continuous change in the local coastline. See *breakwater, coast, fan, sediment.*

Demerara Abyssal Plain A part of the Guyana Basin off the northeastern coast of South America. One side of this plain is the abyssal cone of accumulated sediment brought downriver by the Amazon. The plain is also fed by the waters of the Orinoco and other river systems of northeastern South America. See *Amazon River, Atlantic Ocean, Orinoco River.*

Demospongia A class of the phylum Porifera. This group of sponges is a fairly large one, with about 9,500 species. They are classified by the type of spicule (support structure) they have with some having silicate spicules and others having spicules made of spongin, a rigid protein, while still others have no spicules. See *Porifera, spicule, sponge.*

density The mass of a substance per unit volume. The density of seawater is a function of salinity, temperature, and pressure. The vertical change in density is greater at lower latitudes, (i.e. near the equator), since the temperature of the water varies so much with depth. In colder regions the density of a vertical column of seawater is fairly constant. An increased salt content increases the density of seawater. Thus, the introduction of large quantities of river water or glacial water (icebergs) will decrease the density of the water locally. In areas where the temperature is high and humidity low there is considerable water loss due to evaporation. This increases the density of the water locally. The locale in which this occurs can be quite large—the Mediterranean Sea is one such high-density, high-salt area.

Increased pressure will also increase the density of seawater. However, liquids are not very compressible, and the increase in density due to pressure is therefore small and negligible compared to that caused by salt and temperature effects.

deposits   See *minerals, sediments.*

Deryugin, Konstantin Mikhailovitch (1878–1938)   An outstanding Russian oceanographer and expert on the oceanography of the White Sea. In 1899, while still a student, he made his first trip to the White Sea. He took his doctorate at the University in St. Petersburg (now Leningrad), and remained there as a professor for the rest of his career. Starting in 1920, he was also the manager of the Oceanic Division of the State Hydrological Institute in Leningrad.

Deryugin's first major scientific reputation rested on his work on the classification of aquatic arctic biota. He went on to study the entire ecosystem of the Barents Sea: its biota, shoreline, bottom, geology, hydrology, and chemistry. In 1921 he reestablished the water-sampling station on the Kola meridian, at 33°30′ East Latitude, 75° North Latitude. This station sampled the North Cape Current, whose direction, temperature, and water volume have a profound effect on the quality and quantity of fish in the Barents Sea.

By 1922, Deryugin transferred his scientific interests to the White Sea. This body of water is unique both hydrologically and biologically. The principal reason for its singular composition is the Gorlo Strait, which lies between the White and Barents seas and acts as a biological barrier. The strait does not have a high sill: the barrier is instead tidal.

Deryugin was in charge of the organization of the Pacific Ocean Expedition of 1932–1933. The explorations of the Sea of Japan, the Bering, and the Chukchi Sea brought a "new world of organisms" (Deryugin's comment) to scientific scrutiny.

This able organizer was also instrumental in the direction of more than fifty expeditions, and was a pioneer in the methodology of oceanography. See *Bar-*

ents Sea, Bering Sea, Chukchi Sea, White Sea.

desalination   The process of removing salt and other dissolved minerals from seawater to make it useful for agriculture, industry, and human use. This is an energy-intensive procedure and therefore an expensive one.

The process used in desalination is either distillation (controlled boiling) or reversed osmosis. In distillation, the seawater is boiled and the resulting steam is forced to leave the boiler through a conduit, which is cooled by surrounding air or by water. The dissolved or suspended minerals remain behind in the boiler, while the cooled, condensed water is collected at the delivery end of the conduit as potable water.

A simple distillation apparatus, which is both affordable and portable, uses solar energy to raise the temperature of the water in a plastic boiler. The use of plastic cuts down on maintenance costs, which are considerable if metal is used, since the dissolved materials in seawater are frequently corrosive. The water does not necessarily have to boil as long as the temperature is high, since evaporation will still occur and the distillation apparatus will still work, albeit more slowly. In regions of high insolation there is sufficient light and heat to power a still. The removal of the mineral residue of seawater and its separation is another industrial process. It is also possible to freeze out fresh water, which freezes at a higher temperature than salt water. This is less successful than the evaporation process.

In the process of reversed osmosis, saline water is forced through a membrane while under pressure. This concentrates the minerals and releases potable water. The membrane simulates a cell membrane, which ordinarily allows water flow from regions of high salt and other solute concentration to regions of lower concentration. Commercial membranes are made of cellulose acetate or nylon. Pressure applied to the flow of water pushes desalinated water out of the membrane container. This process may also be solar-powered, and used for collecting minerals.

The electrolytic separation of water as a means for purifying it is economical enough to be used in large-scale operations. By passing a current through a sample of water it can be separated into its component elements, hydrogen and oxygen. See *water*.

detritus   Loose, unconsolidated material that is either finely divided rock or the finely divided remains of animal or plant tissue or both. See *ecosystem*.

deuterium   An isotope of the element hydrogen, also referred to as heavy hydrogen. The atomic mass of this naturally occurring isotope is 2 daltons, as the result of an extra neutron in the nucleus of the deuterium atom, whereas the usual form of hydrogen has a mass of 1 dalton, from the single proton that constitutes the nucleus of its atom. Deuterium forms the same compounds as does the more common isotope of hydrogen. See *hydrogen*.

Devonian Period   A period in the Earth's history from 400 million to 345 million years ago, a part of the Paleozoic era that preceded the Carboniferous Period. The English geologists Adam Sedgwick and Roderick Murchison worked on the classification of Devonian rocks, and named the period for the distinctive rocks and fossils in Devon, England.

The Devonian Period is called the Age of the Sea, since more of the Earth was under water than is now. During the Devonian, the large landmasses of Europe and North America were in contact. The southern continents were all grouped at the equator. The Tethys Sea was the major divider of the landmasses. The face of the Earth changed dramatically in the middle Devonian as the northern Appalachian buildup occurred. Major changes in South America, Asia, and Australia resulted from violent volcanic and seismic activity. The rise of mountains trapped seawater in North American and Siberian basins, while coastal erosion produced

much of the red sediment now found in basement ocean floors.

The biota of the Earth was also rapidly diversifying during the Devonian. While trilobites and grapholites declined and disappeared, conodonts, brachiopods, and cephalopods became numerous and varied. Corals appeared in many new species, taking advantage of the warm, shallow marine environments. There was spectacular diversification of fish in the mid to late Devonian, which explains why the Devonian is perhaps best known as the Age of Fish. Marine forms increased in number of species and number of individuals. Some moved into areas that were occasionally out of water, and eventually developed as lungfish, while some of them ultimately became amphibians. Others moved into brackish and then freshwater habitats.

The end of the Devonian was a sharp break in evolutionary development. Catastrophic extinctions occurred. Several theories, including one of meteorite bombardment, have been advanced as causes for this. Whatever the reason(s), many corals and attendant reef animals and plants ceased to exist during the end of the Devonian, as did many cephalopods and other invertebrates. See *Cephalopoda*.

dewpoint   The temperature at which the air is completely saturated with water vapor. If cooling occurs, the water condenses as dew on land or as fog near coasts and over cold, open oceans.

The dewpoint can be read directly using a smooth metal plate and a thermometer. The temperature at which there is a film of water condensed on the plate is the dewpoint. See *air, condensation, fog, humidity*.

Dezhnev, Semyon Ivanov (1605?– 1673) (also spelled Dezhnyov)   A Russian (Siberian) explorer who made the first transit of the Bering Strait. Cape Dezhnev, the easternmost point of the Chukchi Peninsula in Siberia, which, in turn, is the easternmost point of Asia, is named for him. In English atlases it is called the East Cape. Dezhnev, sailing from the Lena River in a series of voyages, rounded the East Cape in 1648–1649, sailing with F.A. Popov. Their account of this voyage was lost and not found until after Bering's transit of the same Strait. See *Bering, Vitus.*

dhow   A sailing vessel used on the Red Sea and in Arabian waters. The dhow is designed to beat against the wind, particularly in regions of very gentle winds. The usual dhow has one mast, although there may be two, and the masts are lateen rigged, carrying triangular sails. The main sail is very large and intended to catch whatever wind is available.

Dias (or Diaz), Bartolomeu (1450?–1500)   A Portuguese navigator who first sailed around the Cape of Good Hope. The Portuguese empire had established trading positions on the Guinea coast of Africa when the king, Alfonso V, assigned further exploration to his son and heir, Prince John, later crowned King John II. It was then thought that Africa and India were connected by land, and that an overland trek from the West African coast would eventually end at Abyssinia and India. Somewhere between the latter two places, one might find the legendary Christian kingdom of Prester John.

Prince (later King) John (or Infante João) sent Alfonso Paiva, a military commander, overland and Dias by a sea route. Part of Dias' charge was to explore the west coast of Africa south of 15° South Latitude. The flotilla of three ships left Portugal in August 1487. They rounded the Cape of Good Hope in a storm, probably in the first week of February 1488, but did not actually see the Cape. They did sight it on the return, having established the possibility of a sea route to India. The name originally given the Cape by Dias was Cape of Storms. It was possibly King John II who settled on the present name.

In 1497 Vasco da Gama led a large flotilla to India with Dias as captain of one of the ships. The armada sailed too far west and actually sighted the Brazilian coast before rounding the Cape of Good Hope. Dias was drowned near the Cape.

diatom   A group of unicellular algae essential to the support of most marine food webs. These organisms contain a yellow-brown pigment in addition to the essential chlorophyll and possibly xanthophylls or carotenes (yellows and reds) needed for photosynthesis. The outstanding feature of the group is the two-part, transparent, siliceous cell wall. These cell walls are called tests. Diatom tests may be circular (Centrales) or elongated (Pennales). Fossil diatoms are well known, since the cell walls of these animals are virtually indestructible.

The reproduction of diatoms is complicated, since it involves the unhinging of the valves of the cell wall. One half of the wall, with one valve, goes with each new daughter cell. Very recently, it has been discovered that the bacteria that form symbiotic associations with mats of diatoms add nutrients to the ocean by nitrogen fixation. Previously, it had been assumed that only land plants supported nitrogen-fixing bacteria. See *brown algae, food chain, plankton, nitrogen fixation, ooze, test.*

diffraction   The bending or redistribution of light around an obstacle. It is a different phenomenon than reflection or refraction. Diffraction occurs not only with light waves, but also with water waves and sound waves. See *reflection, refraction.*

diffusion   The transfer of matter or energy from a region of high concentration to one of lower concentration. This is the mode of transmission of light through the photic zone of the ocean.

dinoflagellates   Single-celled organisms, less than 1 mm in size, classified as plants or animals since some, but not all, contain chlorophyll. Dinoflagellates have been found in fossils that date back to the Cretaceous Period, and may have existed before that era. These organisms constitute a part of the phytoplankton—

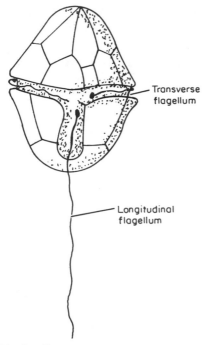

Transverse flagellum

Longitudinal flagellum

Dinoflagellate

those microscopic autotrophs that are the basis of all sea life. They have flagella which provide them with locomotion, and move vertically in response to light.

The dinoflagellates are found in two main groups: armored and naked. The former, the Peridinales, are found inshore. The latter, the Gymnodinales, are found in the open ocean, floating in warm seas. The armor of the Peridinales is a cellulose wall. Some forms are fish parasites, others live symbiotically with corals, specifically the Alcyonacea.

A characteristic of many dinoflagellates is phosphorescence, and their abundance in certain places has given this name to specific locales, such as Phosphorescent Bay (Bahia Fosforescente) in Puerto Rico.

Dinoflagellates "bloom" or greatly increase in number periodically, sometimes with disastrous results, such as the toxic (to marine life) red tides. Other dinoflagellate blooms may also be toxic to shellfish, causing gastroenteritis in the shellfish

and in fish, birds and other predators that eat them, including humans. See *flagella, Protista, red tides.*

dinosaur   A popular name for the giant lizards found on the Earth and in its oceans in the Jurassic, Triassic, and Cretaceous periods of 195 to 65 million years ago. Most of the best known and most spectacular dinosaurs were terrestrial animals. See *ichthyosaur.*

Discovery, H.M.S.   The name of several British ships used in explorations. The first *Discovery* went to Hudson Strait, in a voyage led by George Weymouth (1602); Hudson used the ship to sail into the same waters on his last voyage in 1610.

Another *Discovery* started life as a collier (coal carrier) named the *Bloodhound.* This ship and the *Resolution,* were under the command of James Cook when he sailed on his last voyage in 1788.

The third significant *Discovery* was part of George Nares' polar expedition. The ship had a reinforced bottom designed to withstand ice pressure. It was the model for Scott's ship on his 1901–1904 antarctic expedition. The same ship (*Discovery*) was used as the base for antarctic surveying in 1925–1927, and again to examine the Australian coast. The *Discovery II* (a rather confused designation since it was the fourth vessel of that name) charted antarctic waters immediately after World War II. Much of the work that led to the mapping of the Antarctic Currents and the Antarctic Bottom Water was done by the exploration teams on these ships.

Part of the charge of the "Discovery Expedition" of the 1920s was an attempt to explain the serious population decline of whales in the South Atlantic. It was this expedition that foretold of the possible extinction of several whale species because of overfishing. This problem has worsened considerably since then.

Another significant biological discovery of this research effort was the realization that the Antarctic Convergence is a "biological barrier." Many organisms such as plankton, birds, and fish are present either on both sides of the convergence line, or exclusively on one side or the other. The sinking of the Antarctic Water and the different layers, different temperatures, oxygen, nutrient, and salt content produce very different growing conditions and therefore very divergent populations on the two sides of the line. See *Antarctic Convergence; Cook, James; Nares, George.*

diurnal oscillation A term referring to variations in the planktonic population in the course of a day. Plankton move up toward the surface of the ocean in daylight and down at night. The reason for this migration varies with different species. Some animals migrate for reproductive purposes, others follow their food supply, still others, which harbor photosynthesizing protists, migrate to give the symbionts their daily exposure to the Sun. See *deep-scattering layers, plankton, tides, vertical migration.*

diurnal tide A tide that exhibits one high and one low tide during each tidal or lunar day. See *tide.*

divergence See *convergence.*

diving Descending beneath the ocean's surface. Dives of more than one to three minutes (the length of time a person can hold his breath) require the use of breathing apparatus. The use of the hard-hat diving apparatus, made reasonably safe by the early 20th Century, tethers the diver to an air pump at the surface. While it does provide security and an air supply, the hard hat apparatus encumbers the diver with bulky, heavy equipment and restricts his or her movements. SCUBA equipment, developed in the 1940s, allows a diver to carry his own compressed air supply in portable tanks, thus providing free range down to about 80 m (250 feet). Saturation diving is descent in a submersible which is intended to remain submerged for some extended period.

The physiological effects of diving result from pressure on tissues, and gases. Without an air supply, a diver cannot go below a point determined by his particular physiology—usually about 100 m (300 feet), where the water pressure forces air out of the lungs. Given an air supply, the pressure on the lungs exerted by the inhaled gases increases and this is injurious to lung tissue. Increased pressure within the lung causes collapse. The tiny air sacs, the alveoli, burst as an overfilled balloon would. These "balloons" are enmeshed in a tracery of tiny blood vessels and these too can rupture. An increased nitrogen content dissolved in the blood at about the 100-m mark produces intoxication by obstructing oxygen transport by the blood. This impairs judgment, reflex action, and accurate performance, and is potentially highly dangerous. The exact mechanism of nitrogen narcosis is unknown.

The collection of symptoms that are decompression sickness are exhibited by a diver emerging from compression too quickly. The nitrogen dissolved in the blood bubbles out and symptoms can occur at any time—up to about 18 hours after decompression. The presence of nitrogen bubbles in joints or muscles causes localized pain, hence the common name for decompression sickness—*the bends.* Over 80% of all decompression sickness involves localized pain. Other symptoms depend on the location of the bubbles. In the lungs, they cause labored rapid breathing. In the brain a range of symptoms from headache to vertigo, to unconsciousness or paralysis or death have been documented.

The replacement of some of the nitrogen in the breathing mixture by helium eliminates many of the problems encountered in using nitrogen. Helium is less soluble in human tissues, and therefore makes decompression after a dive easier but not problem-free. Helium does, however, extend the practical work limit of a diver to depths of about 175 m (580 feet).

The pressure of both oxygen and carbon dioxide increases in the body at increased depths, and is potentially dangerous. Increased oxygen in the tissues is toxic, and can lead to convulsions that may be fatal. The most common adverse effect of this is labored breathing. Tingling sensations in the arms and legs, vertigo, nausea, and confusion may also occur. Unfortunately, such symptoms do not occur until several days after the pressure increase itself. Carbon dioxide is retained in greater than normal quantities in tissues at great depths. This produces an effect similar to that of nitrogen. See *Cousteau, Jacques-Yves; marine archeology; nitrogen narcosis; SCUBA; submersibles.*

DNA (deoxyribonucleic acid)   The genetic encoding molecule of all living organisms. Each plant or animal has its own unique DNA. The constituents of DNA, which is a long chain molecule, or polymer, are the sugar deoxyribose, phosphate groups, and the four nitrogen-containing organic bases adenine (A), quanine (G), cytosine (C), and thymine (T). (See diagram.)

The DNA molecule exists in the form of a double helix—a coiled ladder in which the "sides" are the sugar-phosphate strings and "rung" is a base pair. Each sugar is attached to its respective base by a strong chemical bond, whereas the bonds that hold together the base pairs forming the rungs of the ladder are

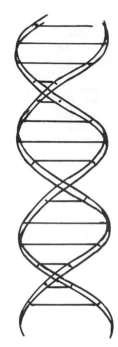

(2) DNA structure (double helix)

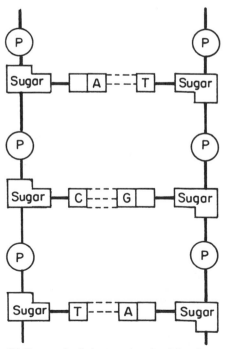

(3) Base pairs between strands of the double helix

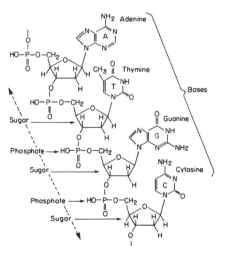

(1) A section of a DNA chain

weak chemical bonds (hydrogen bonds).

The genetic information is carried in the linear sequence of the four types of bases along the DNA chain. DNA replication, which duplicates the genetic information and allows it to be passed on during cell division, occurs as the weak chemical bonds in the base pairs forming the rungs of the DNA ladder come apart and a new DNA chain is polymerized onto each of the two old chains that formed the sides of the original ladder (replication).

By an analogous process, a segment of the genetic information contained in the linear sequence of the bases along the DNA chain is transcribed, or copied, into a complementary strand of messenger RNA (ribonucleic acid). This is then translated into protein in a reaction catalyzed by a large complex aggregate of RNA known as a ribosome. The amino acids used for protein synthesis are brought one at a time to the messenger RNA chain by transfer RNA molecules, each of which recognizes a specific sequence of three bases. The sequence of bases in the messenger RNA is "read" from one end to the other in sets of three. Each such triplet, called a codon, specifies a particular amino acid, and in this fashion the "genetic code" in the DNA molecule is translated into the proteins that make up the skin of an eel or the feathers of a penguin. *See peptide formation, RNA.*

dogfish   A member of one of the most familiar groups of sharks (for example, the Squalidae and Carcharhinidae). The dogfish, which at maturity is about 1 m (3 feet) long, is a well-known biological laboratory specimen. It is fished commercially as it swims in schools following herring. Its oil is used industrially.

The dogfish is ovoviviparous. Its eggs are fertilized internally and hatched about a year and a half later. It is a long-lived animal, specimens having been tagged and then recorded more than 25 years later. *See marine oil, shark.*

Dogger Bank   A submerged sandbank, roughly in the center of the North Sea. It is about 250 km (160 miles) long, quite shallow (13 m or 42 feet), and composed of glacial moraines and outwash. There are submerged peat bogs in the Dogger Bank. This area is one vast commercial fishing ground used by the vessels of several nations. Historically, it is remembered as the site of the largest World War I naval encounter, the battle of Jutland. *See North Sea, Pleistocene Epoch.*

doldrums   *See Intertropical Convergence.*

dolphin   A relative of the whale, belonging to the family Delphinidae. Dolphins are slender and have a beaklike snout. They are the most agile cetaceans. The different species vary from 1.25 to 25 m (4–85 feet) long and range up to 230 kg (500 pounds) in weight. They move in herds that follow and feed on schools of herring or sardine. Tagged specimens have shown that in the wild dolphins live more than 25 years, possibly 35.

The most common characteristic of the dolphin is that it has both fins and flippers. Like other cetaceans, it breathes in great gulps at half-hour or longer intervals, descends rapidly to great depths without compression or decompression effects, and is noted for its familial and community adherence.

The most frequently seen dolphins are the common dolphin and the bottle-nosed dolphin. Both are highly intelligent, and communicate with other members of their group and with humans using a series of high-pitched honks, barks, and squeaks. Dolphins appear to like humans, and seafarers' tales of dolphins saving shipwreck victims seem to have some validity.

Dolphins swim and share food populations with tuna, and if they become enmeshed in the large seines used to catch tuna, will drown. In response to this, dolphin-proof tuna nets have been designed. *See cetacean, echolocation, whale.*

dorsal   Relating to the back or upper surface of a fish or other animal.

dory   A small rowboat-sized craft with high, curving sides. Dories are dropped by fishing boats and used to set lines. See *fishing industry.*

dory fish   See *John Dory.*

Dover, Strait of   A body of water that separates England from the European continent and constitutes the southern boundary of the North Sea and its connection to the English Channel. The strait is about 33 km (21 miles) wide at its narrowest point, and about as long. Since it is in the shallowest part of the North Sea perimeter (60 m or 180 feet deep at most), there is relatively little Atlantic water entering the North Sea through this strait. It is thought that this area was once exposed during the Pleistocene. The strait itself was carved out by the now-drowned ancient Thames river system. The strait will be undercut by the Anglo-French venture which is building a tunnel to connect the road and rail systems of the two nations. See *Atlantic Ocean, North Sea.*

Drake, Sir Francis (1541? or 1543?–1596)   English navigator, explorer and buccaneer; the first Englishman to circumnavigate the globe. Drake first went to sea in 1566 and 1567–1568 with John (Jack) Hawkins, a buccaneer, sailing on slave ships to the West Indies. In 1570, 1571, and 1572 he sailed again on voyages to raid Spanish treasure ships. The last of these was very successful: Drake captured the Spanish treasure convoy in Panama, which made him an international figure.

A syndicate of Protestant gentlemen financed Drake's voyage of circumnavigation. The stated objective was to explore and exploit the western coast of South America. It also served to promote the cause of the English Protestants and to get Drake out of England at a time when the political climate made it important to Queen Elizabeth I to attempt to retain diplomatic relations with Spain even while her privateers were harassing Spanish shipping.

Six ships left Plymouth in December 1577 (the number was quickly reduced to five.) The flagship was the *Pelican,* which was later renamed the *Golden Hind* in honor of Sir Christopher Hatton, one of Drake's patrons, whose family crest included a golden deer. The expedition first moved south along the West African coast and then across the Atlantic at its narrowest point to South America. At this point every narrative about Drake becomes vague. It is known, however, that Drake sailed through the Straits of Magellan in the winter of 1578. His fleet was reduced to three ships, the others being abandoned because they were unseaworthy. Drake then moved up the western coast of South America, raiding Spanish ports and ships as he went. On April 16, 1579 he left Spanish America (Mexico) and made landfall on June 5 in what is now Oregon. Drake named the West Coast of the United States New Albion—the first New England.

He sailed south and commented (unfavorably) on the brown uninviting hills frequently covered with fogs. This seems to have been his sighting of San Francisco Bay.

On July 24, 1579 the *Golden Hind* headed west. Sixty-five days later it made land, probably on the Palau island group in the Pacific, and then went on to Mindanao. The ship then rounded the coast of Sulawesi (Celebes) and struck out into ocean waters from Java. Sailing around the Cape of Good Hope without stopping, it finally anchored in June 1580 in what is now Sierra Leone, on the west coast of Africa. The long run at sea was intended to avoid Spanish or Portuguese vessels. The *Golden Hind* sailed into Plymouth harbor on September 26, 1580. Drake brought back a ship in good condition, and a healthy crew of 58 men, having lost 17 to illness, fights, and desertion.

Although memorable, the voyage did not bring many new discoveries, but was a great example of seamanship.

Drake became a fabled hero in his own lifetime. He continued to harass the Spaniards and died in the West Indies of yellow fever on January 28, 1596.

Drake Passage (Drake Strait) The narrows of the Southern Ocean, probably opened in the late Oligocene epoch; it connects the Atlantic and Pacific and separates the tip of South America at about 50° South Latitude from the South Shetland Islands and the Antarctic Peninsula. It is about 1,000 km (450 miles) wide and has an average depth of about 3,500 m (12,000 feet). Sir Francis Drake, for whom the passage is named, discovered it inadvertently in the Anarctic in the spring of 1578, when blown off-course by a storm.

The Antarctic Convergence bisects the Passage along an east-west line. Water from the Antarctic continent moving north sinks at about 50° South Latitude and then spreads north and rises to the Subtropical Convergence. There it sinks again and moves further north as Antarctic Intermediate Water. It is still recognizable as cold, low-salt and low oxygen-content water. Intermediate Water from the middle latitudes moves south to replace what is lost. Since the Bottom Water comes up from great depths, it is responsible for the great fertility of the antarctic waters.

Because it is so constricted, the Drake Passage is the Earth's greatest carrier of water moving in a west-to-east direction. The total volume of water pouring through this passage into the Atlantic has been variously estimated at anywhere from 75 to 150 million cubic meters per second (20 to 40 billion gallons per second)—a volume hundreds of times greater than the volume of water moved by the Mississippi River, for example. See *Antarctic Circumpolar Current; Antarctic Convergence; Atlantic Ocean; Drake, Sir Francis.*

dredge A floating device used for removing underwater material, and usually held in place by supports that descend to the bottom.

The removal of sediment from canals, rivers, and harbors is vital for navigation and sanitation. The fill has traditionally been used to create new land or build dams, dikes, and breakwaters. Dredges are also used in salvage or to remove commercially significant materials from the sea bottom.

Dredging techniques have been used for centuries. One of the earliest examples of this work is the reshaped landscape of the Netherlands. Dredges and mudmills were common features of the medieval Dutch landscape. Earlier still, dredging was part of the general maintenance work done on the canals of ancient Babylon. When the irrigation canals silted up, the life of the city ended.

Basic modern dredge design involves some combination of mechanical and hydraulic techniques. Dredges are either dippers which operate at the end of a flexible arm—a type of dredge that can operate at almost any depth—or buckets that grab bottom material and are brought up by cables or a continuous feed. The continuous feed or ladder is a line of buckets that digs into the mud and brings it to the surface. Hydraulic dredges liquefy bottom silt and pump it to the surface. See *coast.*

dry dock A shipyard structure used for major overhaul and repair to ships' bottoms. The dry dock is a basin cut into a harbor and separated from the rest of it by a wall. A ship can be positioned over supports and the basin emptied, leaving the ship standing on the supports. After repair and cleaning, the water is allowed to flow back into the basin and the ship is refloated.

Floating dry docks have at least one section that can be pumped out to contain and service a ship. These containment basins can be complex and large enough to float a naval vessel of battleship size. The dry dock is a direct descendant of the primitive earthen walls with which the ancient Greeks held back the sea while standing a vessel on the beach in order to repair it.

duck One of about 90 species of aquatic birds of the family Anatidae, order Anseriformes. Ducks generally have broad bills, short necks, and legs located on the tail-side of their centers of gravity. This results in a waddling type of gait.

The plumage of the male is markedly different from that of the female. Ducks undergo two molts each year, and true ducks have overlapping scales on the exposed skin of their legs. Ducks do not mate for life, as do geese and swans.

Ducks that live in brackish coastal waters or on open coasts such as the eiders, are likely to be divers who pick their food out of the water or off the bottom. Among sea ducks that winter at the seacoast and nest inland are scoters and mergansers. Almost all of these ducks live in the Northern Hemisphere. See *bird.*

dugong A herbivorous Asiatic mammal living in coastal, tropical waters, where it does little more than eat. These slow, fat animals reach a size of about 3.5 m (12 feet) and 270 kg (600 pounds). They have gray-to-brown exteriors, front flippers, no rear appendages, and a broad, powerful bilobed tail. The dugongs have been hunted almost to extinction. The only surviving species, *Dugong dugon,* is in the same family as the manatees, Sirenia. See *Sirenia.*

Dumont d'Urville, Jules-Sébastien César (1790–1842) A French explorer of the South Pacific and Antarctica. He also made contributions to archeology because, while charting the eastern Mediterranean Sea for France, he was part of the venture that brought the Venus de Milo to France (1820). He also participated in a French voyage that circumnavigated the world in 1822–1825.

Dumont d'Urville was commissioned to find the remains of Jean François Galoup, comte de La Pérouse, who disappeared in the Pacific Ocean in 1788. On this reconnoitering mission, he stopped at and charted parts of New Zealand, New Guinea, the Loyalty Islands, Van Diemen's Land (now Tasmania), the Carolines, and Sulawesi (Celebes). He found what he believed to be parts of La Pérouse's wrecked ships in the Santa Cruz islands, and returned to France in 1829.

The careful maps made on this voyage led to the designation of the charted island groups into Micronesia, Polynesia, Melanesia, and Malaysia. (Fiji Islands, Loyalty Islands, Maritius, etc). The expedition was also a scientific success since it returned with samples of plants, animals, minerals, and ethnological artifacts, as well as dictionaries of native languages.

Dumont d'Urville was promoted to captain and again commissioned to explore the Pacific for France. He departed in 1837 for a voyage to the antarctic, hoping to better the record of the English explorer James Weddell, who had gone as far south as 74°15', in 1823. Dumont d'Urville charted the Straits of Magellan and found ice at 63°29' South Latitude, and then drifted east from the Straits to the South Orkney and South Shetland islands. After this he headed west again to Fiji, New Guinea, and Borneo. After taking on supplies the expedition again returned to the antarctic, hoping to locate the South Magnetic Pole. It sighted the Adelie coast of Antarctica in 1840, which it named for the captain's wife. The French expedition and the American one led by Wilkes from 1838 to 1842 each claimed to be the first to have sighted Antarctica. Dumont d'Urville returned to France late in 1841, and died in a train wreck the following year.

See *explorations; Wilkes,* Charles.

Dungeness crab A large, commercially important, edible crab, *Cancer magister,* found in western North American waters. The Dungeness is a decapod found from Alaska to Baja California. It is usually 18 to 25 cm (7 to 9 inches) wide and 10 to 15 cm (4 to 6 inches) long. The upper surface is dark, the belly is yellow. The Dungeness is a relative of the Atlantic rock crab *C. irroratus,* the Pacific rock crab, *C. antennarius,* the Jonah, *C. borealis,* and the common crab of North European waters, *C. pagurus.* All of these are edible; the last is usually even larger than the Dungeness of the Pacific coast. See *crab, Decapoda.*

E

earthquake A rapid movement of rock caused by the buildup of strain created by the movement of one crustal plate against another. When this occurs in water, the displacement is spread both vertically and horizontally in all directions. Since water is a medium that disperses shock poorly, tsunamis (incorrectly called tidal waves) are created.

Earthquakes are associated most closely with specific geographical areas. The rim of the Pacific Basin is a series of deep trenches parallel to the South American coast, the Aleutian and Japanese island chains, and New Zealand. At these trenches, crustal plates are moving together, and as one plate moves below another one (i.e., as it is being subducted and moves into the molten interior of the earth), volcanoes are sometimes formed behind the trenches and eruptions and earthquakes result.

Earthquakes are categorized as either shallow-focus, occurring at a depth of about 400 km (250 miles), or plutonic or deep-focus, occurring at depths of about 750 km (470 miles) depths. Seismic waves of the two basic types spread in different ways.

P waves (primus) are compressional, s (secondus) have more shear quality. They arrive at different times at a distant detector. The time difference depends on the composition of the rock traveled through. See *crust, trench, tsunami, volcano.*

East Australian Current The western portion of the gyre (the circulatory route of surface water in an oceanic region) in the South Pacific Ocean. It forms between the Great Barrier and Chesterfield reefs, and is augmented by warm equatorial water moving west-southwest, pushed by the spring (January–March) monsoon. From April to December this

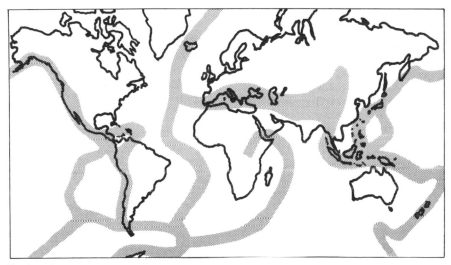

Earthquake zones

warm water comes from the Coral Sea. At its strongest point the East Australian Current moves at an average speed of about 50 km/day (30 miles/day). There is an offshore countercurrent in winter.

South of about 32° South Latitude, the Current broadens to about 200 km (150 miles) wide and disappears into an eddy system. These warm-core eddies are a notable feature of the East Australian Current, and last about a year each. The eddies can coalesce to form new systems.

This current has been called the Southern Hemisphere's Gulf Stream. However, unlike the Gulf Stream or the Kuroshio Current, it disappears rapidly when it moves away from the coast. The Australian continent ends at about 44° South Latitude. South of the continent the water moving west to east is swept along by the Antarctic Circumpolar Current which, because of its size and volume, incorporates any other moving water in the area. See *Antarctic Water, gyre, monsoon, ocean currents*.

east boundary currents   Currents that flow toward the equator on the eastern side of an ocean. These currents include the Canary, California, Humboldt, West Australia and Benguela currents. The east boundary currents are surface currents with water temperatures lower than currents further west at the same latitude. The currents are relatively slow. Upwellings, when they occur, are on the eastern edges of the currents, bringing nutrients to the surface. The result is a large number of both plant and animal species and individuals. See *currents, individual currents*.

East Cape   The easternmost part of the Eurasian continent. It is on the Chukchi Peninsula, and is the eastern boundary of the Bering Strait. The Soviet name for East Cape, Cape Dezhnev, after its discoverer, is becoming internationally recognized. See *Dezhnev, Semyon Ivanov*.

East Greenland Current   Part of the water of the Greenland Sea. This body of cold, polar water of low salinity moves south along the east coast of Greenland, picking up icemelt and water of terrestrial origin. This further decreases the salinity of the Greenland Sea to below 3% in summer. Polar Intermediate Water joins the East Greenland Current near Spitsbergen to form the Greenland Current, a flow of south-moving surface water that pushes a belt of ice before it. See *Arctic Ocean, Fram Strait, ice*.

East Pacific Rise   Or East Pacific Ridge. A broad, relatively continuous formation that runs roughly north-south. It rises 2 to 3 km (1 to 1.5 miles) above the ocean floor and is 2 to 4 km (1 to 2 miles) wide. The average depth of the rise is about 2.7 km (or about 1.7 mile). The East Pacific Rise disappears under the North American Plate at the Gulf of California.

Recent studies have revealed considerable volcanic activity in this region of seabed spread. Chimneylike vents pour hot sulfides into the surrounding water. An ecosystem based on bacteria that survive using sulfur instead of solar energy makes possible a wide variety of organisms that can survive in hydrothermal environments.

Geologic studies of the northern edge of the East Pacific Rise near the Mexican coast reveal recent lava flows, sulfide-rich areas, and significant manganese and iron accumulations. It is thought that similar finds would be made in other parts of the Rise. See *evolution of oceans, Pacific Plate, sulfur bacteria, vent communities*.

East Siberian Sea   A broad continental shelf north of Siberia. It lies between the New Siberian Islands to the west and the Wrangell Islands on the east. The Wrangell Islands are the traditional division between the East Siberian Sea and the Chukchi Sea. North of the East Siberian Sea is the Wrangell Abyssal Zone. The area was explored by the Russian Baron Wrangell in the 1820s.

The East Siberian Sea is quite shallow. The depths at the eastern or deeper end are frequently in the 30 to 40 m (100–130-foot) range. The temperature and salinity of the water depend on river input. Since the Sea is so shallow, there is thick ice on its surface through most of the year. This results in a relative diminution of large animal life. However, there are large fish communities, such as char and salmon, and the birds that feed on them, including cormorants, terns, and gulls. Seals and walruses are found at river mouths and on terrestrial bars in the Sea. See Jeanette; *Wrangell, Baron Ferdinand Petrovich von.*

Easter Island   A small (130 km² or 50 square miles), volcanic, South Pacific Island, lying about 3,300 km (2,300 miles) west of Chile. It is a part of the Easter Island Ridge, which is the Pacific Ocean equivalent of the Mid-Atlantic Ridge. The ridge system of which Easter Island is a part is somewhat unlike the Atlantic Ocean counterpart in that it is asymmetrical, sloping inward toward the South American continent. Easter Island is best known for the huge and unusual statues found on it. See *Heyerdahl, Thor; mid-ocean ridges.*

Echinodermata   An exclusively marine phylum of great age and worldwide distribution. The name comes from Greek and means spiny skin. While this is true for most of the phylum, which tends to exhibit calcareous spines, some species have warts rather than spines.

The echinoderms are classed as follows:

Crinoidea—Sea lilies
Holothuroidea—Sea cucumbers
Asteroidea—Starfish and sea stars
Echinodea—Sea urchins and sand dollars
Ophiuroidea—Serpent stars and brittle stars
Ophiocistioidea—Extinct class of small domed animal usually 7–8 cm in diameter. They lived from Ordovician to Devonian time.

The common features of the Echinodermata are the absence of a head or end of some sort, and radial symmetry, with arms in a pattern of five or a multiple of five. The crinoids are sessile (stationary), and their original ancestor may also have been stationary. The rest of the phylum is generally bottom dwelling, and moves by creeping along, using the arms for locomotion, although some sea cucumbers are pelagic.

Echinoderms have a water-vascular system, in which water enters the body through a valve called the madreporite; in asteroids, this lies on the aboral (upper) surface of the animal. Water is circulated through the highly calcified "stone canals" to the "tube feet"—suction devices on the legs that adhere to the shells of mollusks and pry them open. Other echinoids are filter feeders. See *Crinoidea, sea urchin, starfish.*

Echinoidea   The class of Echinodermata that represents the sea urchins and sand dollars. These creatures are characterized by a rigid covering called a test. See *Echinodermata, sand dollar, sea urchin.*

Echiuroidea (Echiura)   The spoon worms, a small phylum of about 130 species of marine burrowers of worldwide range. Most of the individuals are small, though they range in size from about 1 to 60 cm (0.4 to 24 inches), and most are free living. They share some characteristics with the annelids. Fertilization is external, and the young are nektonic, swimming freely on their own. Unlike the annelids, the majority of echiuroidean species are dimorphic, with distinct male and female members.

These animals have some unique characteristics, the most notable of which is an elaborate, grooved, ciliated proboscis (food-gathering) organ that amasses food particles, either detritus or plankton. The proboscis may be 90 to 100 cm (30 to 40 inches) long, although the body of the average worm is only about 8 cm (3 inches). The proboscis cannot be retracted, but only shortened somewhat. One species of echiuroidean, *Bonellia*, is remarkable because the males are tiny and parasitic on the females. They attach to the female's proboscis while still larvae, and move down into the body. Fertilization in this species is anomalous because it is internal. See *Annelida, nekton, plankton, proboscis.*

echolocation   The ability to determine the shape and size of an object, and its distance, by the echo it makes. The best-known terrestrial animals who find their way about by echolocation are the bats. In the sea, the range of species using echolocation is much greater. All toothed whales, the Odontoceti, use echolocation, as do some baleen types, such as the blue and gray minkes. Some pinnipeds, including the California sea lion, the Weddell seal, and the walrus also echo locate. Echolocation is useful to these mammals because although they have good vision, they frequently find it obscured by a high sediment load in the water; alternatively, they may be beneath ice or below the photic zone, where there is no light.

The sound used for echolocation is produced by several different structures in the animal's head. In the whales it may (or may not) be amplified by the blow hole. The sound that is used is specific to a given species, and of a frequency range that is usually greater than humans can detect. The pattern of sounds is also species-related, and different sounds mean different things. There have been considerable attempts made to investigate the meaning of the sounds used for echolocation.

The sound receiver for echolocation is not always the ear, even in those species that have external ears. The entire skull may be a receiver, and in some whales the lower jaw is an auditory receiver. The mechanism by which the sound that is received is processed into information is conjectural. See *echo sounding, whale.*

echo sounding   A technique used to map undersea terrain. Sound is carried

well by water, a phenomenon observed by the ancient Greeks. In seawater it travels approximately 500 m/sec (or 1600 feet per second). Not until the 20th century, however, was there a reliable method for both making a sound and timing its rebound from the bottom.

By timing the echo from the bottom, the shape of the bottom surface can be plotted, using a number of soundings. Computer enhancement of this plot then gives a good, fairly reliable picture of the bottom to an experienced observer. It is essential to use repeated soundings and to check the results using lead lines and other methods, since plankton and schools of fish will also produce an echo, called a false bottom. The echoes from schools of fish are used by commercial fishermen to locate such schools. See *SONAR, sounding.*

ecliptic   The plane of the orbit of the Earth as it revolves about the Sun. The line through the Earth from the North Pole to the South Pole makes an angle of 25°30' with the ecliptic. This is the declination of the Earth. See *declination, equinox, solstice, tide.*

ecosystem   A term used to describe the interrelationships between all organisms in a given area, and their relationships to the nonliving materials that make life possible. The specific environments in the oceans are divided according to physical conditions of depth, light, temperature, and salinity. In a large part of the ocean, the density of living organisms is low, but excluding insects in their in-

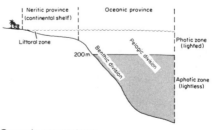

Oceanic ecosystems

credible numbers of species and individuals, the oceans contain a wider variety of organisms and are more densely populated than land areas.

All aquatic life is dependent on the abiotic (nonliving or inorganic) materials present in water. Thus, dissolved calcium salts (carbonates) as well as nitrates, silicates, phosphates, and sulfates must be present, and also a large variety of micronutrients, including nickel, manganese, and copper, as well as the gases carbon dioxide and oxygen. Ultimately almost all life depends upon solar energy, which is used to make sugar molecules by autotrophic (photosynthesizing) organisms. The greater part of this process is carried out by algae, diatoms, dinoflagellates, and nanoplankton.

The living things in aquatic environments are divided into three arbitrary groups: plankton, nekton, and benthos. Plankton are either animal-like, the zooplankton, or plantlike, the phytoplankton. The latter are microscopic autotrophs. Zooplankton includes animal eggs, larvae, protists such as radiolarians and foraminiferans, tiny crustaceans, mollusca, and other small animals. This is subdivided according to the size of the organism. Nekton are swimming organisms, including fish, cetaceans, crustaceans, squid, and octopus. Benthic organisms are bottom dwellers. It does not matter whether the bottom is a tidal pool or an abyssal depth. These organisms are the grasses and large algae, either infauna which live in sediment, or epifauna which cling to rocks, and animals such as crabs, fish, oysters, sea anemones, echinoderms, and corals. Benthic organisms feed on plankton and sediment coming down on them from the surface water, and on each other. Animals which as adults are benthic may be planktonic or nektonic as larvae.

The marine ecosystem is further divided according to depth. The two subdivisions are neritic and oceanic. This may be crudely stated as green water or blue water. The neritic inshore water is green because it contains more phytoplankton and other plants, and is nutrient rich. Nutrients are either stirred up from the bottom or brought by terrestrial runoff. This green-water zone is further divided into the littoral, which is sometimes exposed between tides, and the shallows, where there is always a water covering, albeit shallow. The oceanic region is graded according to depth: epipelagic, mesopelagic, bathypelagic, and abyssopelagic or hadal.

The microscopic organisms in each environment, in addition to serving as photosynthesizers, include the bacteria and detritus feeders, the tiny scavengers that recycle all other living and once-living material and release from it the raw materials that, in turn, are necessary for the photosynthesizers and other autotrophs.

The relationship between competing populations is being explored, with the interests of the fishing industry in mind. Any species in a given area depends upon the food supply and the physical conditions of temperature, light, salinity, mineral content, and upon local predators, as well as the presence of other species competing for food. For some species, competing species are also predators. Herring and mackerel, for example, use each others' larvae as food. Species may also compete for space. Fishing for a dominant population may reduce its numbers to a point at which a subordinate species becomes dominant. This interrelationship of species and their physical requirements is not yet well understood for many commercially fished organisms.

Heterotrophs consist of all the nonautotrophic organisms in an ecosystem. Ultimately, all of them feed on the autotrophs. Herbivores do so directly. The detritus feeders live on dead plants, the carnivores feed on herbivores and on each other. Eventually the detritus feeders and scavengers consume both the carnivores and herbivores. See *algae, estuaries, pelagic community.*

ectoderm   The outer layer of cells of an embryonic organism. This layer in the embryo engenders the nervous system and skin.

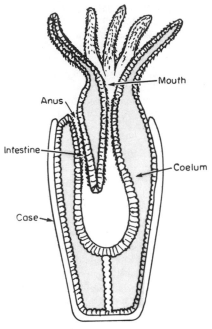

Ectoprocta

Ectoprocta A phylum of bryozoans. There are about 60 species. These colonial animals are found as wrinkled, crackling sheets on the shore. The sheets are masses of calcareous (calcium-containing) or chitinous "boxes" that form the individual organism's living space. See *Bryozoa, chitin, colonial animals*.

eddy A fluid flow that is markedly different from the general movement of a large mass of that fluid. The number and velocity of eddies increase as the velocity of the surrounding mass of fluid (air or water) increases. More eddies and faster-moving ones spin off in rapidly moving water than in a more sluggish stream. The Gulf Stream is a prime example of an eddy-creating system.

Eddies are a means of energy dissipation. Larger eddies form quickly and then break up into smaller ones, which eventually dissipate.

In the ocean, eddies are spun off by wind activity, upwellings, or waves encountering an obstacle. If the water moves around an obstacle, an eddy may form at some critical velocity (this depends on the nature of the obstacle) on the leeward side of the obstacle because the pressure of the fluid is reduced at that point by the rapid movement of the flow. If the velocity of the general flow is great enough, a vortex or whirlpool will form. Whirlpools are permanent eddies. Normally an eddy moves in a particular path. Cyclonic storms are examples of atmospheric eddies. See *Gulf Stream, Straits of Messina*.

eel The order Anguilliformes; elongated, usually scaleless, bony fishes with an average size of about 1 m (3.3 feet). Eels have been caught and eaten as a delicacy since classical times. There are about 500 species. One major family, the Anguillidea, are found in freshwater streams in North America and in the Caribbean, northern Europe, the Baltic, the eastern Mediterranean, and the Far East. These eels breed in the Sargasso Sea.

Another family, the Moringuidae, consists of small burrowing and nocturnal sea eels that are sometimes found in brackish water. The morays, or Muraenidae, comprise about 120 species of tropical eels that are often brilliantly colored and marked with spots or bars of black. They have strong jaws and poison glands, live in rock crevices, and hunt in darkness. The morays are the largest of all eels; one Indo-Pacific species exceeds 3 m (10 feet) in length.

A third family, the Congidae, or conger eels, are also toothed. They are generally under 3 m (10 feet) in length and live in all of the world's warm seas except the eastern Pacific. They live on fish and squid, pouncing on them from rock crevices and other dark hiding places. Congers die upon spawning.

The Ophichthyidae or snake eels live on coral reefs. They bore holes in the sand, tail first, and feed on fish, crustaceans, and copepods. The Nemichthyidae or snipe eels are long-nosed, deep-sea dwellers found at depths ranging from 400 to 4,000 m (1300 to 13,000 feet).

The suborder Saccopharyngoidei are deep-sea organisms. The pelican fish is one. Its whiplike tail is longer than its body. It has many arched teeth and distends its mouth to swallow prey whole. Gulpers are classed as eels by some taxonomists and not by others. These slender, black, 50-cm (1.5 foot) long fish live on plankton or small crustaceans. See *Sargasso Sea.*

eelgrass   Also known as wrack, this term actually comprises a group of angiosperms (flowering plants) of worldwide distribution found in protected inshore locales. Eelgrass belongs to the family Zosteracea. It grows in areas where there is abundant sunlight, and serves as both food and shelter for littoral communities. It is unusual because it lives and flowers while submerged. See *angiosperms, Zostera.*

egret   A long-legged, long-necked heron. Egrets are usually white and have plumes called aigrettes, which play a part in their courtship display. These feathers were popular as hat decoration in the 19th and early 20th centuries. The egret was almost driven to extinction by the feather trade.

The common egret is found in warm to temperate regions worldwide. It is a thin, rangy bird, between 50 and 100 cm (20 to 40 inches tall). The bill is also long and slender, designed for picking at the small fish, frogs, worms, and crustaceans that are the egret's food.

Egret communities usually nest in trees, and they may be found almost anywhere near water. Some species live in saltwater marshes, others range far inland. Cattle egrets are most often found in grasslands, where they are commensal with cattle, picking insects off them. See *heron.*

eider duck   A sea duck related to the scoter. While there are some Pacific species, most eiders are found in the North Atlantic. They are large birds, about 70 cm (28 to 29 inches) long, and have large wings. The head is large and the bill is distinctively shaped and forked.

The male has a white back and a black underside. The female is red-brown with black bands.

Eiders nest in large colonies. Each breeding pair builds a grass-and-seaweed nest on the ground, and the female lines the nest with down plucked from her breast. When the female leaves the nest to fish, she covers the five-egg clutch with these remarkably light, resilient insulating feathers. The down is prized and used commercially, collection taking place after the birds are fledged. Those areas of Norway, Iceland, and Greenland where the eider duck nests are protected. See *duck.*

Ekman, Vagn Walfrid (1874–1954)   A Swedish oceanographer best known for his work in ocean currents. He was a student of physics at Uppsala University and went from there to the International Laboratory for Oceanographic Research in Oslo, Norway. There he worked on several projects, among them the design of devices for sampling and measuring ocean water. Ekman also did the work that explained the "dead water" phenomenon observed by Frijdtof Nansen. This occurrence in fiords and other enclosed areas, such as narrow bays, rendered slow-moving vessels motionless. Ekman explained that this was caused by nearly fresh meltwater spreading over the saltier ocean water and setting up internal waves.

In 1910 Ekman went to the University of Lund, Sweden, as a professor of mechanical and mathematical physics. While there he did the work on currents that made him well-known to oceanographers. This was the culmination of a lifelong interest.

Ekman joined the German-sponsored *Meteor* expedition, and spent five years, 1920–1925, collecting data. The final report of the expedition was not published until after World War II. Ekman, the consummately careful researcher, had been separated from his data and did not work on the project until he had free access to his notes again. See *Meteor Research Expedition, Nansen, Frijdtof.*

Ekman spiral   A mathematically constructed spiral describing the effect of
wind on the ocean. It is named for the
Swedish oceanographer V. Walfrid Ekman, who formulated it in 1905. This
effect is a composite of the force of wind
and the force related to the Earth's rotation, the Coriolis force, which deflects
movement of air or water to the right or
east in the Northern Hemisphere and to
the left or west in the Southern. At the
same time, this deflection is diminished
by friction. The Coriolis deflection causes
circular motion, the friction creates a spiral. At a given depth, which depends on
the force of the wind, the Coriolis and
frictional forces both approach zero. The
layer of water above that depth is the
Ekman layer. It moves as an entity above
the water further down. The details of
how this layer moves and what angles it
makes with the prevailing wind are unclear. The *Ekman transport*—the net effect of water movement in the upper part
of the layer if the water moves seaward—
causes coastal cold-water upwelling, as
evidenced off the coasts of California and
Peru. See *circulation; Coriolis force; currents; Ekman, Vagn Walfrid; El Niño;
Nansen, Frijdtof.*

Ekman transport   The total transport
or displacement of surface water as it
moves with the wind. Excluding the friction factor, the transport is at a 90° angle
to the right of the prevailing wind in the
Northern Hemisphere, and at 90° to the
left of the wind in the Southern Hemisphere. See *Coriolis force, current.*

El Niño   A southeast current in the
southeastern Pacific Ocean bringing warm
water toward the coast of South America
in December or January. Because of its
proximity to Christmas and the tropical
fauna it carries along, it is called El Niño,
the Christ Child.

At irregular intervals of as little as two
or as much as ten years, this benevolent
warm current becomes a very warm current and brings catastrophe. This is part
of an overall phenomenon involving a
change in wind direction, the Southern

Oscillation, and changes in air and water
temperature (SST, or Sea and Surface
Temperature) as well as oceanic circulation. An El Niño event begins to build
about a year and a half before it is evident in South America. The wind in the
western Pacific drops, the sea level falls,
the water temperature rises, and coral
begins to die. The change in the height
of the water level moves warm air and
water into coastal areas and keeps it there
through the winter, deflecting the South
Equatorial Current. This means that cooler
water, instead of moving west across the
Pacific, goes north as far as San Diego,
California.

This bottling up of water of higher
temperature along the South American
coast results in the dying off of plankton,
squid, and fish populations that are
adapted to cold water. With the death of
the anchovy and sardine populations,
hake, herring, and mackerel disappear or
die. The sea birds and seals that feed on
these fish abandon nests and lose their
young. The decaying fish accumulate on
beaches producing huge quantities of
sulfurous gas.

The climatic changes bring torrential
rain and floods to the normally dry South
American regions while producing drought
in the western Pacific. These climate
changes have effects on the weather in
North America as well. The severe winter
and spring flooding experienced in North
America has its origins in the anomalous
wind pattern associated with El Niño.

This phenomenon of destruction has
been chronicled since 1726. Some El
Niño years in the 20th Century have
been 1925, 1930, 1941, 1953, 1957/8,
1969/70, 1976/7, 1982/3. The last was
a major event with associated worldwide
climate anomalies. See *currents.*

Elasmobranchi   A division of the
Chondrichthyes comprising fishes with
plate-like gills, the sharks and rays. The
elasmobranchs have cartilaginous skeletons and no swim bladder. Their buoyancy is controlled by oil storage. They
are hunted for oil, particularly the squal-
ene-manufacturing sharks. See *marine
oil, rays, sharks.*

Electromagnetic spectrum

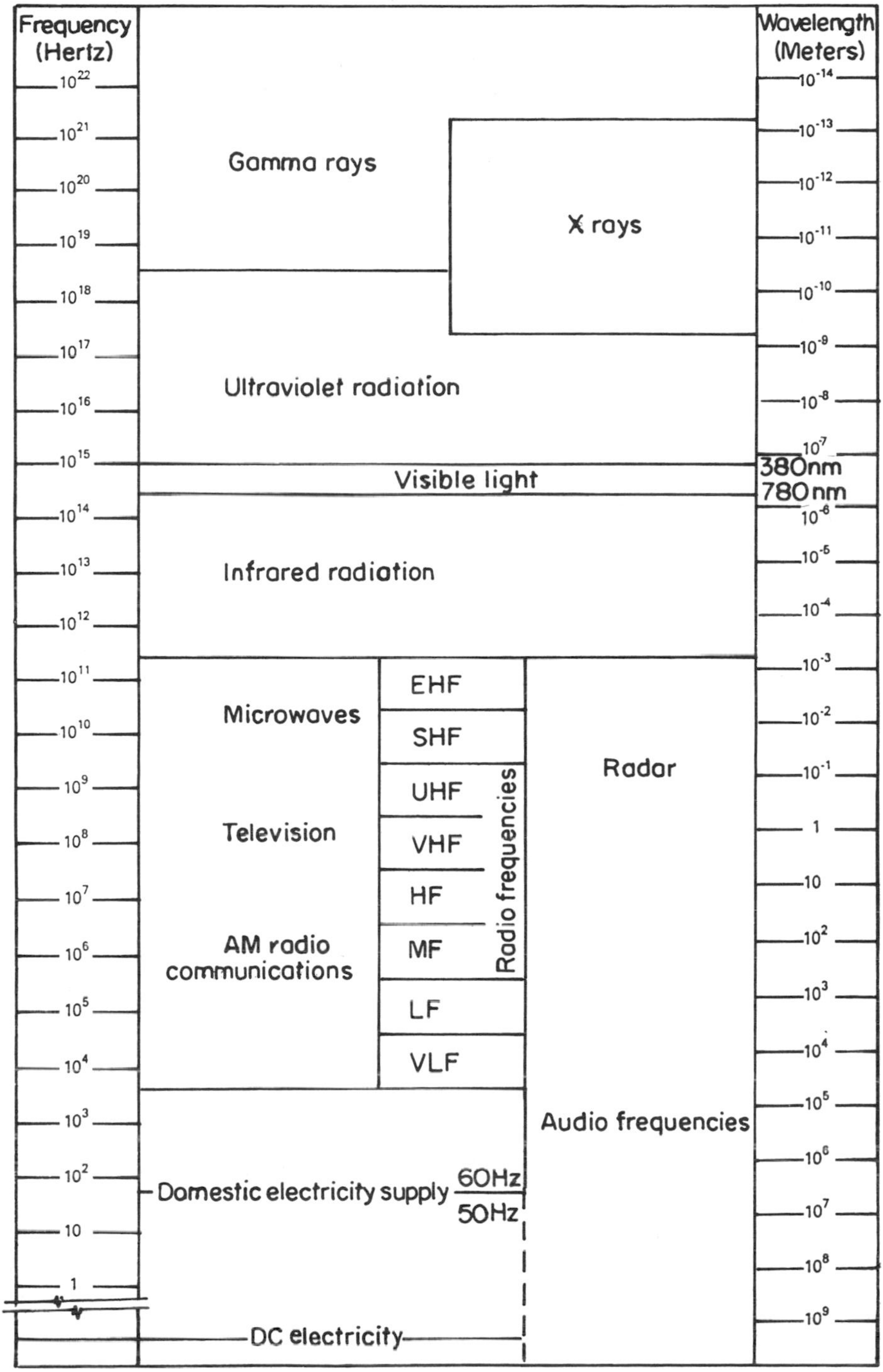

electric eel   A freshwater elongate fish, *Electrophorus electricus,* related to the carp, and capable of giving a strong electric shock with its electric organs. It is not a true eel.

electromagnetic   spectrum   The range of energy emitted by a star or other energy source as waves or particles. The wavelengths in the spectrum vary from about 10 km (6 miles) or more for very long radio waves, to less than 1 billionth of a kilometer for cosmic rays. A wave is represented as a curve about a nodal line. If the wavelength is short, the frequency is great; if the wavelength is long, the frequency is small. Visible light is only a small part of the electromagnetic spectrum.

element   A chemical substance that cannot be decomposed into simpler substances by chemical reactions. All atoms of the element have the same atomic number, and thus the same number of electrons outside the nucleus. See *atom.*

Emperor seamounts   A north-trending line of seamounts (undersea mountain peaks) in the Pacific Ocean. See *Hawaiian Islands, hot spots, seamount.*

endoderm   The innermost layer of cells of an embryonic animal. In the embryo it develops into the digestive and excretory systems.

energy   The capacity for doing work. This transformation is never complete— there is always a loss of energy (entropy). The forms of energy are mechanical, electrical, chemical, thermal, nuclear, and electromagnetic. These are rather arbitrary names because there is some overlap among them. Ocean thermal energy is the energy derived from the circulation of layers of water of different temperatures.

The energy source that is the basis for most but not all terrestrial life is sunlight. See *electricity, electromagnetic radiation, heat, ocean thermal energy, Sun, vent communities, wind.*

English Channel   An arm of the Atlantic Ocean separating England from France. It slopes steeply on the English side to a maximum depth of about 120 m (400 feet). Tidal currents in the channel travel at an average of 1.5 to 3 m/second (5 to 10 feet/second) but at twice this speed in some areas, moving warm water from the Gulf Stream into the colder North Sea, along with a considerable quantity of sand. This is very much a surface-only phenomenon, since there is a high sill blocking entry to the North Sea. The usual wind in the Channel is westerly and strong. See *Dover, Straits of; North Sea.*

Eniwetok   A group of coral islands in the west central Pacific, part of the Marshall islands group. The main island is a volcano whose top subsided. The coral growing on it has produced a limestone cap about 1,400 m (4,000 feet) deep. The structure of Eniwetok has proven Darwin's hypothesis of the origin of coral islands. Darwin postulated the origin of atolls as undersea mountains that subside and form the base for coral growths. The oldest coral on Eniwetok dates from the late Eocene Epoch.

The island was used as an American atomic test site in the 1940s. See *atoll, coral, seamount, volcano.*

Entoprocta   A phylum with similarities to the Rotifera and some to the Bryozoa. Entoprocta larvae sometimes resemble annelids. The adults are small polyps that use tentacles to collect diatoms and other plankton. They are either solitary or colonial. Some species of the class Ascidia (sea squirts), siphynculids, and porifera (sponges) are associated with entoprocts, but the nature of the relationship is unclear. See *Annelida, Bryozoa, polyp, rotifers, sea squirts.*

environment   All of the factors— physical, chemical, and biological—that touch the life of an organism or a community. See *ecosystem.*

Periodic table of elements

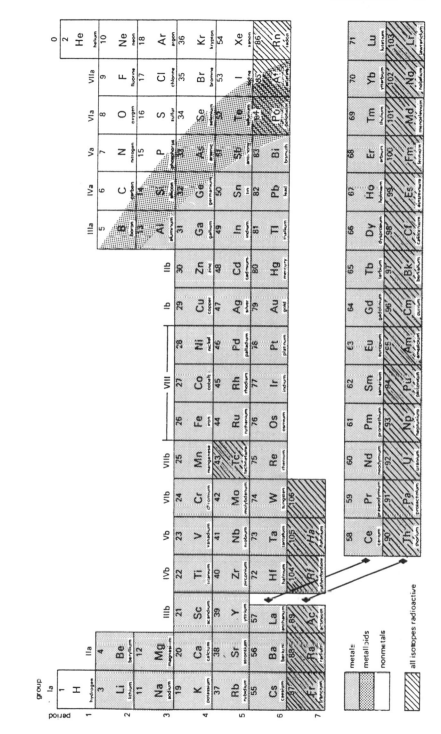

Eocene Epoch   The longest epoch of the Tertiary Period of the Cenozoic Era. This epoch followed the Paleocene and preceded the Oligocene. It lasted about 20 million years, ending roughly 50 million years ago. The name was devised by Sir Charles Lyell and he meant it to represent the "dawn of the recent time," because fossils from it are recognizably like presently found mollusks.

Eocene marine sediment is present in the Alps, the Caucasus, the Himalayas and the Rocky mountains. The sands of the Gulf Coast of North America in which oil deposits are now being exploited were laid down at this time. At the same time, California and the North Pacific coast were undersea, as was much of the Atlantic coast.

Relatively warm seas supported an extensive biota, which continued most of the organisms of the Paleocene Epoch. Birds, teleost (bony) fishes, and siphonate gastropods (snails) thrived, as did echinoderms and nummulites. Plankton also flourished. As the sea level rose, Australia was cut off from other landforms, establishing the "Wallace line." See *cenozoic era; Lyell, Charles;* Nummulites, *Wallace Line.*

epibenthic   A term referring to the region of the ocean just above the ocean bottom and extending to a height of about 3 m (10 feet), and the organisms in it. See *algae, benthos, coral, ecosystem.*

epicenter   The point on the surface of the Earth directly above the focus of an earthquake. See *earthquake.*

epifauna   Animals that live on the ocean bottom, comprising both motile and sessile (relatively stationary) species. See *epibenthos, sessile.*

epipelagic   A term referring to the uppermost layer of the ocean water column, down to a depth of about 200 m (700 feet). See *pelagic environment.*

Equatorial Currents   Currents at the equator that move east to west, pushed along by the trade winds. These currents are fairly shallow and move at 0.5 to 2 knots per hour, or 0.25 to 1 m/second. These currents are separated by a slow, narrower, eastward-flowing countercurrent at about 5° North Latitude. There is also a large, eastward-flowing undercurrent beneath the surface system. When the South Equatorial Current is weak, this undercurrent, which moves at a faster rate (2 to 3 knots/hour or 1 to 1.5 m/second), breaks onto the surface of the ocean. See *currents.*

equinox   The date when the hours of sunlight and the hours of darkness are equal. There are two equinoxes each year, occurring when the sun is directly overhead at the equator at noon. For the Northern Hemisphere, the vernal equinox occurs on or about (leap years make a difference) March 21. The autumnal equinox is September 23, given a day or two variability. The dates of the equinoxes are reversed for the Southern Hemisphere. See *solstice.*

Eratosthenes   (276?–194   B.C.) Curator of the Alexandrian library, mathematician, geographer, and founder of geodesy, the study of the Earth. He believed, as did Pythagoras, that the Earth was a sphere. He measured the circumference of the Earth by a relatively simple calculation: a vertical hole (really a well) at Syene (now Aswan), Egypt, had the sun directly overhead on the day of the summer solstice, June 21. At Alexandria, which was directly north of Syene, the shadow of an upright stick made an angle of 7°12′. By measuring the distance between the two cities, Eratosthenes could then calculate the circumference of a full 360° for the Earth. His measurement was only about 4% in excess of the modern one, despite several errors.

Eratosthenes' map of the known world put Rhodes at the center. His parallels of latitude and meridians of longitude were unequally spaced. The limits of his map were from the Pillars of Hercules (Gibraltar) to the Ganges, and in the north-south span from Taprobane (possibly Sri

Lanka) to Borysthenes, which may be the outflow of the Danube.

Eric the Red (10th century) A Norse sailor and explorer, immortalized by his red hair, who left Europe after a fight in 984 or 985 in which another man was killed. Sailing west, he was heading for Iceland with a party of settlers and was blown off-course to an unknown land. In a prime example of early advertising, he called it Greenland. Eric assumed that giving the new land a good name would entice settlers and encourage them to stay. When the Norse settlers found the eastern coast of Greenland icebound, they moved around Cape Farewell and established a settlement near what is now Godthaab.

Ericsson, Leif (980?–1001) A Norse explorer who is associated with the Viking colony in North America. He had grown up in Greenland and as a young man went on a trading voyage to Norway. The king, Olaf, was surrounded by a court of adventurers who were full of stories of voyages of discovery. The originator of some of these tales was Biarni Heriulfson, who claimed to have seen lands to the west with trees growing on them. This was significant, since both Iceland and Greenland are treeless, as are many of the other islands of the North Atlantic.

In the summer of 1001 Leif Ericsson sailed west. He found one of the islands described by Heriulfson. It is now known as Baffin Island. His party then continued westward. The Norse sagas based on this voyage describe what might be Belle Isle. The next anchorage was most probably L'Anse aux Meadows, Newfoundland. The "vines" of Vinland could have been red currant bushes, which are found at the site. A settlement was established and from the remains seems to have lasted about fifteen years, with continual fighting among the settlers and between them and the "skrallings," or local inhabitants.

estuary The area near the mouth of a river where seawater and fresh water meet.

In the estuary, fresh water draining from the river system dilutes the salt water and tides also push salt water upriver to some extent, depending on the strength of the tide, geologic formations, the offshore currents, and the quantity of fresh water entering the system via the rivers.

All estuaries are geologically "recent". In the last ice age (Pleistocene), with much of the world's water present as ice, river mouths were considerably seaward of their present positions. The present aspect of many estuaries was very different; they were longer river valleys which are now "drowned." Examples of these are the Hudson and the Thames River valleys.

In 1818 the English naturalist John Fleming published a work detailing the mechanism of mixing of waters in a well-mixed estuary, building on the earlier assumption that an "invisible wall" or dam separated fresh water from salt water.

A true, or positive, estuary is one in which more water flows in from rivers than is brought in by the tide. True, or positive, estuaries take several forms. One form, called a salt wedge, occurs when a large volume of fresh water is discharged and flows above the denser ocean water which reaches inshore. There is some, but not much mixing, which results in a sharp difference in the salinity at a given depth. This line of demarcation, the break in the salt level, is the halocline. Some salt water is moved by wavelets into the fresh layer, in a process called entrainment. When rivers seasonally bring great quantities of water and sediment down to the sea and there are no strong offshore currents to disperse this load, deltas form, such as the Mississippi and Niger deltas.

Another type of estuary is formed when there is a strong tide and the river load is more or less the same year-round (this is more likely in temperate zones), the frictional drag of the tide on the bottom serves to stir the water, and the halocline is less steep than in the salt-wedge estuary. In the Northern Hemisphere, fresh water flows to the right as it moves downstream, and salt to the left (in the Southern

Hemisphere the reverse situation prevails), the result of the Coriolis force. The Chesapeake, the Thames, and the Gironde are all estuaries of this classic type.

In a well-mixed estuary system, the tidal range is large and the water in the estuary is thoroughly mixed at high tide. This leaves salt pools and salt marshes when the tide ebbs.

Fiords are deep, glaciated cuts with much freshwater drainage and a sill separating them from open ocean. Some salt water comes over the sill, which is usually less than 100 m deep (300 feet). The fiord is usually deeper than 600 m, and the salt water in it is completely overlain with fresh water. There is very little mixing of the water systems. This is typical of the Canadian Northwest coast, Norway, and New Zealand.

A bar estuary is one built up on a low coast where there is some sediment coming toward the sea, where coastal currents move it around. The Texas coast is typical of such shoals interspersed with inlets whose size and position are highly variable.

Negative estuaries occur in areas where the rate of water loss due to evaporation exceeds that of water inflow from rivers and precipitation. In such a landform, the water flows in from the sea at the surface and out at the bottom.

Because they are partially enclosed, estuaries can become polluted. The pollution may be biodegradable or not; its likelihood of being swept out to sea or rendered harmless depends on both its quantity and the likelihood of the biodegradables in it being attacked by decomposers, sunlight, or sufficient oxygen to degrade it.

The interaction of plants, animals, and their physical world in an estuary is very complex and not at all well understood. Because rivers bring loads of inorganic essential materials, including nitrates and phosphates, turbulence aerates the water, and there is adequate insolation, plants thrive in estuaries. Certain species are arranged in accordance with their tolerance of brackish water, but estuaries are a prime locale for algae of all varieties, from tiny phytoplankton to huge kelps. The fauna living off and on this plant community is also varied and selected for its tolerance to salt water and its exposure at ebb tide.

Each estuary has its own collection of living organisms, which will change with an alteration in any of the factors shaping it. This has enormous significance for the local and ultimately the global ecosystem. See *currents, individual organisms, insolation, marsh.*

Euphausiacea   An order of swimming crustaceans (1 to 6 cm or 0.25 to 1.4 inches) in size. They are found in the open ocean and throughout depths of over 3,500 m (10,000 feet). They constitute about 100 species, and are important links in the oceanic food chain, where they are the principal food of whales. Euphausids look like tiny shrimp; some are transparent, some are red, most are luminescent. This luminescence, which is a product of special photophores, appears on the thoracic appendages, abdominal somites (body segments), or eyes of these animals. Euphausids are filter feeders. See *Crustacea, filter feeders, luminescence, shrimp.*

Eurasian Deep   A part of the Arctic Ocean near the North Pole. It is in the European Basin of the Arctic and adjacent to the Lomonosov Ridge. The water depth averages 4,300 m (13,000 feet). The floor of the Eurasian Deep is a flat, clean plain which tilts slightly upward near the Lomonosov Ridge. The part of the Deep closest to Spitsbergen has the warmest water, since warmer Atlantic water enters the Arctic at that point. See *Arctic Ocean, currents.*

Eurasian Plate   The large portion of the Earth underlying Europe and Asia, extending west to the Mid-Atlantic Ridge. On the east the plate ends in the Sea of Japan, where it meets the Pacific Plate and Philippine Plate. The Java Trench is to the south, and the southern edge of the Plate abuts on the Iran, Arabian, Turkey, Hellenic, and Adriatic Plates, which mark a line of seismic instability. See *continental drift, plate tectonics, individual plates.*

euryhaline A term referring to the capability of an organism to live in environments of variable salinity. See *anadromous fish, catadromous fish.*

eurythermal A term referring to the capacity of some organisms to survive in a wide range of temperatures. See *homeotherm, poikilotherm, temperature of animals.*

eutrophication An accelerated rate of plant growth, particularly the growth of algae, most often associated with fresh water, but which can occur in fairly constricted bays. Algae can "bloom" if there is comparatively little flushing out of an area and a high nitrogen and phosphorus load. The increase in nitrogen, phosphorus, or both may be the result of agricultural fertilizer runoff, manufacturing waste, or the draining of domestic sewage. The last is most noticeable, especially in areas using heavy-duty synthetic detergents. The algae will grow as long as sunlight and temperatures above freezing and below about 25°C are available.

As the algae proliferate, and if there is no large animal population to keep them in check, dense mats of algae form. These choke off the surface and block the water column's access to sunlight. The algae below the surface begin to die for lack of light. The resulting decaying mass deoxygenates the water as it decomposes. The then-poisonous mass makes the water unfit for anything to live in, and there is a rapid die-off of many species.

Eventually the water will become clean again, but unless it is restocked, it will not exhibit the same populations it once had. See *algae.*

evaporation The change of a liquid into a gas. When water evaporates it becomes water vapor. This change requires energy, the natural source of which is the Sun. The warmer the water is, the greater the rate of evaporation. See *condensation, ocean-atmosphere relations, water.*

evaporites Deposits of salts left by the evaporation of a body of water. When water containing dissolved minerals stands, the water evaporates, leaving deposits of gypsum ($CaSO_4 \cdot 2H_2O$), rock salt (NaCl), and a variable mixture of carbonates, sulfates, halides, and nitrates. These salt deposits are formed when either the growth of a reef, or wave action, or shifts in terrain isolate an arm of the sea. The discovery of marine fossils in the same area as salt beds establishes the origin of the evaporites and dates them. See *minerals, sediment.*

evolution The process by which life forms change with time and accumulate successful mutations. There have been many theories in the course of human thought to explain the variety of life found on Earth. For example, Linnaeus, whose system of taxonomic nomenclature is used to describe life forms, believed that the species he saw had been present in the same form from the beginning of time. Earlier, it was held by some that any one species could produce almost any other, thus, barnacles with long necks could be the relatives of geese with similarly long necks.

Erasmus Darwin, the grandfather of Charles Darwin, was a naturalist and contemporary of Georges-Louis Leclerc, comte de Buffon. In 1794 he developed the theory of "gradual transformation." Lamarck, curator of the natural history museum in Paris, believed that the creation of a new species as a response to some environmental stimulus was possible, and that the characteristic difference in that new species was heritable. Lamarck's great critic and rival, Georges Cuvier, was a catastrophist. He believed in a scheme of three great epochs, each of which was devastated by some calamity. The first was that of great fish and reptiles, the second that of small mammals, and the third that of the giant animals: the hippopotami, mastodons, elephants, and the like.

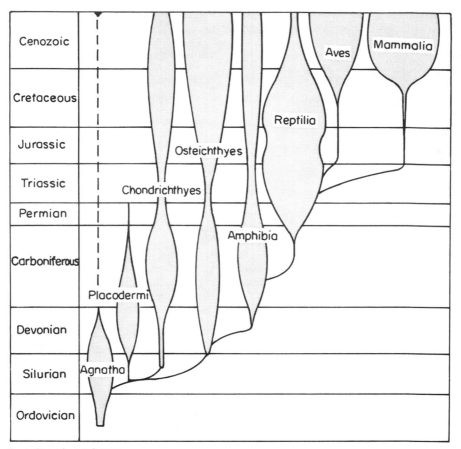

Evolution of vertebrates

The idea of gradual transformation created scandals. The idea of change in species became linked in the respectable (and essentially English-speaking) world with the excesses of the French revolution.

The idea of natural selection was not original with Darwin. Any animal breeder knew how to develop a particular characteristic. However, the theory of natural selection that best explained all the evidence was proposed independently by Wallace and Darwin. It provoked intense controversy and an immense literature. This outpouring still continues, as much effort is devoted to theories of the causes of catastrophic extinctions.

Although Wallace and Darwin ex-plained natural selection, it was not until the work of Gregor Mendel and his concept of the gene was appreciated that the mechanism for this phenomenon was understood. According to this, each individual has a finite possibility of producing a slightly altered chromosome—a change in the DNA of the individual. If this change is not life-threatening, and is viable, the descendants of this individual organism will also have this genetic alteration. Thus, in a given population, many versions of a characteristic are possible.

If the population is geographically isolated, the various separated, related species will continue to change slowly along ever-divergent lines. If, however, the environment changes, only some of the versions may survive the changes. Fairly frequent changes in environment will either lead to extinction or to a rapidly evolving organism, as an individual of one or another genotype (specific genetic makeup) is favored by chance and circumstance.

The aspect of the environment that favors the survival of an individual of a particular genotype can be any one or a combination of a vast array of factors, such as the temperature of air or water or the nutrient value or salinity of water. Alternatively the organism may change color, rendering it more (or less) noticeable by predators (or prey), or it may develop (or lose) an enzyme system that made digestion of a particular food possible (or impossible). A migrating species, (i.e., one that is new to one area but not so different that it would produce sterile matings) may introduce still greater variety into a gene pool and again increase the variability of a population.

Current evolutionary theory explains extinction in terms of wide divergence. If a particular group of organisms becomes so specialized that only a narrow band of conditions will support its existence, a small change in the environment can be fatal. Thus, if a small predator population is *wholly* dependent upon a particular prey, a sharp decline in the prey population (due to any one of many variables) will eliminate the predators because they will be unable to get to enough food for a long enough period to maintain a reproducing population.

Current areas of evolutionary research involve the relationships between organisms, their lines of development and the similarities (or differences) in their structure, and the protein structure and enzyme systems of organisms. See Agassiz, Jean-Louis Rodolphe; Buffon, Georges-Louis LeClerc, comte de; catastrophism, Cuvier, Georges; Darwin, Charles; fossils; Lamarck, Jean-Baptiste Pierre Antoine de Monet, chevalier de; Linnaeus.

evolution of the oceans The currently accepted theory of the evolution of the oceans holds that at one point in the Earth's geologic history there was

only one supercontinent—actually a consolidation of several fragments—called "Pangaea," which was centered around the equator. As Pangaea broke up, the water of the primordial ocean also split into the separate oceanic areas, the Atlantic Ocean being the oldest. Pangaea broke into a distinct northern segment (Laurasia) and a southern segment (Gondwanaland) during the Jurassic Period, about 195 to 135 million years ago. By 150 million years ago, the central Atlantic had spread to about one-third its present width. Europe began to move away from North America, leaving remnants (terranes) on the Carolina mountains. Bits of North America are in Scotland. By the Lower Cretaceous Period, the Iberian Peninsula had moved counterclockwise, creating the Bay of Biscay. By 60 million years ago the last of the large Atlantic basins formed. This created the Norwegian Sea and separated Greenland, a North American fragment, from Europe. At the same time, all the continental mass was moving northward and opening the water channels, making possible water circulation between tropical and polar water.

Gondwanaland fragmented about 140 million to 130 million years ago. The Indian landmass split away from Australia and Antarctica somewhat later; the usual figure is 110 million years ago. An island chain formed at the Ninety Degree East Ridge (90°) and acted as a barrier between the deep eastern basin of the Proto-Indian Ocean and the shallower western part. The eastern basin is now the Wharton Basin. The Ridge continued to act as a barrier, and by 53 million years ago came close to the Mid-Indian Ocean Ridge, which once passed north of Australia, but as that continent drifted northward, the Ridge appeared to its south. This opened the channel between Antarctica and Australia, making polar water and Indian Ocean water freely exchangeable.

The Pacific Ocean dates to the time when India moved north of Antarctica. About 100 million years ago the Pacific Ocean area was composed of four large plates whose borders are now not well-defined. They were the Antarctic, Pacific, Kula, and Farallon plates, each moving away from the others. Trenches or subduction zones consumed much of this oceanic crust. The ridges that seamed this Proto-Pacific Ocean were connected to those of the Indian Ocean, which was opening at the same time.

Between 80 and 60 million years ago, the Tasman Sea developed, pushing Australia and New Zealand apart. By 55 million years ago Australia began to move north and away from Antarctica. One of the four plates that formed the area, the Kula Plate, which lay between Siberia and the Pacific coast of North America, all but totally vanished into the Aleutian Trench. The Farallon Plate disappeared about 27 million years ago; it fragmented into the remains of the Gorda, Cocos, and Nazca plates. See *Atlantic-type margin; continental drift; Cretaceous Period; Jurassic Period; mid-ocean ridges; plate tectonics; trench; Wegener, Alfred Lothar; individual oceans; individual plates.*

Ewing, William Maurice (1906–1974)

An American geophysicist and the first director of the Lamont Doherty Laboratory of Columbia University. He was instrumental in the reexamination of Alfred Wegener's theory of continental drift, and explained it by the idea of seafloor spread, or creation of new bottom at the mid-ocean ridge. Ewing took the first seismic recordings in open seas (1935) and the first deep-sea photographs in 1939. He was one of several scientists who suggested that earthquakes and the undersea rifts associated with oceanic ridges were linked. One of his major publications was *The Floor of the Oceans: The North Atlantic* (1977). See *continental drift; Wegener, Alfred Lothar.*

explorers and explorations

An explorer is defined as one who travels in search of geographic or scientific information. However, the Europeans of the 15th century and those explorers that followed, until the 18th century, were impelled to travel great distances for what some have called the causes of gold,

God, or guns. We know their names but they were not the first ocean explorers, nor were they unique. Polynesians had travelled vast distances in open ocean, navigating by stars and returning to particular island groups. Their voyages were most certainly not unidirectional. Chinese sailors navigated successfully as far as Africa, making regular trips. Arab mariners and traders had well-established routes along the African and Indian coasts.

The first name we know in connection with exploration is that of Necho, the Pharaoh who financed a Phoenician circumnavigation of Africa. Herodotus, in the 5th century B.C., produced a map of the world that stretched from Gibraltar to India. The astronomer-geographer Pytheus was also an explorer. He sailed north to Iceland in 325 B.C. using celestial navigation to arrive at a measure of latitude. Eratosthenes measured the Earth and developed the concept of latitude and longitude. Ptolemy, in the 2nd century A.D., continued and expanded on these works.

The Romans were users of geographic works, but were not particularly interested in exploration for the sake of science. After the fall of the Roman Empire, the center of exploration shifted to both the east and west. Brendan and the Norsemen in the west continued to travel and chart new lands; Arabs in the east did the same.

With the Renaissance and the renewed interest in the world in general came the potential of new trade. That and the technological advances of better ships, better navigational instruments, and more accurate maps made long voyages possible. These were undertaken both for national honor and for profit. This was the age of da Gama and Columbus.

These early navigators were followed by traders and, in the 18th century, by the mainly scientific explorers such as Cook, LaPerouse, and Bougainville. Naturalists accompanied the geographers, thus reviving a practice of the ancient Greeks. Banks, a botanist, and Darwin, a natural scientist, were on voyages of exploration as "scientists on board."

By the 19th century it was recognized

by all maritime nations that maps and soundings of all parts of the world were important. This led to a renewal in various arctic interests and the voyages of the Rosses. Surveying techniques were developed by Maury and others.

Increasing information led to the need for consolidation and classification of data. The need to know and to provide as much information as possible as a public service to all scientists led to the *Challenger* Expedition around the world in 1879, and then the Polar Explorations of Nansen in the early 20th century. The *Meteor* Survey was the major work of the 1920s, although its report was not published until the 1950s.

Exploration in the second half of the 20th century has produced images sent to Earth from space, photography from the deeps, and the development of free diving. New genera unseen and unknown before 1950 are now classified. Yet we still know more about the moon than about the Earth's ocean deeps. There is much left to learn. See *history.*

extinction   The end, or disappearance of an organism. Throughout geologic time, many taxonomic families and groups have become extinct. At least four—some researchers say five or more—distinct periods saw large-scale or mass extinctions. The disappearance of a taxonomic family was an ongoing process that lasted about a million years.

Under normal extinction patterns, fewer than eight families per million years disappeared. The major variations in this occurred around the ends of the Ordovician, Permian, Triassic, and Cretaceous periods. At the end of the Cretaceous up to 19.3 extinctions per million years were noted.

When calculating the number of extinctions, allowances have been made for those organisms, both plant and animal, that are rarely preserved. Shells and skeletons become fossilized, but soft tissue usually disappears without leaving a trace.

There have been a number of theories to explain extinctions. It is no coincidence, however, that the major breaks occur at the ends of geologic periods. The original 19th century geologic time scale was constructed on the evidence of large-scale change, and geologic periods ended with more or less abrupt (give or take a million years) changes.

Salinity changes, the deepening of ocean basins, and meteor impact have all been proposed as causative agents of extinction. Changes in the salt content of oceans would kill off marine organisms if there were suddenly less salt, and would poison freshwater species if the salt content increased. If ocean basins became deeper, marsh organisms would be drowned or littoral life-forms become stranded on dry beaches. Meteor impacts might have showered the Earth with dust and changed the weather much as a large volcanic eruption does now. The anomalously large iridium layer corresponding to the late Cretaceous extinctions may be an example of this phenomenon.

Changes in the polarity of the Earth are still another possible cause of extinctions. Some researchers have also argued that the level of extinctions and emergence of new groups has slowed because the more adaptable—meaning more highly evolved—organisms are better able to withstand environmental shock. Therefore they are more long-lasting or durable than those in past ages that were felled by small but, for them, significant changes.

F

Fahrenheit, Gabriel Daniel (1686–1736) A German meteorologist and instrument-maker renowned for his work on the measurement of temperature and the design of a thermometer scale. Fahrenheit studied physics in the Netherlands, Germany, and England, and was elected to the Royal Society in 1724.

In devising his thermometer scale, Fahrenheit used ice and salt to establish the lowest temperature he could make. He called this 0°. His second point of reference was that of a normal human body. He then divided the interval by units of twelve. On this basis the freezing point of water is 32° and the boiling point is 212°. The Fahrenheit temperature scale is no longer used in most of the world, England and the United States being the two major exceptions. The world's entire scientific community uses the Celsius scale.

Possibly more important than his design of the thermometer scale was the experimentation Fahrenheit did using liquids other than water as indicators of temperature. His attempts at low-temperature work led to the use of various alcohols as indicator liquids, and alcohol is still used in this manner. See *Celsius, temperature.*

Falklands Current A cold polar current flowing parallel to the eastern South American coast; it meets the Brazil Current at about 35–40° South Latitude. See *Atlantic Ocean, Antarctic water, Brazil Current, explorers and exploration.*

Falkland Islands A group of islands on the continental shelf off the southeastern coast of South America, less than 300 km (about 200 miles) west of the Straits of Magellan. They are windswept, rocky, treeless, small peaks perched on a very old Devonian landmass that is analogous to that of the South American mainland. The islands are inhabited by over a half million sheep and several thousand people, mostly of British descent. The islands are claimed reluctantly by Britain and vigorously by Argentina, the latter claim being based on an older claim by the Spanish crown. This unresolved series of claims and counterclaims led to a war between Great Britain and Argentina in 1982. See *Antarctic Ocean, Devonian Period, explorers and explorations.*

Falklands Plateau An undersea plateau east of the Falkland Islands and at a considerably greater depth than the ocean floor surrounding the islands. It ranges from 2,000 to 3,000 m (or 6,600 to 9,900 feet). Steep escarpments separate it from the trench east of South Georgia Island and the Scotia Ridge to the south. See *Atlantic Ocean, Antarctic water.*

FAMOUS French-American Mid-Ocean Undersea Study. In 1974 this expedition explored the axis of the Mid-Atlantic Ridge. See *Mid-Atlantic Ridge, mid-ocean ridges, ocean floor.*

fan A geologic formation at a river or canyon mouth. It is the characteristic spreading pattern of a river-borne load of sediment.

Faroe Islands A group of islands directly north of Great Britain; they are a part of Denmark. There are 17 inhabited islands and many small, uninhabited islets in the Faroe group. The group is one of a series of volcanic peaks with high, steep cliffs and very irregular coastlines.

The climate is mild due to the surrounding warm water of the North Atlantic gyre (water circulation). There is almost daily rain and fog. The dominant native animals are birds, notably the puffins and eider ducks. Because of the constant wind, the islands are essentially treeless. The first colonists of the Faroes were Irish monks, who came about 700 A.D. The Vikings arrived about a century later. See *eider ducks, gyre, puffins.*

Fast Ice   Sea ice that is frozen onto a shoreline and is, therefore, stationary. See *floe, ice, iceberg.*

fathom   An old measure of water depth, now defined as 1.83 m (6 feet). The word is of Danish origin and means outstretched (arms). It is supposed to be the length from left middle finger to right middle finger of a large man standing with his arms outstretched.

fault   A fracture zone in the crust of the Earth. Displacement of blocks of crust occurs at faults. A fault block is a piece of crust, usually rectilinear. A fault block that is pointed down is called a graben; one that is tilted up is called a horst.

fauna   The animal life of a place at a particular time. See *ecosystem.*

feather star   See *Crinoidea.*

felucca   A small sailing vessel of ancient design. These craft were once quite common in the eastern Mediterranean; they are now almost exclusively Nile River craft. They usually have two masts—infrequently there is only one—and lateen sails. The smaller feluccas had rowers.

fiddler crab   A member of the genus *Uca;* a true crab with its first cheliped, or large walking leg, highly developed into a pincer claw. This oversized right claw is a feature only of the male; females have symmetrical legs. The modified leg is used in courtship and mating. See *backshore, chela, crab.*

Fiji   A group of southwest Pacific islands at about 15° South Latitude and 180° West Longitude. Fiji is composed of over 200 islands, of which fewer than half are inhabited. The population is Polynesian and Melanesian. The islands are an important resort area and also produce agricultural products.

Geologically the Fiji islands are interesting because they are not the usual atolls of volcanic and coral construction, but have continental basement rock. See *Fiji Plate, islands.*

Fiji Plate   A small but separate part of the Earth's crust, originally attached to Antarctica. It is therefore "continental" and its various large islands are a composite of continental igneous and sedimentary layers, with adducts of volcanic eruptions and coral encrustation. The smaller islands in the Lau group are truly oceanic and form a tail which gives the entire Fiji Island group a "lambchop" configuration.

The Fiji islands rest on the Plate, which is divided into two areas. The North Fiji Basin (Pandora) is about 2000 m (6,600 feet) deep and circumscribed by the Melanesian Border Plateau, the north-south line of the New Hebrides, and the Hunter Ridge. The South Fiji Basin is a triangularly-shaped area with an average depth of about 4,000 m (12,000 feet), pointed at New Zealand. The "tail" of the Fiji "lambchop" is separated from the almost true north-south line of the Tonga-Kermadec Trench by the Tonga-Kermadec Ridge. The slippage of the Plate is northwest in the South Fiji Basin and southwest in the Pandora Basin. See *Pacific Plate.*

filter feeders   A variety of animals living mostly on the ocean bottom or in the plankton, whose feeding mechanisms are a filter, a means of creating a current running through the filter, a scraping device to move food from the filter, and a water exit.

The filter's function may be accomplished by a mucus net put out by the animal and ingested along with anything it has trapped. The filter itself may be made of cilia, setae (and in crustaceans) antennae or antennules, and may be part of gill (mollusk) or maxilliped (crustacean leg) structures. The cilia, or setae, may beat in concerted waves to create a current which moves detritus or plankton past the net to entrapment. Either the cilia or setae or maxillipeds then deliver the item to the mouth.

The filter will absorb particles of an acceptable size and reject those that are too large. In the sponges and sea cucumbers the entire body acts as the filter, water inlet, and water exit. The baleen whales, which are the largest filter feeders, have blowholes that expel the water ingested along with the krill that is filtered out of the water. See *benthos, cilia, detritus, krill, plankton, setae, sponge.*

Finland, Gulf of   A 400 km (250-mile)-long arm of the Baltic Sea between Finland and the Soviet Union. Its maximum width is 130 km (80 miles). The depths range from 8 m (26 feet) to 100 m (290 feet). The Gulf is separated from Lake Ladoga, which was once a part of it, by the Karelian Isthmus. The eastern end of the Gulf is quite narrow, being only about 20 km (12 miles) wide.

The bottom of the Gulf is sandy and has many bars, which, along with the heavy winter ice, make navigation difficult and perilous. The Gulf receives the outflow of several rivers, the largest of which is the Neva, the river that runs through Leningrad. See *Baltic Sea.*

fiord (or fjord)   A waterway in a rocky or mountainous region that may be currently glaciated or have been covered by moving ice in the past. The coastal cliffs are deeply infiltrated by seawater. Fiords are characteristic of the Alaskan, Chilean, Norwegian, and southern New Zealand coastlines. See *estuary, glacier.*

fish   Aquatic animals of the superclass Pisces, which breathe by means of gills. The jawless fish, the Agnatha or Ostracoderms, are sometimes put into a phylum of their own. Fossil agnathans are the evidence of the earliest vertebrates; they were well-established in the Ordovician Period (500 million years ago). Agnathans had, in addition to bony skeletons, formidable bony armor plate.

The living species of fish represent a considerable portion of all the members of this superclass. The classes and orders of fish are as follows:

Class Agnatha
    Order Cyclostomata
        Lampreys and hagfish
Class Placodermi (known only as fossils)
        These creatures had both jaws and armor. They have been found in some very ancient rocks
Class Chondrichthyes (cartilaginous fish)
    Subclass Elasmobranchii
        Order Selachi
        Sharks and dogfish
    Order Batoidea
        Skates and rays
    Order Holocephali
        Chimaeras
Class Osteichtyes (bony fish)
    Subclass Actinopterygii (ray-finned fish)
        Superorder Chonrostei
        Most of these fish are fossils only. The sturgeons, distinguished by their ganoid (tile-like) scales, are living examples of this group.
        Superorder Holostei
        Most of this group is also extinct; the mudfish are living specimens.
    Superorder Teleostei (spiny-finned fish)
        About 95% of all living fish are in this group of about 20,000 species.
    Subclass Choanichthyes (lobe-finned fish)
        Order Crossopterygii (stalked fins. Most species extinct)
        Suborder Rhipidistia
        Suborder Coelacanthini
        Order Dipnoi (lungfish)
        These animals have swim bladders that act as air reservoirs. A major evolutionary advance in these fish is the nose-to-mouth connection. This "minor" improvement makes possible the transition to air-breathing.

Although fish exist in an incredible variety of forms, sizes, colors, and habits, they share several unifying structures. One is their covering; their skin is covered by colorless, dermal scales. The color comes from the dermis above and below the scales. Fossil scales are osmoid, either bony or ganoid. The extant ray-finned fishes also have ganoid scales. These contain ganoin, a hard, translucent material similar to tooth enamel. The arrangement of the scales in ganoid fish resembles floor tile, being edge-to-edge as opposed to overlapping. Sturgeon scales have no ganoin but the pattern is ganoid. *Placoid* scales are present in the Elasmobranchii; each scale has a rounded element with a spine or dentine projecting toward the tail. *Ctenoid* scales are thin, translucent, and overlapping, with comblike projections on the free edge; their lines of growth vary with species and season. *Cycloid* or roughly circular scales are found on some primitive fish; this type of scale is present earlier in geologic time than is the ctenoid scale.

The digestive system of fish is fairly simple. A mouth is the passage for food, and water used in respiration. A short esophagus leads to the stomach and digestive glands. The liver in some species is unusually large—a phenomenon leading to much speculation but as yet no concrete answers. The anus (or cloaca in some species) is anterior to the tail structure.

Respiration is conducted by the gills. These are made up of filamentous epithelial tissue arranged in tiny parallel lamellae in the interior gill. The exterior gill is covered by ectoderm. The elasmobranchii have gill slits as the external opening, the teleost fish have an operculum. This gill cover acts as a one-way valve. Water rushing into the gill is passed through the lamellae, which remove oxygen from it and deposit carbon dioxide and gaseous nitrogenous waste into it, after which it passes out of the mouth. The removal of most of the waste nitrogen from the body by the gills explains why fish have rudimentary kidneys, if any.

Swim bladders may be respiratory organs, but usually arise as an offshoot of the digestive system. They are the equilibrating hydrostatic structures that make change of depth possible without the animal experiencing decompression problems (i.e., the bends). The mechanism by which air is introduced into the swim bladder is not known, nor is there a satisfactory explanation of why fish or other diving organisms without swim bladders can manage rapid changes in pressure without experiencing illness.

The reproductive mechanisms in this vast group of water creatures vary enormously. Most fish are either male or female, but some are hemaphroditic and some change sex during their lifetimes. Most produce vast numbers of externally fertilized eggs with little yolk supply. This means rapid hatching and many fry that do not live to grow up. Other fish, such as many elasmobranchs, have internally fertilized eggs that develop either from a large yolk, (the ovovivaparous fishes) and produce live young after a long development, or have a placenta-like organ that nourishes the embryos (the viviparous fishes).

The circulatory systems of fish are analogous to those of other vertebrates. They have two-chambered ventral hearts and closed circulatory systems, and have red blood cells that contain hemoglobin as an oxygen-carrying substance. Oxygen can also diffuse directly into the blood from the gills and through the skin. This may explain why there is not very much hemoglobin in fish blood. It constitutes about 2% of the blood in most teleosts, as compared to about 15% in humans.

The sensory reception of fish is good. While there is probably no sense of taste because the palate is too bony and the tongue lacks papillae, the sense of smell is excellent. The senses of hearing and sight also seem good, since the receptor areas for these senses in the brain in all varieties of fish are well developed. The lateral line of fish acts as a receptor for vibrations.

The generalized fish body plan has an oval shape with fins attached laterally and a tail at the end. The tail in teleost fish is a symmetrical one, its upper and lower halves, being of equal size. In cartilaginous fish the upper lobe of the tail is larger than the lower one. The fins of fish are most often paired appendages of dermal origin (as is the tail). Thus, the pectoral and pelvic fins are paired. The dorsal, anal, and ventral fins may or may not be present in any particular species, but when present are usually unpaired. See *color, lateral line, parasites, schooling, individual fish.*

fishing   The catching of marine or freshwater fish and other organisms for food or sport, individually or on a large scale.

Currently the annual worldwide commercial catch is over 75 million metric tons. With some improvements this can be extended to 100 million metric tons. This is considered to be an upper limit, since many fishing grounds are being used to the maximum. A few nations catch the great bulk of fish. In terms of tonnage taken, Peru ranks first, Japan second, and the U.S.S.R. third, with about 25% of Japan's total. In order to prevent overfishing, a number of recent treaties involving two or more North Atlantic nations have been signed with the aim of maximizing the catch while not depleting the breeding stock.

Fishing methods, whether mechanized or not, follow a few basic patterns. Inshore or coastal fishermen rarely exceed the 5 km (3 mile) limit. They do not trawl, and they take most of the known food fish and shellfish, including shrimp, crabs, lobster, herring and the related pilchards and anchovies, cod and its relatives, the pollack and haddock, mackerels, tuna, salmon, flatfish, hake, and mullet. The fishing techniques they use depend on the size of the operation. Pole and line are still used in fishing from small boats. Small tuna are also taken by hand.

The large factory ships or large fleets set purse nets which encircle fish, particularly schooling fish. Drifting long lines bearing multiple hooks are used for tuna. Bottom dwellers such as cod, flatfish, and rays are taken with bottom lines. The major part of the crustacean catch is brought up by trawlers using conical bag nets.

Finding fish can be an art; they do not always appear in the same places. Schools move in response to water temperature, food supply, salinity, wind and water currents, and other yet unknown factors. They are spotted by the likelihood that they will be in a certain place at a certain time of year, and more scientifically by using echo sounding, bioluminescence, or the observation of accompanying species. The last technique is based on the frequent observation of tuna with accompanying porpoises, for example, or of birds on the surface following herring schools. See *mariculture, nets, seines, individual fish and shellfish species.*

fish oil   Oil obtained from the liver of fish, principally cod and sharks. The main commercial use of fish oil is as a dietary supplement, since it is a good source of vitamins A and D. While this use has declined, the oils are still used in the manufacture of paints and coatings. See *marine oil, oil.*

flagella   Tail-like organelles on a cell surface, which move rapidly and usually rhythmically. They are structurally similar to cilia, the difference being one of size and number: flagella are larger, stronger, and present in smaller numbers than cilia. Structurally, both flagella and cilia are bundles of muscle fibrils arranged in distinct packages which divide longitudinally when the cell or organism bearing these organelles divides by fission or reproduces asexually.

Flagellate motion involves a bending backward of the flagellum which both undulates and rotates about a lateral axis. The body of the cell then rotates and moves forward in response to this motion, with an action analogous to that of a screw propeller. See *cilia, dinoflagellates, plankton.*

flatfish   Fish with both eyes on the same side of the head, such as sole, flounder, or halibut. The eyes and skin of flatfish, however, differ from species to species. The larval flatfish has one eye on either side of the head, but as the larva matures, one of the eyes migrates so that both are on the same side. Depending on the species, either the left or the right eye dominates and stays in place, while the other eye migrates. There are other changes in flatfish during maturation. The scales on some flatfish may become cycloid (smooth and rounded) on the underside and ctenoid (with sawed edges) on the superior surface. The swim bladder also tends to disappear on maturation. See *fish, flounder, scales.*

Flinders, Matthew (1774–1814)
An English explorer and navigator, he served on a voyage to Australia with George Bass. They became friends and colleagues and surveyed southern Australia together. Flinders circumnavigated Australia in 1802. The voyage was a very difficult one and his ship, the *Investigator*, almost did not survive it. Most of the crew died and Flinders was never healthy afterwards.

Flinders did very careful work determining the extent of compass deviations due to iron components in the ship. His precise calibrations of compass error and deviations due to nearby ferrous material were vital after the advent of metal ships.

Flinders was to have returned to England after his voyage on the *Investigator,* but he was shipwrecked and captured by the French on Mauritius, where he remained a prisoner until 1811. He finally returned to publish *A Voyage in Terra Australia* in 1814.

FLIP   Floating Instrument Platform, a structure used for making acoustical soundings in the open ocean. This equipment was designed by the Scripps Institution of Oceanography in La Jolla, California, and looks like an oil rig. It must be towed to its research site. Once it is on site its ballast tanks are flooded and the platform flips to its upright and functioning position. See *oceanographic equipment, Scripps Institution.*

floe   A relatively flat piece of floating ice. Ice floes vary in size from quite large (over 1 km in diameter) to smaller than 25 cm (less than 1 foot). See *glacier, iceberg, pack ice.*

flora   The plants of a given locale at some defined time. See *ecosystem.*

Flores Sea   A body of water separating the islands of Flores and Sumbawa from Sulawesi (Celebes) to the north. The Makassar Strait is also north of the Flores. The bottom of this small sea is composed of four distinct regions. The west is a submerged plain with depths of 500 to 1,000 m (1,500 to 3,000 feet). The Postiljon and Pasternoster islands are seamounts that have accreted coral rising up from this plain.

Another feature of the Sea is the Flores Trough, parallel to the Sunda Shelf. It has a maximum depth of over 5,000 m (16,000 feet). The ridges south of South Sulawesi Sea are marked by islands, of which Salajar is one.

It is believed that the Flores Sea is the result of downwarping and subsidence of continental areas. The focus of seismic activity in Indonesia, a region where earthquakes are well known, is in the Flores Sea. See *Pacific Ocean, seamount, seismic activity, subduction.*

Florida Bay   A body of water at the southern end of the Florida peninsula, its southern boundary is the string of small islands called the Florida Keys. The western end of the bay opens into the Gulf of Mexico and is delimited by the last of the Keys, the Dry Tortugas, which are west of Key West. However, Key West, at about 81° West Latitude, is sometimes referred to as the last key.

The Bay and the keys are of geological interest. The bay is quite shallow, and calcium carbonate precipitation in it is an ongoing process that has continued throughout geologic time. This is a present-day example of the process that has produced the limestone beds found in various places on Earth and used by paleontologists and geologists interested in the age of such formations. The keys are exposed coral reefs which formed in the last ice age (Pleistocene) and are still growing by coral accretion. See *Caribbean Sea, coral, Gulf of Mexico, Gulf Stream, limestone.*

Florida Current   The large volume of water leaving the Caribbean and moving northeast at an average speed of 2 to 4 knots/hour through the constriction between the Pourtales Terrace and Cuba which forms the Straits of Florida. The Straits, which separate the Atlantic and the Gulf of Mexico, are quite narrow and only about 800 m (2,500 feet) deep. As a result, the warm Florida Current flows into the Atlantic Ocean with great force. This current forms part of the Gulf Stream. See *Caribbean Sea, Gulf Stream.*

flounder   (Pleuronectiformes)   A widely-distributed marine flatfish taken for food. Flounder are bottom dwellers whose eyes are on one side of the body. There are two families of flounder: right-eyed (Pleuronectidae) and left-eyed (Bothidae). A right-eyed fish has the right eye migrate. The shape of the head is also deformed. Both have free-swimming young. The right-eyed types, about 100 species, are found in north temperate and polar waters. They include the halibut, plaice, winter flounder *(Pseudopleuronectes americanus)*, and starry flounder *(Platichthys stellatus)*. All of these are fished commercially.

The Bothidae comprise about 200 species inhabiting warmer water, and most of these are smaller than the right-eyed types. The sand dab, California halibut, turbot, and summer flounder *(Paralichthys dentatus)* are among the species belonging to this group that are fished for food. See *fishing, flatfish, halibut.*

flowmeter   An instrument for measuring the velocity of water. A flowmeter usually has a wheel, propeller, or rotor whose revolutions per minute can be transmitted to a monitoring station as a measure of the velocity of the water which causes the rotor to turn. Flowmeters are designed to be attached to ocean bottoms or to a stationary object, although some float on a buoy. They may be anchored at some height above the bottom when the object is to determine the velocity of ocean currents at varying depths. See *currents.*

flying fish   (Exocoetidae)   A small fish (usually less than 45 cm or 18 inches) with large winglike pectoral fins, found in tropical and semitropical waters. Some, such as the California flying fish, which looks like a four-winged flying object, also have enlarged pelvic fins. Flying fish build up speed in the water and break out of it using their fins as gliders, while the still-submerged tail acts as a propeller.

fog   A meteorological phenomenon that occurs when the water content of the air is at saturation. Fog usually occurs when warm, moist air contacts cold air or water. The resulting temperature drop forces the equilibrium between water vapor and water in the direction of liquid water. This is condensation. If the water condenses around dust or other particles in the air, fog results. Fog is an almost constant feature of some regions, notably polar areas of air or water convergence. See *polar convergence.*

food chains   Interlocking dependencies of organisms on each other and on their environment. Algae and other chlorophyll-containing organisms, utilizing solar energy, extract minerals from the water, perform photosynthesis, and store the resulting carbohydrates. They are in turn consumed by other organisms, which are prey to larger animals or, on dying, are consumed by scavengers and bacteria. Intermediate steps in this process involve commensal organisms—those which live together without seeming to affect each other directly—and symbiotic organisms, which do affect each other directly and favorably. Such symbiosis occurs with the sea-anemone and the fish that live among its tentacles and lure other fish to the anemone as food.

The predator population, then, is di-

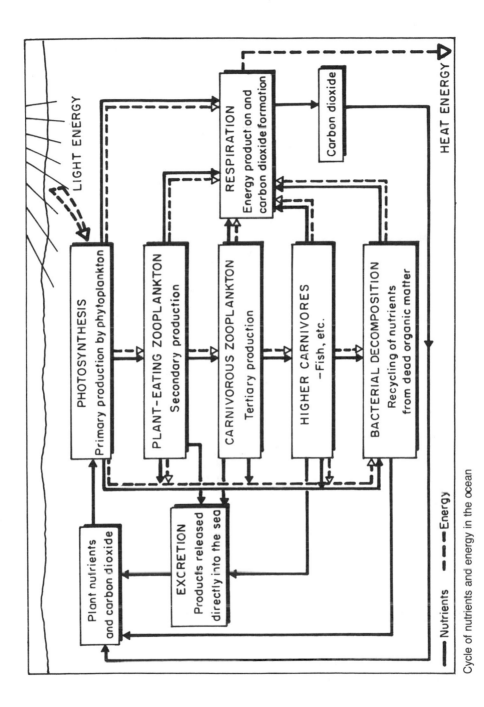

Cycle of nutrients and energy in the ocean

—— Nutrients   ---- Energy

rectly affected by the population it feeds on. Great population fluctuations can occur after natural disasters, earthquakes, weather changes, El Niño, or as the result of human activity, such as overfishing or pollution with oil spills or chemical dumping.

A completely different series of interlocking relationships exists on ocean bottoms at the edges of crustal plates, where hot, sulfurous vents create an environment totally dependent on sulfur as the first element in the food chain. See *coral, ecosystem, El Niño, estuary, sulfur bacteria, vent communities.*

food webs   The interrelationships and intersections of food chains found in a particular place at a particular time. In studying food chains one examines a linear relationship between the plants and animals in the chain and between the animals themselves. In food webs, which are much harder to unravel, the interconnections are frequently not linear or simple. For example, both tuna and sharks eat herring. While the size of the herring population will affect both tuna and sharks, will the shark and tuna populations affect each other? See *ecosystem, food chain, population.*

Foraminifera   A very abundant order of almost exclusively marine protists. Most of them have a test, or shell, either of chitinous or calcareous material. The tests fall to the bottom as the organisms die.

The tests are either single chambered, the chamber expanding as the animal grows, or multichambered, with a series of larger spaces that the animal moves into when it becomes cramped. The walls between the chambers are perforated. The tests are usually colored: Globigerina is often red-brown, but may be pink or dark orange. Other foraminiferan tests are yellow or brown.

There are two types of tests: either a large proloculum (or initial living space) or a megalosphere. The union of two zygotes, in animals whose reproduction is sexual, produces a smaller proloculum. The animal then grows by asexual fission,

and there are a number of asexual generations before the production of zygotes.

Foraminiferans have been incorporated into rock since the Cambrian period, and appear as chalk or limestone. The Egyptian pyramids are faced with limestone produced by fossil foraminiferal products. The ooze of ocean bottoms, ranging from continental shelves to depths of about 4000 m (13,000 feet), consists of unconsolidated foraminiferal tests.

The Foraminifera extant in the Eocene and Oligocene eras were large, measuring several inches across, but most of the specimens found today are quite small, with some tropical exceptions. Some foraminiferans have commensal (not mutually beneficial, merely parallel) relationships with algae. The relationship with other species, however, is symbiotic.

The living foraminiferans are bottom dwellers, some crawl around, others are sessile. Their food consists of bacteria, diatoms, and other algae. In spite of the vast number of Foraminifera, their life history is not well understood. It is believed that foraminiferans can withstand adverse conditions by forming spores, but this is not certain.

Since they are so plentiful and ubiquitous, (about one third of all sea bottom sediment consists of discarded foraminiferan tests) the specific species of these animals are used as index fossils. See *carbonate, chitin, diatom, Globigerina, Protista, symbiosis, test.*

Forbes, Edward (1815–1854)   An English naturalist, Forbes was the scientist-on-board the H.M.S. *Beacon,* commanded by Thomas Graves, which charted the Mediterranean. In the 1840s, prior to the voyage of the *Beacon,* Forbes had studied the marine animals of the English coasts and documented the different groups of species that occur at various ranges of depth. His *Beacon* voyage allowed him the chance to extend this work to warmer and deeper waters.

Forbes divided organisms into eight groups with regard to depth and the type of bottom they inhabit. He noted, quite accurately, that not all sea bottoms are capable of supporting the same ecosystem. Some are literally undersea "deserts." He noted that the deeper his sample came from, the less abundant was the array of organisms. He theorized that plant life disappeared at a depth of 100 fathoms (200m or 660 feet). He extrapolated this to a "zero point" for animals too, concluding that any remains found on deep bottoms died higher up and sank.

This idea of an "azoic" zone at about 300 fathoms (600m or 2,000 feet) so enthralled the scientific community that in spite of evidence to the contrary in existence at the time of Forbes' publication of his work in 1844 (e.g., John Ross's starfish brought up from a depth of 800 fathoms (1,600m or 5,300 feet) in 1818, and Aimé finding animals at 1,800m or 900 fathoms), the idea of an azoic zone hung on. Even the work of a *Beacon* officer named Spratt, who continued Forbes' research, and a refutation by David Forbes, Edward's brother, in 1869 did not completely dispel this comfortable but erroneous idea. See *benthic ecosystems, hadal zone.*

foreshore   That part of the shore between high and low water—the intertidal zone. See *beach, intertidal zone, littoral.*

fossil   Evidence of an organism's past history: it may be the original organism, a mineral deposit in the interstices of the organism's skeleton or soft tissues, a mineral replacement of the organism's tissues, or an impression of the entire organism or its tracks. Since living things decompose quickly, as opposed to mineral replacement, which is a slow process, there are very few fossils relative to the numbers of individuals that lived. The fewer hard parts an organism had, the less likely there is to be a fossil of it at all.

Early attempts to explain fossils ranged from scientific to theological. The ancient Greek, Xenophanes, and the Chinese of the 1st century B.C., both recognized fossils as remains of once-living plants or animals. The Chinese by the 6th century were trying to classify the "stone fish" they knew. These ideas were not revived in the West until after the Renaissance.

Although canal builders regularly turned up fossilized remains and wrote accurate descriptions of them, as did the natural historians of the time, this was not widespread information. Leonardo da Vinci described fossils in 1500, as did Bernard Palissy (1510-1590) in France later in the century, but the official position of all religious authority in Europe forbade any scientific explanation for them. [The official Western religious view of a six-day creation certainly did not allow for any creation to arise and disappear in that period.]

In 1706, the governor of Massachusetts, Thomas Dudley, wrote to Cotton Mather describing a mastodon tooth as the remains of a giant killed by Noah's flood. Twenty years later another theologian (Johannes Beringer, in Germany) viewed fossils as instruments of God (or Satan) sent to test men. Beringer gave accurate accounts of both real fossils he had found and those his students had "baked" for him as practical jokes.

Eventually, as religious authority lost control of the scientific community, the study of fossils increased. Also, large public works such as the construction of canal networks turned up more and more fossils.

By the 18th century and the work of Linnaeus and then Cuvier (1769–1832), it was established that these remains of formerly living things could be fitted into an almost continuous pattern of living organisms, with some related to organisms currently existing.

With the growing interest in stratigraphy (the study of the Earth's strata) in the 18th and 19th centuries, fossils became identified with distinct eras in Earth history. The fossilized flora and fauna of each epoch (i.e., the fossil record) gave a clue to successive stages of life's development. Through the work of William Smith (1769–1879), Cuvier, and Charles Lyell (1797–1875), fossil analysis became, and still is, the stratigrapher's tool

ERAS	PERIODS	YEARS x10^6	Angiosperms	Gymnosperms	Pteridophytes	Bryophytes	Bacteria, Algae, Fungi	Protozoa, Porifera Simple Metazoa	Myriapods	Insects	Molluscs	Echinoderms	Fish	Amphibians	Reptiles	Birds	Mammals
CENOZOIC	Tertiary	65															
MESOZOIC	Cretaceous	135						Ammonites									
	Jurrassic	195															
	Triassic	230											Trilobites				
PALEOZOIC	Permian	280															
	Carboniferous	355															
	Devonian	405															
	Silurian	440															
	Ordovician	510															
	Cambrian	590															

The fossil record

for identifying unknown strata. Although various 19th century natural scientists (even Charles Darwin!) were skeptical about the fossil record, eventually an extended fossil history for mankind became accepted. See *evolution, individual organisms, individual eras and periods of geologic time, palaeontology.*

fracture zone   A long, narrow, fresh break in the ocean floor, with sheer drops of more than 1km. In the Atlantic, fracture zones are more-or-less perpendicular to the Mid-Atlantic Ridge, where the north-south orientation of the line of the Ridge is offset by numerous fractures. The East Pacific Rise has no corresponding single ridge, but more spectacular fracture zones. Of these, the Clipperton, Clarion, Murray, Molokai, and Mendocino fractures extend for thousands of kilometers perpendicular to the edge of the North American Plate.

Magnetic anomalies are offset by the fracture zones. This is to be expected from the theory of seabed spreading. Fracture zones are not usually sites of volcanic activity unless at the edge of a plate.

While the fractures in the Atlantic and Pacific oceans usually run east-west, those in the Indian Ocean run north-south or north northeast-south southwest, demonstrating the complicated series of shifts of continental plates in that area. See *magnetic anomaly, mid-ocean ridges, ocean floor, plate tectonics.*

Fram   The ship on which the Norwegian explorers Frijdtof Nansen and Otto Sverdrup drifted across the Arctic Ocean in the late 19th century. The name means forward. This ship was built according to Nansen's specifications; it was a three-masted schooner and had a rounded hull that enabled it to rise onto pack ice and not be crushed by it. Nansen and 12 others were frozen into the ice near the New Siberian Islands on June 14, 1893. They drifted with the ice, hoping to cross the pole, thus proving the nonexistence of an arctic landmass. They missed the pole but proved their point. The *Fram*

drifted with the ice for three years before it broke free in 1896 near Spitsbergen. It returned to Norway on April 16, 1896, eight days after Nansen. Nansen and a companion had left the ship on skis, hoping to reach the pole. They got as far as 86°14' before turning back.

Later, Roald Amundsen used the *Fram* on his expedition to the South Pole. The ship is now in a museum in Oslo, Norway. See *Amundsen, Roald; Ekman, Vagn Walfrid; Nansen, Frijdtof.*

Fram Strait   A body of arctic water separating the east coast of Greenland from Spitsbergen. Cold water of low salinity moves south fairly slowly along the surface of the strait. This comprises the upper 30 to 50 m (90 to 150 feet) of water in the strait and is the exit for a great part of the Eurasian Basin runoff. The very distinct layers of arctic water make mixing almost impossible, particularly with the warmer, more saline intermediate water that moves north.

The Fram Deep is a very small cut in the Fram Strait, with reported depths of over 5,000 m (16,000 feet). It has not been thoroughly explored, however. The bottom of the strait is highly varied, with seamounts rising from it. One seamount comes to within 730 m (2,400 feet) of the surface pack ice. See *Arctic Ocean, Fram, pack ice.*

Franklin, Benjamin (1706–1790)
An American polymath who made contributions to oceanography among many other sciences. Although he did no systematic work in the field, his observations and comments on significant phenomena led to systematic work by others. Thus, Franklin was the first on record to refer to the Gulf Stream as a "river in the ocean." He also observed the phenomenon of internal waves, which are waves created by motion at an interface or boundary where saline water is overlain by fresh water, such as in an estuary of the fiord type. Internal waves have only recently been given extensive study, and are important in the understanding of water movements.

"Pouring oil on troubled waters" is an ancient cliché. Franklin tested the validity of this statement on a pond. Oil on the windward side was spread by the wind over the surface of the pond, and its waves were indeed stilled. If the waves on the pond were wind-created, the layer of oil (it would have to be a very thin film) would cut the aeration of the water that creates the foaming crests of waves. This is due to the surface tension of the oil particles and their lack of adhesion to water particles. The smaller the body of water involved, the more likely it is that the film of oil will completely cover the surface, and the more effective it will be in stilling wave cresting. See *Gulf Stream, waves.*

Franklin, John (1786–1848?)

An English naval captain and explorer, the nephew of Matthew Flinders, the explorer of Australia. Franklin was in the Royal Navy from boyhood; he had been at the battle of Copenhagen in 1801. Since he showed talent as a surveyor, he was chosen to accompany Buchan on the first of that captain's voyages to the arctic in 1818, which charted western North America. By 1821 Franklin had been promoted to captain and was introduced as a member of the Royal Society.

A second voyage to North America and the Northwest Territories began in 1825. From 1836 to 1843, Franklin was in the Southern Hemisphere as governor of Tasmania. In May of 1845, as commander of the *Erebus* and accompanied by Francis Rawdon Crozier, captain of the *Terror,* he sailed into the arctic to find the Northwest Passage. The ships were last seen afloat on July 26, 1845. After the ships became icebound, the surviving crews attempted to walk south from King William Island to a settlement. None survived. Captain (later Admiral) Francis Leopold McClintock, on one of the many exploratory expeditions financed thereafter to find Franklin, found the log of the expedition, the last entry dated April 28, 1848. It is unclear whether Franklin made the last entries. The expedition had found the Passage but didn't know it.

A small cemetery on Beachey Island holds the remains of three crew members who died in January 1846, when the expedition was still well-equipped and in reasonable health. While the graves were known for years, they were untouched until 1981, when Dr. Owen Beattie of the University of Alberta opened the grave of John Torrington. Beattie believes that Franklin lived at least a year after the foundation of the cemetery. The Inuit (Eskimo) who witnessed the end of the Franklin party thought them mad. They may have been acting in a bizarre manner from lead poisoning brought about by eating food preserved in badly soldered cans.

free diving   See *diving, SCUBA.*

fresh water   Water which is drinkable and of almost no salt content. Fresh water enters the ocean most regularly at river mouths. Some fresh water is introduced into seawater through artesian springs under the sea. Water of relatively low salinity is also produced as glaciers and icebergs melt. The result is a specialized ecosystem in bays and estuaries, where the salinity is very different from that in open ocean. See *estuary, salt marsh.*

frigate   A class of naval ship. In the days of sail it was a three-masted ship that carried 24 to 38 guns on one gundeck. Frigates were more maneuverable than ships of the line. The name frigate was revived in World War II, and is now used to mean a smaller warship of multipurpose capability.

frigate bird   Also known as the pirate bird, a member of the family *Fregatidae,* order Pelicaniformes. The alternate name is the result of the bird's habit of stealing food from other birds while in flight. The frigates are large birds and almost entirely black. The females, which are larger than the males, have white undersides. The frigate has albatross-sized wings with a span of over 2 m (7 feet). It is almost constantly in flight. The birds nest in trees on warm coasts on all continents. Both parents care for the single chick. Frigates live on fish and small crustaceans, which they skim out of the water. The male frigate bird has a spectacular courtship display, inflating a large, bright-red, featherless neck. See *bird, pelican.*

fringing reef A reef attached to an island or a continent. The seaward side may be submerged and, therefore, a navigational hazard. See atoll, coral, island.

Frobisher, Martin (1535?–1594) An English explorer who made three voyages west in search of the Northwest Passage. The first expedition sailed in 1576; it explored the Davis Strait and other points in the Canadian north. Frobisher brought back minerals, sure that he had found gold. He was deceived by mica and pyrites, but made a good enough case for the commissioning of a second voyage. In 1577 he sailed into Frobisher Strait and the next year returned to the arctic again. This time he commanded a fleet of fifteen ships. They charted the waters they explored, but found nothing of commercial interest, and the project was abandoned.

The last voyage marked the end of Frobisher's career as an explorer. He later sailed with Drake (1585) to harass the Spanish and then took part in England's defense against the Spanish Armada. In 1590 and again in 1592 he was in the West Indies, attempting to capture the Spanish treasure fleet. In 1594, he sailed for France to defend the port of Brest against Spanish attackers, and was killed in the battle.

front The leading edge of an air mass. Both polar and tropical air masses move into the temperate regions of the Earth from west to east. A warm front is a warm air mass moving into a cold one. A cold front means that cold air is invading a body of warm air. In either collision the

warmer air rises above the cold air. Thus, the water vapor in the warm air condenses as it rises into colder regions of the atmosphere, and precipitates as rain or snow. See *polar front, weather.*

frustule The silaceous shell of a diatom. See *diatom.*

fucoxanthin A red-brown pigment present in and characteristic of brown algae. It masks the chlorophyll in these plants. See *algae, Fucus.*

Fucus With *Sargassum,* it constitutes the Phaetophyte, or brown algae. *Fucus* grows in temperate intertidal zones. Specimens of 2 m (7 to 8 feet) long and more are known. The holdfast cells (anchoring cells), midribs, air receptacles, and conceptacles (reproductive cells) of this alga are examples of cellular differentiation. In *Fucus* a diploid organism produces oogonia or antheridia, which in turn produce eggs or sperm. There, when they unite, produce the zygote that grows into a new plant.

Sargassum is usually a tropical plant. It looks like a vascular plant because it is branched. It is algal and a truly floating species that has pillowlike structures to buoy it up. It grows luxuriantly, sometimes forming dense mats. *Sargassum* in turn supports an ecosystem of crustaceans, mollusks, small fish, and plankton that live on the plants or hide in the branches, feeding on the alga or on each other. See *algae, Phaetophyta.*

fulmar A gull-like oceanic bird. The Procellariidae (order Procellariiformes) includes petrels and shearwaters—scavengers who eat everything from dying fish to carrion and garbage. As a result, they are called "stinkers." Fulmars are usually white or grayish white. Their habitat is near the Earth's poles. The great fulmar of the antarctic, *Macronectes giganteus,* is about 90 cm (3 feet) long and has a wingspan of over 2 m (6.5 feet); most fulmars, however, are closer to half that size. See *petrel, shearwater.*

Fulton, Robert (1765–1815) An American inventor best known for his invention of the steamboat. In his first career Fulton was an apprentice jeweller and then an artist; he finally settled on engineering. His first success was a submarine, the *Nautilus,* whose development had been supported by a grant from the French government in 1806. Although it dove, neither the French or the English or the American navies were interested in it.

Fulton's attempts to power vessels with steam were more interesting to private backers. He displayed a steam-powered craft on the Seine in 1803, and by 1807 ran his famous steamboat, which used an English engine, up the Hudson River in New York. By 1815 he had constructed an oceangoing paddlewheel steamboat, the *Fulton,* this time for the U.S. Navy.

Fundy, Bay of A large inlet on the North American Atlantic coast, lying in a northeast direction separating the Maine and New Brunswick coast from Nova Scotia. It is about 240 km (140 miles) long and about 75 m (250 feet) deep. At its head end, the Bay splits into a forked tail, with Chignecto Bay to the north and west of the Minas Basin.

Because of its shape and orientation, the Bay of Fundy, and particularly the Minas Basin, exhibits some of the highest tidal ranges in the world. The standing wave is over 2 m (7 feet), and the tide range is greater than 17 m (56 feet). The tide enters the Bay along the south shore and leaves via the north shore, creating an eddy within the bay area itself. Periodically the idea of utilizing the enormous force of the moving water to generate electricity has been explored but there has yet been no commercial exploitation of this energy source. See *river bore, tidal range.*

Fungi Essentially land plants that have no chlorophyll. They live on dead plant tissue. The origin of fungi is unknown, but one theory is that they developed out of filamentous seaweeds, whose structure and growth pattern they duplicate. See *algae, heterotrophs.*

Fury and Hecla Strait A short, barely navigable stretch of water in the Canadian arctic, it is part of the boundary of the western end of Baffin Island. Hudson's Bay is south and east of the Strait. The Strait is part of the so-called southern route of the Northwest Passage. See *Northwest Passage.*

G

Galapagos Islands An island group isolated in the Pacific Ocean about 1,000 km (650 miles) west of Ecuador, of which the Galapagos are a part. The group consists of 14 large and many smaller volcanic outcroppings near the mid-ocean ridge where the Pacific Rise branches off into the California Rise and the Easter Island Rise.

Darwin visited the Galapagos Islands as part of the explorations made while he traveled on the *Beagle*. He was not the only naturalist to observe the highly differentiated biota on the islands. Many have noted the exotic animals: the tortoises, iguanas—both terrestrial and marine—sea lions and penguins.

The islands are unusual since they lack coral. They are temperate due to the cold Peru Current that touches them in spite of their position near the equator.

Recent geologic exploration of the Galapagos has shown that some of them formed under the sea while others have limestone layers indicating coral accretion at some time in the geologic past. Espinola (or Hood) Island has an eroded undercone with pillows of younger lava above. This is evidence of a much greater age for these islands than was previously theorized. The erosion means that at some time in the past the island was above sea level, making its age about 3 million years.

Another instance of the changing face of the sea bottom in this area is the Galapagos Spreading Center, a site of hot springs which bring calcium and magnesium into the sea water. This region is part of the Mid-Pacific Ocean Ridge. See *Darwin, Charles; mid-ocean ridges; Peru Current; volcano.*

Galilei, Galileo (1564–1642) An Italian mathematical physicist who discovered the laws of falling bodies and the parabolic motion of projectiles. Galileo was perhaps the outstanding physical scientist of his time. Although he did not invent the telescope, he made one and used it to observe objects on Earth as well as in the skies, including sunspots and the phases of Venus. His use of the telescope to plot many new stars led to experimental refutation of the "perfect Aristotelean" cosmos with the stationary Earth at its center, and brought Galileo into conflict with the Church. His publication of the *Dialogues on the Two Chief World Systems—Ptolemaic and Copernican* led to his being considered a dangerous heretic. His last published work, *Discourses Concerning Two New Sciences . . . ,* was published in 1638 in Leiden, the Netherlands.

Galileo spent his last years working on the use of the pendulum in clockmaking, thereby laying the foundation of precision timekeeping.

galleon A redesign of the carrack. John Hawkins, an English navigator, is credited with the modification of that large commercial vessel to create the galleon. The change removed the large forecastle that had caught the wind, forcing the ship to leeward (away from the wind). The elimination of the large bow structure made the ship more maneuverable and better able to tack, or sail into the wind. See *carrack.*

galley A warship of the Mediterranean; galleys existed before 3000 B.C., and were finally retired in the 18th century. The typical Greek galley was about 5.5 m (18 to 19 feet) wide and about 33 to 34 m (130 feet) long. The oldest examples had a line of rowers on each side.

Later there were biremes with two banks of oars and triremes with three. The long oars of the upper decks of the biremes (two-tiered galleys) and triremes were manned by several rowers for each oar. It is thought that the rowers were freemen or pirates in the early days of the galley; from classical times onward they were most often slaves. Sails were used on the mast and there was most often only one square-rigged mast. The oars were the chief means of propulsion, making the vessel very maneuverable. The principal weapon of the galley was the ram—a metal-sheathed protrusion from the bow. After the use of gunpowder became widespread, small cannon were also mounted on the galley's bow.

Gama, Vasco da (1460?–1524)

A Portuguese navigator who made three voyages to India and established Portugal as a world power. He first fought for the Portuguese king John II, and was involved in the capture of a French fleet. John was succeeded by Manuel in 1495. The latter, like his father, was interested in funding voyages of exploration that would lead to the opening of a route to the Indies that would allow Portugal to bypass Moslem intermediaries. Da Gama and his four-ship squadron sailed from Portugal on July 8, 1497. They were accompanied by Dias, who was en route to the Gold Coast (now Ghana). They all spent several weeks on the Cape Verde Islands and finally came round the Cape of Good Hope in early 1498. Da Gama stopped in Mozambique, where the expedition found much treasure and stories of Prester John, the mythic king of a Christian kingdom. A characteristic stone pillar was left to claim the land for Portugal. The group left the African coast in April and made a 23-day run across the Indian Ocean to Calcutta (then called Calicut).

The trade component of this voyage was a failure. There were Moslem traders on site in Calcutta whose goods were much superior to those of the Portuguese, who had come with what they had used on the West African coast. The

Indians were unimpressed. Again leaving a pillar, the Portuguese left for a long and difficult voyage home. Much of the crew died of scurvy and the remainder limped into port in July of 1499.

After Cabral's expedition, a small factory (trading post) had been left in Calcutta, protected by a garrison. They were overrun and slaughtered. When the news finally drifted back to Lisbon, da Gama was sent to India as an avenger. Sailing in 1502, he stopped again in the Cape Verde Islands and then on the coast of East Africa where he captured an Arab trading vessel, killed the crew, and then proceeded to India. There he again captured several vessels and killed their crews. He returned to Portugal in October 1503.

After this punitive expedition, da Gama took an inactive role in Portugal's seafaring ventures. He was the king's advisor on Indian affairs, but largely spent his time as a country gentleman. In 1524, the administration of the reestablished factory in India having fallen into chaos, da Gama was sent to reorganize it. He died, presumably of overwork, in December 1524, in Cochin, India. See *Cabral, Pedro Alvarez; Dias, Bartolomeu*.

Gambia Abyssal Plain

A long valley perpendicular to the Mid-Atlantic Ridge and to the African coast. The Cape Verde islands are north of it and the Sierra Leone Rise is to the south. See: *abyssal plain, Atlantic Ocean, Cape Verde islands*.

gamma rays

See *atoms, electromagnetic radiation*.

Ganges River

The most important North Indian river (2,500 km or 1,550 miles) long. However, it is not the longest river in India: One of its tributaries, the Brahmaputra (2,900 km or 1,800 miles), is longer. The combined system drains a territory of about 4.5 million hectares or 422,000 square miles. The area consists of clay, sand, and silt interspersed with peat bogs. There is evidence of forestation in earlier times. The abundant river water of the Ganges comes from Himalayan meltwater and the summer monsoon rains.

The mouth of the Ganges at the Bay of Bengal opens into a series of shifting channels. The eastern fringe of the delta is swampland. To the west, the Sundarbans (the word means beautiful forest) is a wildlife preserve. See *delta, Indian Ocean.*

gannet   A fish-eating sea bird found in arctic to temperate waters. It is the largest northern sea bird. The gannets are variously classified as *Morus,* or with the boobies as *Sula.* They are in the order Pelecaniformes. The adults are white with black wing feathers and "black eyes." They produce a single egg and the chick reaches adulthood in 3 to 4 years.

Crowded colonies of gannets are a feature of sea cliffs in Canada, Greenland, and northeastern Europe. See *bird.*

gaper clam   See *horse clam.*

gar   A mainly North and Central American freshwater fish, although some species are found in brackish and salt water. Gars breathe air some of the time.

Gars have descended from the Eocene epoch; their fossils are found both in the Americas and in Europe. Their long slender bodies, behind small heads and sharp, strong beaks, are encased in an armor of ganoid scales. The alligator gar of the southern United States grows to lengths of 3 m (10 feet) and more. Technically the gar is edible, but few people try to eat it.

Unrelated species, such as the needlefish and its relatives, are sometimes mistakenly called gars.

GARP   The abbreviation for Global Atmospheric Research Project, an international group created to map and follow meteorological phenomena. It also tracks the effect of these occurrences on the world weather system. See *weather.*

gas   One of the three states of matter; the others are solid and liquid. Gas exists in sea water, which dissolves the various components of air. Surface water is saturated with oxygen, nitrogen, carbon dioxide, and the rest of the gases in air because of its contact with air.

As water moves, its gas content is affected by the temperature encountered at different latitudes because the temperature determines the solubility of gases in water. Additionally, subsurface waters are usually less well aerated because less gas can dissolve in water that is under greater pressure.

Respiration and the decomposition of organisms tend to consume more oxygen than the subsurface water layer can generate. This means that subsurface layers of water will have less oxygen dissolved in them than will surface water, regardless of the pressure they are under. This in turn limits the kinds of plants and animals that may be found at a given layer of water. See *BOD, carbon dioxide, nitrogen, oxygen.*

Gaspé Current   A current that moves St. Lawrence River water around the Gaspé Peninsula in Quebec, Canada. The current skirts the southern shore of the Gulf of St. Lawrence before merging with the Labrador or Cabot Current. The combined flow moves through the Cabot Strait. The Gaspé current is strong in summer and weak in winter, when the river and its tributaries are icebound. See *Cabot Current, Cabot Strait, Labrador Current.*

Gastrotricha   Microscopic round worms, comprising both freshwater and marine species. The head and ventral surface have cilia, and the worms glide in the water. Some are pelagic living in the open ocean and feeding on diatoms, bacteria, or protozoa. They have a sucking pharynx like that of the nematodes. See *Nematoda.*

geochemistry   The study of the chemistry of the Earth, and particularly its crust, waters, and atmosphere. The chemistry of the Earth affects the ocean, since a number of ions are present in both sea water and fresh water. Most often the ions in sea water come from terrestrial runoff; others are introduced by undersea volcanic activity or from vents in the mantle at the mid-ocean ridge lines. See *crust, manganese, mid-ocean ridges, ocean floor.*

geochronology The dating of events on earth by analysis of the radioactive elements in sea water and corals. Several radionuclides are useful in the dating process. Uranium is a terrigenous element brought to the sea by river water. Thus, its incorporation into corals and the tests of foraminiferans dates those structures. The isotope carbon 14 is present in greatest quantity in the oceans, and is useful in assigning an age to fossil shells.

When examining a core sample of ocean bottom and assigning a date to it, allowances must be made for the problem of turbulence. Ocean bottom sediments may be disturbed by bottom currents, the movement of animals, or shifts in the bottom as a result of volcanic activity. Any and all of these may seriously disturb the bottom sediment. See *radioactivity.*

geology The study of the history of the Earth and its life as recorded in rocks. Geologic stratifications were well known by the early 19th century. Since marine fossils were found on land, some early naturalists assumed the reverse would be true and that land fossils would be found on the sea bottom. Not until the *Challenger* voyage reports were published, however, was it generally accepted that this was not the case. However much sea levels had changed in the history of the Earth, the ocean basins were never land areas.

Still, the oceans do change with time. Very obvious events in oceans are rapid erosion and land slippages. Both were noticed by the cable engineers who found breaks and burials of cables even in the absence of volcanic activity. The activating force in the movement of the sea bottom is the underlying movement of the Earth beneath the water. This was very poorly understood for a long time. Another misunderstanding of the age of ocean basins and the rapid events that occur in them, resulted in the ignoring of Wegener and his arguments for continental drift as a long-term, ongoing event. His theory was not considered valid in spite of the striking similarities of the edges of continental shelves and the relationships between the Earth's land masses.

Results of fairly recent dredging that revealed the Earth's youngest rock at the mid-ocean ridge and thickest crust at the continents, together with evidence of the movement of the plates and mantle, have totally changed the geological picture of Earth. See *abyssal floor; catastrophism; continental margin; island arcs; Lyell, Charles; mid-ocean ridges; Murchison, Roderick; plate tectonics; trenches; Wegener, Alfred Lothar.*

Gesner, Konrad (1516–1565) A Swiss theologian and one of the "Encyclopedic Naturalists." Gesner was a protégé of Zwingli, the Protestant reformer. He was attracted to medicine and studied that subject along with theology and ancient languages. He practiced the latter two professionally.

Gesner's five-volume *History of Animals* was begun in 1551. He worked on it for the rest of his life, but the massive collection was not published until 1587, well after his death. Gesner attempted a total reclassification of all animal species. The book was so complete that it achieved instant scientific success and remained a major reference work into the 18th century Enlightenment. It was considered a foundation work by Cuvier, the French expert on paleontology of the late 18th and early 19th centuries, who sought to classify species by their fossil remains. See *Belon, Pierre; Buffon, Georges-Louis Le Clerc, comte de; Cuvier, Georges-Leopold Dagobert, baron de; Rondelet, Guillaume.*

ghost crab See *sand crab.*

giant clam A clam of tropical Pacific and Indian Ocean waters *(Tridacna)* that measures over 1 m (3.3 feet) in diameter and very occasionally closes its shell on divers who manage to step into it.

Because it is so large, the clam is hunted for its shell. One very large specimen is the holy water font in the church of St. Geneviève de la Sorbonne in Paris. See *clam, Mollusca, sea monster.*

giant crab *Macrocheira Kaempferi,* a spider crab found in waters near Japan. The habitat of this creature is usually between 50 and 300 m (150 to 1,000 feet) deep. It is the largest known arthropod, the length of the crab from one claw tip to another frequently measuring more than 4 m (13 feet). It weighs about 18 to 20 kg (40 to 45 pounds). See *crab, sea monsters.*

Gibraltar The limestone top of a granite formation, rising sharply at the southern tip of Spain at the western end of the Mediterranean Sea. The highest point is 450 m (1,350 feet), and the formation is a natural bastion. Because it commands the strait between Europe and Africa and is easily defended, Gibraltar has been a fortified bastion throughout recorded history. See *Mediterranean.*

Gibraltar, Straits of A body of turbulent water separating North Africa from Europe (Spain) at the western end of the Mediterranean. Only 13 km (8.5 miles) wide at the narrowest point, the Straits are very ancient, apparently having existed when the Mediterranean was much larger than it is at present. The depth of the sill separating the Atlantic from the landlocked Mediterranean controls the water circulation in the latter. Cold Atlantic surface water of relatively low-salinity spills into the Mediterranean and warm, relatively high-salt water leaves the Mediterranean and enters the Atlantic. See *Atlantic Ocean, Mediterranean, sill.*

Gilbert, Humphrey (1539?–1583) An English soldier, navigator and relative of Sir Walter Raleigh, Gilbert was the leader of an English movement for the colonization of the Americas, which resulted in the English annexation of Newfoundland.

His interest in the New World began with the proposal for expeditions to find the Northwest Passage. In 1578 he was given a charter entitling him to plant a colony in America, so long as it was not in the dominion of any other Christian prince. He sailed in November 1578 with seven ships, but could not control his men; some turned pirate and the others drifted back to England. After a second attempt, which left Plymouth in June of 1583, Gilbert arrived in Newfoundland and claimed it for England. At the end of August, this expedition began moving south, but several ships were wrecked near Sable Island (Nova Scotia) and the rest turned east in a storm. Gilbert's ship, the *Swallow,* went down near the Azores. There were no survivors.

gill A respiratory organ of fish, mollusks, crustaceans, and some worms. A gill is constructed of finely divided, featherlike tissue that is well-supplied with blood. The gill is often enclosed in a supportive or protective structure made of bone, shell, or chitin. In teleost fish the gills are covered by an operculum (covering flap).

The finely divided structure of the gill makes possible maximum contact of the blood in the gill with surrounding medium. The blood brings carbon dioxide to the gill, and this gas is passed from the blood to the surrounding water. At the same time, oxygen diffuses from the water onto the gill and into the circulatory system. In many organisms the water is forced over the gill by muscle action; respiration in gilled organisms is not a passive process. See *carbon dioxide, chitin, respiration.*

glaciers Slowly moving masses of ice accumulated either on mountains or in polar regions. Glaciers are found in areas of heavy precipitation where warm, moist air or warm water meets cold air or water. Ideal sites near oceans are in Alaska, Antarctica, Greenland, and New Zealand. More than 95% of glacial ice is polar.

Glaciers reach an equilibrium between accumulation or growth and ablation or decay. In polar regions this occurs at sea level. Glaciers move slowly; their movement is lubricated by a slush layer under the ice and, barring surges, they will move downhill eventually to sea level, or toward the coast, and into the water as icebergs. Polar glaciers are extremely stable, with little overall change of volume. The most rapidly moving glaciers are in western Greenland, and produce the largest number of Northern Hemisphere icebergs.

Glacial surges are more often than not mountain occurrences. They are of uncertain cause but are very impressive. In a surge, a glacier can move up to 15 m (60 feet) in one day. See ice, iceberg.

glass sponge A sponge with siliceous spicules (support structures) of the classes Hexactinellida and Hyalospongiae phylum Porifera. On drying, the skeleton resembles glass. See Hexactinellida, Hyalospongiae, sponge.

Globigerina Shelled protozoans. Their shells are the single predominating sediment that covers much of the ocean bottom. There is a continuous layer of this deposit in all oceans north of the Antarctic Convergence.

The tests of Globigerina are about 30% calcium carbonate. This compound usually dissolves at depths greater than 2,000 m (6,000 feet). Some anomalous Globigerina deposits exist at greater depths, where the waters are oxygen deficient.

John Murray, in the Challenger expedition's samplings, brought up Globigerina. He correctly surmised that it was a surface-dwelling organism whose test was deposited on the seafloor after the death of the organism. He also correctly explained the absence of deposits of tests at great depths. See Challenger, H.M.S.; Foraminifera; Murray, John; ocean bottom.

Glomar Challenger A drilling ship built in the United States, in operation since 1968. Although primarily an oil rig, it has more recently been used to take core samples. The use of this vessel for commercial drilling, however, depends on the cost of petroleum. If the price of crude oil is high, the high cost of the Glomar Challenger's operation can be justified. During periods of low prices for crude oil, drilling for oil from deep ocean bottoms, is not economical and the ship is used to take scientific samples. See Scripps Institution.

goby A small perciform carnivorous fish. The gobies are a group of about 800 species distributed in all waters. The greatest number of species are in tropical oceans. About 85% of all species are in the family Gobiidae.

The crystal goby (Crystallogobius nilsonni) is found in Europe. It is about 10 cm (4 inches) long and transparent. It is one of the largest species of goby. Gobies may be as tiny as the Philippine goby (Pandaka pygmea), which is only 13 mm (0.5 inches) long.

Many gobies are scaleless. Gobies also lack a lateral line, but do exhibit prominent spots of color. Some gobies are commensals, living near or on large fish and picking parasites—usually copepods—off their hosts. They attach themselves to the larger animals by means of a suction plate that is a modified pelvic fin. Other species attach themselves to rocks or other features of the ocean bottom with the same plate. See benthos, commensal relationships.

Gondwanaland The archaic southern continent that resulted from the breakup of the still older supercontinent Pangaea. Gondwanaland then broke up again to form South America, Africa, Australia, Antarctica, and India. The name is based on the Gondwana region of India. See continental drift; plate tectonics; Wegener, Alfred Lothar.

goniatites A group of extinct clams (genus Goniophora) that appeared in Silurian seas and continued into Devonian times. Their shells have a distinctive heart-shaped outline and a network of radial growth lines. Modern clams have annular-ring-type (once a year) growth lines. See clam.

Gorda Plate A small segment of the Earth's crust under the Pacific Ocean. This plate abuts the western North American plate north of the San Andreas Fault. It has not yet been forced under the North American Plate by the pressure of the much larger Pacific Plate. See *crust, Pacific Plate.*

gorgonians Also known as "horny corals," this group of sea animals includes the sea whips, sea feathers, and sea fans, all of which are related to the soft-bodied corals. Both the soft and horny corals are in the subclass *Octocorallia,* with eight-part symmetry.

Although gorgonians are found in all seas, they are most prevalent in warm waters at maximum depths of about 1,000 m (3,300 feet), but usually in much shallower waters. These sessile (stationary) animals attach themselves to a firm surface by means of a basal plate or stolons, constructed of a calcareous, (calcium-containing) mass of spicules in a horny matrix called gorgonin. These basal plates in turn serve as anchorages for hydroids, brachiopods, and copepods.

Most gorgonians are both graceful and attractively colored: reds, yellows, orange, and various shades of purple are the most common colors. The coral used in jewelry is a gorgonian.

Some gorgonians produce toxic compounds, usually terpenes, as chemical defenses against predators or other animals that would compete with them for the same planktonic food supply. See *chemical defenses, coral, defense mechanisms.*

Gorlo Strait An arctic waterway separating the Barents and White seas in the U.S.S.R. See *Barents Sea; Deryugin, Konstantin Mikhailovitch; White Sea.*

graben A valley formed by rock that is downthrust between two parallel faults in the Earth's crust. Also called a rift valley. See *crustal block, plate tectonics.*

gradient wind A theoretical wind based on the combination of the Coriolis force and atmospheric pressure. This calculation is useful in predicting the wind acceleration. It is vital to meteorology because it provides some basis for predicting the force of wind in a particular storm such as a hurricane. See *Coriolis force, storm, wind.*

Grand Banks A submerged continental margin east of Nova Scotia and separated from it by the Laurentian Channel. Newfoundland is northwest of the Banks, where the average depths are between 40 and 100 m (130 and 330 feet). The Banks are roughly triangular in shape, with Newfoundland forming one angle of the triangle. The northernmost point on the Banks, Flemish Cap, might be thought of as another angle, but it is shallow, and separated from the rest of the Banks by much deeper waters (over 1,000 m, or 3,300 feet).

A part of the remains of the Gulf Stream reaches as far north as the Tail of the Banks, where the cold, low-salt Labrador Current moves south along the coast and into the area.

The Grand Banks are a prime fishing ground and the source of much of the North Atlantic's cod and herring catch. Despite fluctuations in the fish population, overfishing, worries about pollution, international competition, and fluctuation in the cost-to-profit ratio of the industry, this region is still paramount in its tonnage of several commercially caught fish species. See *cod, herring, fishing industry, Saint Lawrence, Gulf of.*

granite Igneous rock that is largely feldspar and quartz. It is both light in color and less dense than basalt. The continental masses are largely granitic at their lowest levels.

grapholite An extinct marine colonial organism that had a chitinous exterior. The grapholites may be ancient relatives of the hemichordates. They first appeared in the Cambrian Period and were most prevalent in the Carboniferous. Their remains indicate their existence for more than 250 million years, and they are found in limestone deposits worldwide. See *chitin, colonial animals, Hemichordata.*

gravimeter An extremely sensitive spring balance that will deflect with minute changes in gravitational force.

After World War I, V. Meinesz made several voyages during which, using gravimeters, he charted belts of gravitational anomalies. The combination of his charts with the seismic readings of the areas gives a picture of the shapes, geologic features, and foldings of the seafloor. Trenches show negative anomalies, island arcs show positive ones. See *magnetic field; Meinesz, Vening; trench.*

gravity wave A wave in which gravity is the force that restores equilibrium. These waves are of relatively long wavelength, being greater than 1.75 cm (0.5 inches). See *wave.*

gray whale A relatively small (about 15 m or 50 foot) long, slender baleen whale. It is usually gray to gray-black, and splotched with white. The grays are found most often closer to shore, as opposed to out in the open ocean. Their habitat is the northern Pacific. See *whale.*

Great Australian Bight The wide (1,100 km or 700 mile) indentation in Australia's southern coast; a part of the Indian Ocean. The Bight and the arid coastline it touches were first explored by Pieter Nuyts in 1627, and charted by Matthew Flinders in 1802. See *Australia; Flinders, Matthew.*

Great Barrier Reef A coral reef that lies east of Australia, stretching from the Torres Strait south of New Guinea (Irian) to the Tropic of Capricorn. It is more than 1,750 km (1,100 miles) long and the world's largest reef. The Great Barrier reef is really a reef system, not a single entity, since it encompasses channels, islands and smaller reefs. The islands of the system's western boundary have navigable channels between them and the mainland. These islands are not all just sandy bars, but vary in both size and elevation. The coral accretion to the reefs grows eastward and provides an incredible series of undersea landscapes, habitats, and exotic animals.

Geologically, the entire region was once a terrestrial landscape of undulating country with ranges of small hills. The Queensland plateau sank—most likely a geologically recent event, and the coral upbuilding on the drowned plain then began. The entire area is one of continuous change, with more upbuilding by coral than there is subsidence. See *Barrier Reef, coral, Coral Sea, Fiji, island.*

great circle The largest circle that can be drawn on a sphere. When the sphere is the Earth, the meridians of longitude and the equator are all great circles. The great circle is of navigational importance since the shortest distance between any two points on the Earth's surface is the great circle that both of those points lie on.

While the theory of great-circle sailing has been known and understood since classical times, it was not technologically possible until the 19th century. Earlier sailors had to be content with inadequate charts and chronometers, which could only give estimates of longitude, as well as the vagaries of wind and current. See *navigation.*

Great Lakes Five large lakes in North America which collectively form the largest body of fresh water on Earth; they cover an area of about 245,000 km^2 or 95,000 square miles. Because of their size, they are tidal and also subject to storm surges, as is the ocean. The lakes descend in elevation from Lake Superior, the westernmost, to Lake Michigan, Lake Huron, Lake Erie, and Lake Ontario. The last empties into the St. Lawrence River.

The lakes were formed in the Pleistocene Epoch (about 4 million years ago); the exact mechanism of their formation is explained by a number of somewhat differing theories. The present lakes are remnants of larger and more extensive bodies of water. They have had their present conformations, however, since the last retreat of the glacial ice, about 10,000 years ago. See *tide.*

Greece *See marine archeology.*

green algae (*Chlorophyceae*)
Algae in which the chlorophyll is not
masked by another pigment. They include Caulerpa, or sea grapes, found
near corals in warm temperate-to-tropical
seas. These plants do not have discrete
cells, but a multinucleate protoplasm; *Ulva,*
Sea lettuce, found in temperate waters;
Enteromorpha, flattened tubules found
in temperate waters; Cladophora, algae
with branching filaments; and numerous
planktonic species.

Nowhere in open ocean waters are the
green algae dominant. They are overwhelmed by the brown species, since
algae with more pigments than chlorophyll alone are more efficient in utilizing
the light they receive. See *algae, blue-
green algae, brown algae.*

greenhouse effect A blanket term
used for describing a worldwide heating
trend linked with a general rise in the
level of atmospheric carbon dioxide.
During the last 200 years the increased
use of fossil fuels such as coal and oil,
and particularly the industrial use of these
fuels, has increased the atmospheric carbon dioxide content by about 10%. Concurrently, the mean temperature has increased by 0.2°C. At the same time, the
amount of airborne pollution generated
in large part by aircraft and automotive
exhausts, vapors, and aerosol propellants
has increased. These gaseous, long-lasting additions to the atmosphere prevent
insolation while trapping infrared radiation close to the surface of the Earth. The
total effect is one of blanketing and raising the temperature in spite of the fact
that preventing insolation by itself would
cause the temperature to fall. See *carbon
dioxide, pollution, temperature.*

Greenland-Scotland Ridge An east-
west ridge in the North Atlantic. It forms
a sill that directs one branch of the Gulf
Stream-warmed Atlantic Ocean waters
between the Shetland and Faroe Islands.
See *Atlantic Ocean, mid-ocean ridges,
Norwegian Current, sill.*

Greenland Sea A body of water that
lies east of Greenland, north of Ireland,
west of Jan Mayen Land, and south of
Spitsbergen. The Sea is divided by the
western Jan Mayen Rise (with minimum
depths of about 1,000 m or 3,300 feet)
into the Iceland and Greenland basins.
The latter is the larger and deeper, with
maximum depths of 2,800 m (9,200 feet)
in the Iceland Basin and 4,800 m (16,000
feet) in the Greenland Basin. The Sea
has been referred to (only by its ad-
mirers) as an arctic Mediterranean.

Since Greenland and Scandinavia are
being actively separated by the mid-ocean
ridge, there are active volcanoes under
the sea producing lava. When arctic air,
which is cold and dry, and polar maritime
air, which is warmer and moist, meet
over the sea, there is heavy precipitation.
Most of this falls as snow.

The water system in this area is a
complicated one. There are four distinct
bodies of water: *Polar water,* which is of
low salt content and cold (−1.85°C),
forms the Greenland Current and moves

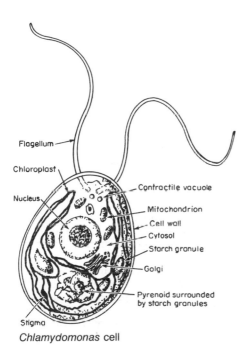

Flagellum

Chloroplast

Nucleus

Contractile vacuole

Mitochondrion

Cell wall

Cytosol

Starch granule

Golgi

Pyrenoid surrounded
by starch granules

Stigma

Chlamydomonas cell

south; *Intermediate water,* which is derived from the Norwegian Current, is saline (relative to polar water) and of higher temperature, averaging about 5°C; *Norwegian deep water,* which is colder and less saline than the intermediate water; and *North Atlantic water,* which comes from the Irminger Current, is 8°C or higher, and has a salt content of at least 3.5%. It exists near Iceland and overlies the Polar water.

Spring and summer runoff further decreases the salinity of the Greenland Sea. The glaciers of Greenland produce icebergs which almost totally cover the Sea's surface in April. See *currents, glaciers, iceberg.*

green turtle   A large pelagic reptile, this tropical animal is in danger of extinction. The adults have been hunted extensively in the Caribbean Sea, and increased use of beach areas has demolished the green turtle's nesting sites. Since the female does not produce eggs every year, and is faithful to her particular native beach, the population has diminished markedly in the 20th century. See *reptile, sea turtle.*

groin   A man-made rocky structure built perpendicular to a beach to inhibit the movement of beach material and to prevent beach erosion. Groins extend from the landward side of the highest high water and continue into the water. They are usually placed in a series of parallel lines, which as a unit encourage the deposition of sand and other beach materials, thus expanding the beach. Care must be taken in the placement of the groins, or the growth of one beach may result in the erosion, undercutting, and ultimate destruction of another. See *beach, jetty.*

grouper,   Any of a large number of species of fish belonging to the family Serranidae, order Perciformes. Two of the most numerous genera are *Epinephelus* and *Mycteroperca.* Groupers live in warm seas in all oceans. They are large, heavy-bodied animals that are usually dark brown or green. Some species are either flecked with red or are completely

red. Some of the Caribbean varieties can vary their color.

Although the groupers are hunted as food fish, some species manufacture toxic substances and are dangerous.

Experimental stations in the Mediterranean have been marginally successful in attempting to raise grouper, but while the juvenile fish can be reared in a lagoon or pool, the adults need considerably more room. The average fully grown adult is about 2 m (7 feet) long and weighs over 200 kg (440 pounds). See *fish.*

growth   The increase in numbers of individuals in a population (members of one species), or the enlargement of an individual. The first depends on the existence of favorable conditions; the second also involves hormonal control without which there can be no differentiation of cells. Growth also encompasses the series of transitions whereby an egg divides in specific ways to produce different tissue and systems in the organism.

grunion   A small (15-cm or 6-inch long) silvery fish *(Leuresthes tenuis),* found in Pacific Ocean waters off the California coast and known for its unusual spawning behavior. The eggs are deposited and fertilized in sandy beds at the highest high-tide mark in spring (April and May). The next high spring tide washes the hatchlings out to sea. When the fish spawn, they are present in such numbers that they can be scooped up by hand. See *fish, tide.*

grunt   One of about 175 species of edible fish related to the snapper. Grunts are usually tropical and small; the average individual is 20 to 22 cm (8 to 10 inches) long. Most grunts are elongated and slender, but a few species are deep-bodied. The color varies from white to black, with yellow, blue, striped, or mottled species also known. The inside of the mouth is often red.

Grunts make noise, from which they derive their name. They grind their teeth and resonate the sound through their swim bladder, thus producing their characteristic sound. See *snapper, swim bladder.*

guano   The excrement of sea birds. In areas of large bird population, notably the islands on the western coast of South America, the guano is commercially mined for fertilizer because its nitrogen and phosphorus content is very high. In some places that have little rain, notably coral reefs, the calcium carbonate of the coral is altered by contact with quantities of bird excrement and becomes more crystalline and sometimes chemically changed. See *El Niño*

Guinea Current   A current found on the west coast of Africa, that changes its location with the season, being more northerly and broader in summer than in winter. The Current moves east, bending along the southern edge of the west African "bulge" as it moves into the Equatorial Countercurrent.

The path of the Guinea Current is well documented. By 1850 it had been noted that sailing ships moving west in the current were held back about 60 km (40 miles) a day. Although it does vary somewhat with season, the current can be generalized as a warm, highly saline, and relatively oxygen-poor body of water. See *Equatorial Currents*.

gulf   A large coastal indentation caused by the downwarping or subsidence of continental shelf areas or a rise in the sea level. The latter occurred most often after the melting of the Pleistocene ice sheet about 10,000 years ago.

Gulf Stream   A current in the North Atlantic, first reported by Ponce de Leon (1460–1521). Its point of origin was assumed to be the Gulf of Mexico, based on Ponce de Leon's statements and those of other Spanish navigators. It is really the northern and then western swing of the North Atlantic Current. In the early 19th century, British explorations, using Humboldt's observations, rightly attributed the high seas and resulting current of the West Indies to the trade winds.

In the 1840s, Alexander Dallas Bache began a long study of the Gulf Stream, thus following a tradition established by his great-grandfather, Benjamin Franklin.

In the 1880s, Lt. J.E. Pillsbury of the U.S. Coast Survey measured the currents of the Stream from an anchored ship, using Benjamin Franklin's phrase "a river in the ocean" to describe it.

The Gulf Stream separates the warm, salty Sargasso Sea from the cold, less saline inshore water. It moves north along the edge of the Blake Plateau at about 2.0 to 2.5 m second (or 4 to 5 knots/hour). This huge stream is 800 m (2650 feet) deep. As it rushes along, it brings with it sudden storms, deep blue water, and phosphorescent displays at night.

About the latitude of Cape Hatteras, North Carolina, the Gulf Stream leaves the coast and changes character. It broadens and deepens to about 1000 m (3,300 feet). The popular concept of the Gulf Stream as a warm current is true only of a part of this huge flow.

North of Cape Hatteras the Gulf Stream flows over the continental rise, and is deflected eastward by the Labrador Current. It also spreads out over the abyssal plain and increases greatly in width. There are large meanders and eddies that form, vanish, and reform. Countercurrents form on the right (south) of the Stream. By the time the system reaches the Grand Banks of Newfoundland, the Gulf Stream has lost about 20% of its volume.

The path of the system and its direct effect on Europe is not clearly understood. Although the North Atlantic gyre brings warmer water into the waters of northern Europe, it is not clear whether this water is of "Gulf Stream" origin. See *Bache, Alexander Dallas; Caribbean Sea; currents; Equatorial Currents; Florida Current; Franklin, Benjamin; Hatteras.*

gull   A long-winged, web footed sea bird of the family Laridae with a relatively heavy body and strong, slightly hooked beak. Gulls are best known in the Northern Hemisphere. About 30 to 40 species live in temperate to polar regions. Gulls migrate but are less likely than terns to travel long distances to their winter grounds. Gulls are scavengers: they eat insects, crustaceans, mollusks, worms, fish, the eggs of other birds, and garbage.

Some commonly known gulls are:

Common Name	Scientific Name	Distinguishing Feature	Habitat
Black headed	*L. ridibundus*	Red legs	Iceland, Europe
Bonaparte's	*L. philadelphia*	Black bill/red legs	North America
California	*L. californicus*		Western North America
Franklin's	*L. pipixan*		North + South America
Glaucas	*L. hyperboreaus*	White/pink legs	Hawaii + Mediterranean
Great black-headed	*L. marinus*	1.6 m (63-inch) wingspan (largest gull)	Polar
Herring	*L. argentatus*	Gray, fleshy legs (most common gull)	Northern hemisphere
Kelp	*L. dominicanus*	Black head	Southern hemisphere
Laughing	*L. atricilla*	Black head/red bill/red feet	Maine to South America
Little	*L. minutus*	Smallest, 60 cm (24 inches)	Europe + North America
Pacific	*L. pacificus*		Australia + Tasmania
Ringbilled	*L. delawarensis*		North America inland
Sooty	*L. hemprichi*	Dark brown	Indian Ocean

The following look like gulls but are not in the Laridae.

Ross's	*Rhodostethia rosea*	Pinkish white	Siberia, seldom in North America
Sabine's	*Xema sabini*		Arctic
Swallowtailed	*Cregrus furiatus*	Unusual tail	Galapagos

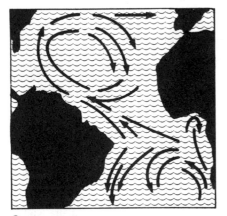

Ocean gyres

guyot   Also called a tablemount. A flat-topped seamount rising from the sea floor. These structures were named in honor of Arnold Guyot, a 19th century geologist, by their discoverer, Harry Hess. See *seamount; Hess, Harry.*

gyre   the roughly circular path of water circulation in the open ocean. The North Atlantic gyre comprises the Gulf Stream, the North Atlantic Current, and the Canaries Current. The return flow is the Equatorial Current. Although this is probably the best defined gyre, gyres exist in all oceans. See *currents.*

H

habitat The space occupied by an individual or group of individuals, including everything in that space both living and nonliving. Typical marine habitats are mangrove swamps, marshes, estuaries and mudflats, intertidal zones, and benthic (depth of the ocean) and pelagic (ocean surface) zones. Within each habitat are specific variations due to the temperature, light, the substratum, and other physical variables. See *benthos, estuary, intertidal zone, polar biome.*

hadal zone The name used for the ecological zone at ocean depths greater than 6 km (20,000 feet). There is no light at this depth; the water pressure is over 600 atmospheres and can be as high as 1,200 atmospheres in ocean trenches. Both the temperature and salinity of hadal zones are fairly constant; the salt content is about 3.47%. The bottom itself is composed of red clay of terrigenous origin, and radiolarian ooze. The dominant animals are omnivorous crustaceans, echinoderms, polychaetes, and mollusks. Present in very small numbers, these animals feed on each other, or on detritus and the mud-and-detritus-eating organisms that exist even under these difficult conditions. See *benthos, pelagic zone.*

haddock A large North Atlantic fish of the cod family. The haddock resembles the cod but has a different coloring, usually dark brown or green, with a noticeably darker lateral line and white or creamy undersides. Haddock are about 84 to 95 cm (30 to 35 inches) long, and are bottom-dwelling carnivores. Like the cod, they comprise an important component of the commercial fishing catch. See *cod, fish.*

hagfish Primitive fishlike animals that, like the lampreys, have no cartilaginous skeleton, no jaws, and belong to the class Agnatha. Also like the lamprey, which is a freshwater animal, hagfish have sucker mouths with many teeth. Hagfish are cold-water organisms that live in burrows and feed on invertebrates, detritus, and dying animals. See *Cyclostomata, fish.*

hake A fast-swimming carnivorous fish that is caught commercially off the Canadian coasts and in the eastern Pacific near New Zealand. Hake are sometimes classed as comprising a family distinct from the cod, the *Merlucidae.* Another genus also commonly referred to as hake is *Urophycis.* See *fish.*

Hakluyt, Richard (1552–1616) An English collector of travel diaries. His famous work was *Principal Navigations, Voyages, Traffic and Discoveries of the English Nation,* written in 1589 when Hakluyt was the chaplain of the British ambassador to the French court.

Hakluyt continued to collect and publish accounts of the sea after the publication of his great work. He was convinced of the patriotic necessity of exploration, and continued to write extensively to support this position. The Hakluyt Society, a British group founded in 1874, was named for Richard Hakluyt as a memorial. His original twelve-volume work, by then much expanded, was included in the society's publications of original texts of travel logs. See *explorers and explorations.*

Haldane, John Scott (1860–1936) An English physiologist who, in conjunction with J.G. Priestley, studied the effect of the partial pressure of carbon dioxide on the control of breathing, a body of research that was published in 1905. Haldane then carried out research on respiration under stress, in mines and caissons. He worked out the method of slow decompression still used in all ocean descents, and particularly in ascents from ocean depths. His work was essential to the design of ventilation systems for diving equipment and submarines. See *diving*.

half-life   The length of time required for half of some substance to disappear. If the substance disappearing is a radioactive isotope of some chemical element, then the half-life is that time necessary for half of the atoms of a given sample to decay and form another element by losing alpha and beta particles and gamma radiation. Depending on the isotope in question, the half life can range from fractions of a second or billions of years. Since the half-life of radioactive isotopes is a constant, the rate of decay of these isotopes can be used to measure time. This is the basis of radioactive dating.

The concept of half-life is also used to measure the disappearance of pollutants, particularly those that are not easily attacked by naturally occurring enzyme systems. See *gamma rays, radioactive dating*.

halibut   A large, right-eyed flatfish of the genus *Hippoglossus*. At maturity the fish is either brown or dark green, with a light underside. Halibut may be larger than 2 m (6 to 7 feet) and weigh over 300 kg (700 pounds). However, since the halibut currently caught commercially are considerably smaller than this, there is the possibility that halibut is being overfished in the north Atlantic.

Other genera are also known as halibut. They are usually smaller than *Hippoglossus*. The California halibut *(Paralichthys californicus)* may have eyes and color on both sides, but is nevertheless related to the left-eyed flatfish. Its range includes Japan and Indo-Pacific ocean water. See *flatfish*.

halocline   A layer of water in which the salt content changes rapidly. It forms a boundary between regions of different salt content.

halogens   The elements fluorine (F), chlorine (Cl), cromine (Br), and iodine (I). At room temperature and atmospheric pressure fluorine and chlorine are gases, bromine is a liquid, and iodine is a solid. These elements are important as ions (F^-, Cl^-, Br^-, I^-) in seawater; the chloride ion is the most abundant and noteworthy of the four. See *bromine, chlorine, element, iodine, water*.

hammerhead shark   A type of shark named for the bizarre shape of its head: a grotesquely flattened T-shape with an eye and nostril at each end of the T. Hammerheads live in warm seas and have a worldwide range. They feed on fish, other sharks, rays, and skates. Since they range into coastal waters and will eat anything, they are potentially dangerous to humans, but prefer a fish diet. Hammerheads are sometimes fished commercially for their skin and oil. The common hammerhead, *Sphyrna zygaena*, grows to lengths of 4 m (over 12 feet) or more. See *shark*.

harbor seal   The common or spotted seal *(Phoca vitulina)*, frequently found near shores or in harbors, particularly fishing harbors, of the Northern Hemisphere. Male harbor seals are much larger than females, with the average male most often being about 2 m (7 feet) long and weighing 120 to 130 kg (250 to 270 pounds). The young are dirty white or gray, and darken to dark grey or black as they mature.

The harbor seal is a social creature, easily tamed, and a voracious feeder on fish, squid, and crustacea. See *seal*.

harp seal   A migratory seal *(Papophillus groenlandicus* or *Phoca groenlandica)*, found in the Northern Hemisphere. It is a far-ranging animal that is a strong swimmer. The breeding grounds of the harp seal are Newfoundland, Greenland, and the ice shelves of the White Sea. The adult is a grayish yellow to brown-black color. Harp seals have strong, light markings, which are more distinct on the males than on the females. Males are about 1.75 m (6 feet) long and weigh about 180 kg (400 pounds). The harp seal is killed for its fur. The young are white for about two weeks after birth, and it is the fate of these pups, who are clubbed to death by hunters, that has stirred conservationists and humane societies to force sealers to agree to quotas and less ghastly killing methods. See seal.

hatchet fish   Fish belonging to two groups; a freshwater group related to the carp, and a marine group akin to the salmon and trout; and belonging to the family Sternoptychidae. The range of the marine group is from warm to temperate waters, and from depths of 200 to 1,000m (650 to 3,300 feet). They are flattened, deep-bodied, small animals with slender, shapely tails. The overall length of the marine hatchet fish is most often less than 10 cm (4 inches). Many hatchet fish have light organs on their heads. All of the hatchet fish are carnivorous. See light organs; salmon.

Hatteras   The Hatteras Abyssal Plain is part of the floor of the northwest Atlantic Basin east of the continental shelf of North America. The Bermuda Rise forms its western boundary.

Cape Hatteras is a sandbar forming a promontory on Hatteras Island, North Carolina. The shallows are a navigational hazard. It is at this point that the Gulf Stream, which has moved north along a line about 150 m (600 feet) off the coast, bends east and meets the cold coastal Labrador Current from the north. The cold air over warm water and the cold water meeting warm water at Cape Hatteras create both water and atmospheric disturbances that produce storms, fogs, high tides, and swift, short-lived eddies. See Atlantic Ocean, Gulf Stream.

heat   Energy that passes from one body to another, because of the temperature difference between them. Heat may be generated by sunlight, chemical or nuclear reactions, or by the conversion of electrical or mechanical energy into heat. Heat waves are part of the electromagnetic spectrum. See energy.

Heezen, Bruce Charles (1924–1977)   An American oceanographer associated with the Lamont-Doherty Laboratory from its inception in 1949. Heezen was involved in many of the 20th century's discoveries about the sea floor. He and Marie Tharp mapped the North Atlantic bottom and then all the world's oceans. He participated in the discovery of the role of turbidity currents, proposed the meteorite explanation of tektites, and was co-author of The Face of the Deep, a book on undersea photography. Heezen's most significant work, done with William Maurice Ewing of the Lamont Doherty Laboratory was the elucidation of the ridge-rift valley complex that is the most continuous feature on the Earth. In plotting the ridge profiles, Tharp noticed a dip at the centerline of the ridge. The plotting of undersea earthquakes coincided with these rifts. See Ewing, William Maurice; Lamont-Doherty Laboratory; mid-ocean ridges.

Heliozoa   A subclass of the Protozoa having helical form. Most heliozoans are freshwater species. Almost all of these organisms form scaly shells, or tests, with numerous projections; the animal resembles a watery, microscopic version of a dandelion head. See Protozoa, test.

helium   An inert chemical element that occurs as a gas in the Earth's atmosphere, of which it comprises 0.0005%. Helium is used in high-pressure deep-diving equipment because its solubility in blood is much lower than that of nitrogen, and its use can therefore prevent nitrogen narcosis or "the bends." See air, diving, element.

Hellenic Plate   A small section of the Earth's crust which bears Greece and the Ionian islands. It is being pushed against the Eurasian Plate by the African Plate. Seismic activity is frequent at the edges. See crust, plate tectonics.

Hemichordata   A phylum of marine invertebrates that look like worms and feed on detritus. The name was proposed because at first examination, the protruding mouthpart of these animals was taken for a notocord. The classification of the hemichordates is a subject of continuing controversy. They are solitary or colonial, have a three-part body consisting of a proboscis, collar, and trunk, a rudimentary nervous system, and an open circulatory system.

There are three known classes of hemichordates. The *Enteropneusta,* or acorn worms, is the best known. These have worldwide distribution in intertidal zones, with a few deep water species, and range from 3 to 250 cm (1.5 to 100 inches) long. They are solitary burrowers and some are spectacularly colored in brilliant oranges, reds, and yellows. A second class, the *Pterobranchia,* tend to live at greater depths, and are sessile and aggregative or colonial. The third class of hemichordates is the *Planctosphaeroidea,* known only from large, transparent larvae.

Henry the Navigator (1394–1460)
The patron of Portuguese navigation, Prince Henry, the Infante Dom Henrique, was the third son of John I, king of Portugal.

As Governor of the Algarve, Henry established an information center at Sagres. He had the first telescope-equipped observatory in Portugal, and assembled an international court of geographers and mathematicians.

Henry's interest and support led to voyages of exploration to the Madeiras, the Azores, and the coast of Africa. The interest and support of his court and the concentration of knowledgeable people made these voyages of exploration possible. Henry continued to inspire the Portuguese in feats of navigation, and after his death voyages were dedicated to his memory. See *explorers and explorations.*

herbivore   An animal whose diet consists largely or totally of plant material. See *carnivore, heterotroph.*

hermaphrodism (hermaphroditism)
Either a change of sex or sexual flexibility. While hermaphrodism is quite common in invertebrates (e.g., mollusks, barnacles), it is unusual in vertebrates.

Some deep-sea fish, however, can produce both eggs and sperm. Change of sex occurs in coral reef fish, such as wrasses, parrot fish, groupers, and damselfish. Most of the changes are from female to male.

It is thought that the change in sex increases the possible number of offspring and is therefore an evolutionary advantage. Since large males predominate, a female that has grown large has a reproductive advantage if it becomes male.

hermit crab   A type of crab that uses an empty snail or whelk shell as the protective shield for its body. As the crab grows, it changes the shell for a larger one several times during its life. The two families of hermit crabs, *Paguridae* and *Coenobitidae,* are most often found on sandy bottoms of shallow water in almost any warm to temperate region. Some hermit crabs have complex commensal relationships with sea anemones: when the crab moves from a smaller shell to a larger one, it takes its anemone along to its new home. A few species of hermit crabs are terrestrial, and some of these use plant stems or coconut shells as protection. See *commensal relationship, crab, molt.*

heron   A fairly large, long-legged wading bird that lives in shallow water and feeds on fish, crustaceans, and amphibians. The range of herons is worldwide, but most species are tropical. The great blue heron of North America, *Ardea herodias,* has a wingspan that exceeds 1.5 m (5 feet). The slightly smaller gray heron, *Acinerea,* is typical of Europe. See *bird, estuary.*

herring   Members of the numerous *Clupeidae* family, herring are silvery-blue fish that move in huge schools. Their diet is plankton, and in turn they provide the food for other fish and mammalian predators.

The common or Atlantic herring, *Clupea harengus,* and *C. pallasi,* the Pacific species, account for the largest single fish catch taken commercially. Herring produce vast numbers of eggs which hatch at varying times ranging from December to July, depending on latitude and temperature.

The adult fish are 10 to 40 cm (4 to 16 inches) long, and are mature at about 3 to 4 years. They have a long lifespan: tagged individuals older than 20 years have been found. Several related species of fish are also called herring, including sprats, alewives, summer herring, lake herring, and shad. See *anchovy, sardine.*

Hess, Harry H. (1906–1969) American geologist and president of the American Geological Society, Hess was part of the MOHOLE project and the person in charge of the choice of drilling sites. Hess discovered that the ocean bottom was composed of rock younger than that on the continents. The age of the bottom also varies, he found, being younger close to the mid-ocean ridges and increasingly older farther from them. In order to explain the origin of this younger rock, Hess assumed that the sea floor was spreading. This theory was eventually proven by F.J. Vine and D.H. Matthews, of Cambridge University, England, who, by means of a chronology in stone that constituted a permanent record of changes in the magnetic field of the Earth, verified that the sea floor was spreading out from the mid-ocean ridges. The deposition of new sea floor reflects the magnetic orientation of the Earth. It is a pattern of stripes, each corresponding to the position of the magnetic poles at the time of the deposition of that part of the sea floor. See *mid-ocean ridges, MOHOLE.*

heterotrophs Organisms that cannot manufacture their own food from inorganic material; animals, bacteria, and fungi. Heterotrophs ultimately depend on autotrophs for their energy supply (usually in the form of food). In most cases this means that they digest organic material made by autotrophs or other heterotrophs. See *autotrophs, carnivores, detritus feeders, ecosystem.*

Hexactinellida Glass sponges, class of the phylum Porifera. They have radial, six-way symmetry and their spicules (support structures) are silaceous, or glasslike. There are over 450 species. See *Porifera, sponge.*

Heyerdahl, Thor (1914–) A Norwegian anthropologist and adventurer who showed that it was possible for primitive, albeit skilful, navigators to travel over vast distances. In 1937, Heyderdahl went to the Marquesas, where he developed his theory of a Polynesian migration from South America.

World War II interrupted his attempts to prove his theory, but in 1947 he was ready to try it. Sailing with five other men, he tested his hypothesis and proved that it could be done. A log raft named for the sun king of the South American indians—*Kon Tiki*—drifted and sailed from Peru to the Tuamoto islands in less than four months. Heyerdahl's book about the voyage, *The Kon Tiki Expedition,* made him a popular success.

Subsequent voyages in 1953 to the Galapagos and to Easter Island in 1958 continued his demonstrations of the seaworthiness and possible navigational skills of so-called primitive people. The next endeavor involved a reed boat, the *Ra,* with which Heyerdahl tried to establish the possibility of an Egyptian origin for American Indians, but the *Ra* sank. The second *Ra* finally crossed the Atlantic from Morocco, and landed in Barbados in 1970. See *Easter Island.*

high water High point of the high tide at a particular place. See *tide.*

Hirudinea The leeches, a class of annelids with few body segments and no setae. They are parasites on vertebrate fish, both cartilaginous and teleost.

These hermaphrodite worms have sucking devices at either end. Most of the coelom is taken up with gonads. The single eggs are laid in cocoons and deposited in the water. The adults live on the hosts' blood.

Leeches have been used for centuries, and are still being used in some countries by physicians to bleed patients. A constant blood flow is insured by the leech, which produces an anticoagulant that keeps the patient's wound open. The anticoagulant compounds produced by leeches are being explored for possible pharmaceutical applications.

While the leech is a parasite, it also has a commensal relationship with the bacteria that live in its gut pockets. These bacteria digest the blood that the leech ingests, feeding both themselves and the leech. See *Annelida, coelom, commensal relationship, hermaphrodite, parasite, setae.*

Holocene Epoch The younger portion of the Quaternary Period and the most recent time in the history of Earth, it follows the Pleistocene. The significant development of the Holocene was the melting of the great sheets of glacial ice. This resulted in the springing back of large parts of continental landmass, such as Scandinavia, and a dramatic rise in the sea level. This rise in sea level obliterated the land connections between Alaska and Siberia, Europe and England, Sicily and Italy, Sri Lanka and India, and many other neighboring locations.

The beginning of the Holocene, about 10,000 years ago, was marked by increasingly warm weather worldwide. It was also wet: areas of Africa, India, and the Arabian peninsula that are now desert were arable. The temperature declined, then rose once more from about 1000 to 1250 A.D. It then declined again about 1300 to 1800 A.D. and has been rising since. This latest rise may be normal or may be due to the world's increasing use of fuel or a combination of these factors. See *greenhouse effect.*

Holocephali A subclass of bottom-dwelling carnivorous fish related to sharks. Many of these fish have light-emitting organs. Like sharks, they can concentrate urea in their body tissues, and it has been suggested that they use this as a flotation mechanism. See *chimaera, light organs.*

homoiotherms (homeotherms) Warm-blooded animals. Birds and mammals are homoiotherms. Homoiotherms have much greater flexibility in selecting their environment but require much more insulation and much more food than do the poikilotherms (cold-blooded animals). The alternative to a constant and abundant food supply is hibernation or migration. See *poikilotherm, temperature of animals.*

hook A spit or narrow peninsula, the end of which bends toward the land. The Hook of Holland and Cape Cod, Massachusetts, are examples.

horse clam A large, edible mollusk, most often about 20 cm (8 inches) long, and 3 to 5 cm (1.5 to 2 inches) wide, found on the west coast of North America from British Columbia to Baja California. The horse clam burrows into the sand to depths exceeding 1 m (40 inches).

horse latitudes Atmospheric belts of high pressure found in both the Northern and Southern hemispheres at about 30 to 40° latitude. These regions are known for good weather and light winds. In the Southern Hemisphere, where there is more water and less land, the high pressure belt is more complete than that in the Northern Hemisphere. See *atmosphere, intertropical convergence, wind.*

horseshoe crab An arthropod, belonging to the class *Merostomata;* but not a true crab. These animals have a three-part body consisting of the prosoma, which is the largest segment and includes the head; the middle segment or opisthosoma, which is the heavy shell; and the spiny tail or telson. Like lobsters, horseshoe crabs have compound eyes. They have numerous legs for movement, for capturing prey, and for feeding. Their circulatory and digestive systems are also similar to those of the lobster.

Horseshoe crabs spawn on beaches in the summer, at high spring tide; the next high spring tide then washes the larvae out to sea. Their growth is a complex series of body changes and moltings, and the horseshoe crab does not reach adulthood until it is 10 or more years old. The best known genus is *Limulus.* All five existing species are ancient, having existed in their present forms for 200 million years or more. See *Arthropoda*.

horst A segment of crustal rock that lies between two parallel faults and is thrust upward. See *crustal block, graben, plate tectonics.*

hot spots Theoretical centers of volcanic activity beneath crustal plates. The idea of hot spots was formulated in an attempt to explain the existence of seamounts in areas where there are no volcanoes. According to this idea, as a plate moves across a hot spot, volcanic eruptions occur and a seamount forms if the plate is an oceanic (and therefore thinner) plate. The seamount can then become the basis for a coral atoll.

The island areas that are being studied in an attempt to establish the validity of the hot spot theory are in the Pacific between Midway and Hawaii. The Hawaiian island chain, according to the hot spot theory, is the result of the movement of the Pacific Plate across a mid-Pacific hot spot.

One explanation for chains that have different directions in the Pacific is that there are "lost plates." These different alignments of Pacific crust mean that a crust swept across a hot spot in different directions during past ages, pushed by plates no longer there, but leaving island chains in place to mark the passage.

The hot spot theory is also used to explain the existence of some Atlantic Ocean islands, notably the Atlantis Seamount and Ascension Island. See *continental drift, evolution of oceans, plate tectonics.*

hot springs Thermal springs with temperatures of approximately 100°F (40°C) or above. They occur under water and on land. They are frequently in areas of volcanic activity or where fossil water (water that has been trapped in rock for years) makes contact with hot magma (liquid rock). See *artesian water, Red Sea, thermal spring.*

Huang Ho (or Hwang Ho or Huang He) The Yellow River, China's second longest and northernmost large river,

flowing over 4,800 km (3,000 miles) from Tibet to the Gulf of Chihli in the Yellow Sea. It is the world's leading river in sediment carried, hence its name. The approximately 2,000 km^2 (1,100 square miles) delta of the Yellow River grows rapidly as a result of the enormous sediment load deposited. The course of the river has changed repeatedly in recorded time (since the 6th century B.C.).

Since the river frequently shifts, the delta is a lobed one, as sediment buildup goes in one direction for a number of years, after which the direction of buildup then reverses. Disastrous flooding along the Yellow River has been well documented in the past, and not surprisingly, the River is also called China's Sorrow. Because of the torrent of water and the silt it carries, the Huang Ho has produced a vast rice-growing plain which supports China's huge population. Irrigation dams and hydroelectric installations have also been built along the river. See *Yellow Sea.*

Hudson, Henry (1565?–1611)

English navigator and explorer of North America. Hudson sailed for English interests on three of his voyages, in 1607, 1608, and 1610. He is best remembered—certainly in North America—for his voyage in 1609 for the Dutch East India Company.

On his first voyage Hudson's employer was the Muscovy Company. He sailed north to Spitsbergen (now Svalbard) in search of a Northeast Passage to China. The next year he tried to sail around the arctic island group and found the way blocked by ice. On his third voyage, for the Dutch, he tried to find a sea lane through the North American landmass. He sailed his ship, the *Half Moon*, into the magnificent bay discovered by Verrazzano, and then sailed up the river now named for him. He realized that it was not a strait leading to the Pacific, and returned to Europe.

This third voyage was the basis for the Dutch claim to territories in North America. Hudson's other expeditions were the English rationale for claims to Canada.

Hudson's last voyage was commissioned by the British East India Company and the Muscovy Company as a joint venture. He left London in April 1610, sailing on the *Discovery*, to explore a northern river in North America whose outflow was very voluminous. It was hoped that the volume of water from that river would indicate a water passage to the Pacific. The *Discovery* entered Hudson Bay, the sailors hoping that the sheer size of the waterway was a good sign. Unfortunately they moved south and entered the embayment known as James Bay, where the severe winter overwhelmed the crew. They mutinied and set Hudson, his son, and several still-loyal crewmen adrift on June 22, 1611. When the *Discovery* returned to England, the leaders of the mutiny were not aboard; they had been killed in a fight with the Inuit (Eskimos). See *Northeast Passage, Northwest Passage.*

Hudson Canyon   A canyon extending southeast of the mouth of the Hudson River to the edge of the continental shelf in the North Atlantic. The canyon is about 90 km long and terminates in a fan valley, formed by Hudson River sediment, at the bottom of the continental slope. The area has been explored for possible oil or gas deposits, but these have not so far been of commercial interest. See *Atlantic Ocean, canyon.*

Hudson Strait   A strait linking Hudson Bay with the Atlantic Ocean via the Labrador Sea. The Strait is about 700 km (440 miles) long and ranges from 75 to 210 km (40 to 150 miles) wide. It is clogged with ice most of the year. The first exploration of the Strait was attempted by Martin Frobisher in 1578, in an early attempt to find the Northwest Passage between Europe and Asia. See *Frobisher, Martin; Hudson, Henry; Northwest Passage.*

Humboldt, Friedrich Heinrich Alexander von (1769–1859)   A German explorer, geographer, and naturalist. Humboldt was influential in a number of fields, and had sufficient money and energy to explore all of them. His

training as an engineer strengthened his interest in natural science. His first book, published in 1790, was on geology. Humboldt also worked in chemistry, designed mine-rescue equipment, and in 1795, on a trip to the Alps, realized the relationship of altitude to climate and plant zonation with the plants and animals found at higher altitudes similar to those found at higher latitudes. Geomagnetism was also one of Humboldt's interests, and gave rise to his work toward standardization and precision in geographic measurements.

In 1798 Humboldt went to Spain to "measure" the country with sextants, chronometers, thermometers and barometers. He then obtained permission to "measure" the American colonies of Spain. Landing in Venezuela in July of 1799, he conducted extensive exploratory voyages into the tropical forests and sent home plants, made thousands of observations, and collected data on magnetism, meteorology, climate, geology, minerals, oceanography, zoology, and ethnography. Some of his travels were very hazardous. Traveling in native canoes, on packhorses, and on foot, he moved through the basins of the Orinoco and Magdalena rivers and proved the connection of the latter with the Amazon River.

Humboldt was well received in his own time. He visited the United States in 1804 and appeared at the American Philosophical Society. He later went on to do experiments with Gay-Lussac in Paris. In 1829 he again embarked on a grand voyage of exploration, this time through Siberia.

While not a great innovator, Humboldt was a "connector" who recognized the links between various fields of study. See *currents.*

Humboldt Current   See *Peru Current.*

humidity   A measure of the quantity of water vapor in the air. It is reported as grams/kilogram (g/kg) of water. See *absolute humidity, air, atmosphere, fog, relative humidity.*

humpback whale A relatively short, 12 to 15 m (40 to 50 feet) long, but stubby baleen whale, *Megaptera modosa* or *novaeangliae,* with a hump on its back. It is black or sooty brown above and cream or white on the underside. This whale has long, thin flippers and some few hairs on its head; its throat and underside are deeply ridged. The humpback breeds in summer in warm waters, migrating there from the polar regions. It has been hunted almost to extinction. See *baleen, whale.*

hurricane A tropical cyclone in the western Atlantic Ocean. In the western Pacific the same kind of storm is called a typhoon. The storms develop around the Equatorial Convergence after the summer solstice in each hemisphere (June 21 or December 21). Thus, the hurricane season is July to September in the Northern Hemisphere and January to March in the Southern.

One theory of hurricane genesis is that the storms are set off by the westerly winds invading the doldrums. The winds circulate around the low-pressure center (counterclockwise north of the equator, clockwise south of it) in a swirling column that rises to a height of 15 km (10 miles). The entire system moves west at a velocity of 5 to 14 m/second (18 to 54 km/hour or 30 to 90 mph). Not all hurricanes develop to their full potential, however.

The accuracy of predicting hurricanes has increased greatly, making protective measures more meaningful. Cloud seeding with silver iodide crystals has been attempted in an effort to promote rain and break the rush of the wind in these storms, and has had sufficient success to warrant further experimentation. See *doldrums, intertropical convergence, storm.*

Hutton, James (1726–1797) A Scottish geologist and chemist who originated the theory of Earth's development called uniformitarianism. This theory postulates that while millions of years were required to develop the Earth, the processes were identical with current ones. First a law apprentice, Hutton changed careers and became a medical student and finally a manufacturing chemist. He prospered and retired to be a gentleman-farmer, later moving to Edinburgh, then an important center of scientific study.

Hutton presented his theory of the formation of Earth's layers to the Royal Society of Edinburgh in 1785. He published some of his work in two volumes in 1795; some remained unpublished at the time of his death. John Playfair condensed and edited Hutton's notes, and it was this version that became a foundation stone in the study of geology and of the history of the Earth. See *Lyell, Charles.*

hydrocarbons Simple organic compounds (molecules) composed only of the two elements hydrogen and carbon. The smallest and simplest hydrocarbons are gases such as

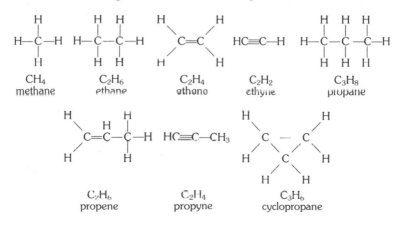

CH_4 methane	C_2H_6 ethane	C_2H_4 ethene	C_2H_2 ethyne	C_3H_8 propane

C_3H_6 propene	C_2H_4 propyne	C_3H_6 cyclopropane

Hydrocarbons can, however, be much larger than these molecules, and quite complex. As a general rule, the larger the molecule (and the more carbons and hydrogens it contains) the higher the boiling point. Molecules with chains of five to fifteen carbons are liquids at room temperature. Hydrocarbons whose carbon chains are 25 or more atoms long, are waxy solids. Very long chains of hundreds or thousands of carbon atoms (i.e., polymers) are plastic materials, such as polyethylene.

Hydrocarbons in the form of oil are of interest to marine scientists. Some algae store a food supply in the form of oil; other organisms use hydrocarbon molecules to build other organic molecules, notably sugars and amino acids. Oil, when spilt from tankers or other vessels by accident or as the result of a shipwreck, presents a serious threat to marine and beach organisms. See *amino acid, polymer protein, starch, sugar.*

hydrologic cycle   The path by which water moves in nature from one physical state to another, and from one place to another. Water is present on land in large aggregates in lakes and rivers, as well as being dispersed in soil and entrapped in rock. It exists in the atmosphere as water vapor and is the dominant feature of the Earth's oceans. About 97% of the world's water is in the oceans. An estimated 75% of the fresh water—only 3% of the world's entire water supply—is tied up in glaciers and polar ice.

Water vapor in the atmosphere con-denses and appears as dew or precipitates as rain, snow, or hail. Liquid water and ice or snow evaporate (i.e. water changes from the liquid or solid state into a gas, or vapor). The evaporation rate is greater if the water is fresh than if it is salt because the dissolved salt slows down the rate of evaporation. The movement of air, in the form of wind, increases it and humidity decreases it.

The movements of water from place to place or from one state (gas, liquid, solid) to another are physical changes. At no time in this process does the water molecule, H_2O, change. See *water, weather.*

hydrothermals   Vents of mineral-rich water, common at the the ridge lines along the edges of crustal plates. Hydrothermal vents are present in the Pacific Ocean on the East Pacific Rise, in the Red Sea, and in the Caribbean Sea. See *hot springs, vent communities.*

Hydrozoa   A class of coelenterates, small aquatic and chiefly marine animals, that includes *Hydra* and the Portuguese man of war. Most species have both polyp and medusa forms. See *Cnidaria, medusa, polyp, Portuguese man of war.*

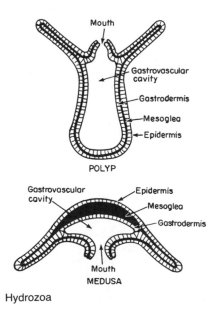

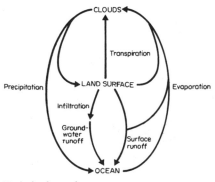

Hydrologic cycle

Hydrozoa

ice Frozen or crystalline water. It forms in regions of low temperature. Terrestrial ice is glacial, and if it enters the sea it forms icebergs. If the temperature is low enough, seawater will freeze despite its salt content. At a salt concentration of 3.5%, the freezing point of water is $-1.91°C$.

The formation of ice crystals segregates the salt in seawater. Salt doesn't fit well into the crystal lattice (the crystal structure) of ice, and so it remains in the water, increasing the salinity of the water and further depressing the freezing point. The more saline, denser water sinks toward the bottom, and this brings up less salty water, which then freezes.

Sea ice forms as needle-shaped hexagonal crystals that first accumulate as slush and then in sheets. The thin sheets are dispersed by wind and waves as pancakes. The pancakes coalesce and form floes. The faster the ice forms, the more salt is trapped in it: the salt content of ice varies from 0.4 to 1.5%. As the ice ages, the salt trickles out, so that by the time a floe is a year old it is almost pure ice with no salt. Some floes are almost permanent and ice stations have been put on them. See *fast ice, floe, glacier, iceberg, pancake ice.*

Ice Age A term coined by the 19th century naturalist Louis Agassiz, for the epoch when most of Europe north of the Alps was covered by glacial ice. See *Agassiz, Jean-Louis Rodolphe; evolution; Pleistocene Epoch.*

ice shelf A thick, floating platform that is to a glacier what a delta is to a river. Glacier ice and snow deposit on this shelf, which grows seaward until it is stopped by a combination of warm air or water, or ocean currents. In sheltered areas, an ice shelf can exist for millenia. The Ross Ice Shelf in Antarctica is the only example of such a shelf. It is firmly attached to the sea bottom, and has been extensively drilled in studies of ice formations of previous geologic times. The Ross Shelf is huge, having roughly the same area as all of France.

The structure of an ice shelf provides a theoretical basis for fiord formation as a result of glacial action. As the glacier continues to grow and depress the land at the edge of the sea, a wall of ice forms and grows from the sea bottom up. See *glacier.*

iceberg A large mass of floating ice. Ordinarily a river brings water from landforms to the sea. A glacier is a frozen river bringing ice to the sea. At the shore, large blocks of ice fall into the water, forming icebergs. The process is called calving, and is noisy. Just as a liquid river carries its greatest load in spring, so does a glacier. The greatest production of icebergs occurs in April in the Northern Hemisphere and in October in the Southern. See *glacier.*

ichthyosaur One of several extinct marine reptiles. Ichthyosaurs flourished in the Mesozoic era, which ended about 65 million years ago. Good specimens have also been found in deposits of the Jurassic period. The average ichthyosaur was about 3 m (10 feet) long, and looked something like a porpoise. It had a very sleek outline and a strong, propelling tail. The animal had large eyes and impressive jaws lined with numerous teeth. Some fed on mollusks, others on fish. See *Mesozoic Era, reptile.*

igneous rock A product of volcanic activity, it is formed from rapidly cooled magma and can be characterized by its crystalline appearance. Some of the crystals of igneous rock can be quite large.

IGY (International Geophysical Year)
A period of worldwide geophysical research that began on July 1, 1957 and ended December 31, 1958. More than 70 countries participated in the research, which included glaciology, seismology, geodesy, and many aspects of oceanography.

Several important projects were carried out simultaneously during the IGY. One was to explore the variability of solar radiation and sunspots, others collected information on cosmic radiation, the nature of the upper atmosphere, the nature of ice in glaciers and ice sheets, and the interpolation of this data with weather fluctuation. These projects used orbiting artificial satellites for the first time.

In terms of oceanography, the IGY yielded more new data on the oceans than had been produced since the HMS *Challenger* voyage of 1872. Charting was done using a variety of sounding devices. Undersea landslides and turbidity currents were studied, undersea volcanic eruptions were noted, and their output charted. Ocean currents were mapped, including subsurface currents, which have profound effects on all life on Earth and move vast quantities of water over great distances.

The exploration of the oceans during the IGY produced the evidence of continuous mid-ocean mountain ranges. Continuing research was one very vital legacy of this period of international research. See *H.M.S.* Challenger, *currents; Ewing, William Maurice; Heezen, Bruce Charles; magnetic pole; mid-ocean ridges.*

Indian Ocean
An ocean south of India, west of Australia, and east of Africa. Its origin traces back to the Mesozoic era, with the breakup of Gondwanaland into South America, Africa, Australia, Antarctica, India, and Madagascar. The large islands in the Indian Ocean—Madagascar, Sri Lanka, Suqutra [or Socotra]—are all continental.

The Indian Ocean has a surface area of about 76 million km² (33 million square miles). Unlike the Atlantic and Pacific oceans it has only a few seas: the Persian Gulf/Gulf of Oman and Red Sea/Gulf of Aden combinations in the northwest, the Arabian Sea to the west, the Bay of Bengal on the northeast, and the subdivision of the Bay of Bengal, the Andaman Sea. Since its division from the Pacific is indistinct, one can include in the Indian Ocean the Arafura and Timor Seas, the Gulf of Carpentaria, and the Great Australian Bight. From east to west the Indian Ocean lies roughly between 140° and 20° East Latitude.

The Indian Ocean is the site of a three-plate juncture. The ridge lines—the marks of the juncture—are the outstanding ocean-bottom features. The Chagos-Laccolive Plateau runs along the mid-ocean ridge. It is a north-south feature, resembling an inverted Y, which breaks into two legs: the Southwest Indian and Mid-Indian ridges. The Southwest Ridge runs into the Atlantic-Indian Ridge at the Mozambique Fracture Zone. The Mid-Indian Ridge merges with the Southeast Indian Ridge at the Amsterdam Fracture Zone. The 90° East Longitude Ridge runs north-south, east of the fan of the Ganges River down to the Southeast Indian Ridge, and is at right angles to the Broken Ridge and Diamantina Fracture Zone, which skirts southern Australia. There are several aseismic ridge systems in the ocean: the Madagascar, Seychelles, Mauritius, Agulhas, Kerguelen, and West Australian.

The 90° East Longitude Ridge is thought to be the evidence of India's move north before its collision with Asia. The ridge is the scour line marking the subcontinent's movement. The Mascarene or Madagascar plateau, which is northeast of Madagascar, includes the Seychelles Islands and Mauritius. These islands have "continental" bottoms, and their crust is thicker than the ocean crust. This leads back to the Atlantic myth, in that some authorities believe that this "microcontinental" region may be leftover, unconsolidated pieces of what was once Gondwanaland, the primitive southern continent. Deep core samples taken during the international Indian Ocean Exploration in 1965 have shown that the continental breakup of Gondwanaland was not instant; the separation of Madagascar from Africa dates to the Jurassic.

Winds in the Indian Ocean are a function of the monsoon pattern. In the Northern Hemisphere in winter, the Asiatic high pressure means a northeast wind blowing north of the equator. Between 35° South Latitude and the Intertropical convergence, the winds are southeast. These winds take over in the northern summer, in the form of southwest monsoons bringing rain. Circulation to some extent follows the wind pattern and the West Wind Drift. The salinity of the Indian Ocean is highest near the Persian Gulf and lowest at the Bay of Bengal. The surface water in the Ocean is of local origin, but the intermediate and deep water is of polar origin. Since there is no connection with the north polar regions, Antarctic water is found almost exclusively in the Indian Ocean. Some Greenland water has been sampled south of Africa. See *Antarctic water, evolution of oceans, Ganges River, islands, mid-ocean ridges, Pacific Ocean.*

Inland Sea   Also known as the Seto Naikai, a sea that separates the Japanese main islands, Honshu, Kynshu, and Shikoku. It has a very broken coastline and many small islands. See *Sea of Japan.*

insolation   The quantity of solar radiation that arrives at some given area on Earth. It is measured in calories per square centimeter, and varies with the several factors of latitude, type of surface (land, grass, snow, water, etc.), and wind. The type of surface is the most unpredictable variable in determining insolation, since reflecting surfaces such as ice and water bounce much of the Sun's radiant energy back into the atmosphere. This reflectance is further dependent on the incident angle of the Sun's rays.

interface   A surface forming a boundary or demarcation between two different bodies, substances, or phases. For example, there are interfaces between water layers of different saline concentration or different temperature. See *halocline, internal wave, thermocline.*

internal waves   Waves caused by the flow of water over an irregular surface, or sharp atmospheric differences in a small geographic area or a rapid tide flow. Internal waves or tides, can occur in bodies of water where different layers have different densities. They are noticeable on the surface as a moving slick or a glassy area that is not wind-driven. These waves are usually 7 to 15 m (20–50 feet) high, but in deep water can be ten times that high. They damp out to zero at the water bottom and at the surface.

Benjamin Franklin observed internal waves, and created them by carefully moving a glass of water. The roiling of the liquid below the surface was obvious in the transparent container.

Frequent internal wave activity does occur in some oceanic areas. This phenomenon has been offered as a rational explanation for some otherwise inexplicable losses of ships at sea. See *wave.*

intertidal zone   Also known as the littoral, it is a part of the seashore usually under water at high tide and exposed at low tide. See *tide.*

Intertropical   Convergence   Also called the Equatorial Convergence. This is the term used for an area in both air and water at or near the equator. It changes in latitude with the season, a change that is most obvious in the Indian Ocean. Because hot air rises, thunderstorms are produced as water vapor, which is carried aloft and then cools rapidly. These storms, which are typified by heavy rains, alternate with the calm surface winds that are a noticeable feature of the area called the doldrums. See *Antarctic Convergence, doldrums, weather.*

iodine   An element present in seawater as iodide ion (I^-) and iodate ion (IO_3^-), in concentrations of about 0.06 mg/liter.

It is commercially extracted from seawater. Some organisms incorporate iodine in metabolically important compounds such as the thyroid hormone thyroxin and its various analogs. See *element, halogen.*

ion An atom or group of atoms that has an electrical charge resulting from the gain or loss of one or more electrons. See *atom, electron.*

Iran Plate A crustal region south of the Eurasian Plate and between the Turkish Plate to the west, Arabian Plate to the south, and Indian Plate to the east. The Iran Plate moves east-northeast and overrides the eastern edge of the Eurasian Plate. It is in turn pushed northward by the Arabian Plate approaching from the southwest. See *crust, plate tectonics.*

Irish moss A dark red alga, *Chondrus crispus,* also called carrageen. It is usually 5 to 25 cm (2.75 to 8 inches) long, and grows around rocks of the North Atlantic on both the European and American coasts. Carrageenan is a gelatinous material commercially extracted from carrageen, or Irish Moss, and used as an emulsifying agent and thickener. See *algae.*

Irish Sea A body of water that separates Great Britain from Ireland. Its southern end is arbitrarily divided from the English Channel and the Celtic Sea between England and Brittany by the St. George's Channel. The northern end of the Irish Sea is the channel between Bangor, Ireland, and the Mull of Galloway, Scotland. The two large islands in the sea are the Isle of Man and Angelsey.

The Irish Sea is not very deep. Its depth averages less than 100 m (330 feet), with a maximum of 175 m (576 feet) just off the coast of Ireland.

While the terrain on the surrounding shores is old (Paleozoic) the sea bottom is Triassic. The rising sea level at the end of the most recent glacial activity flooded large areas in the eastern parts of Ireland.

A fanciful story has placed leprechauns and "drowned villages" in the area that was once flooded. See *English Channel, Paleozoic Era, Triassic Period.*

Irminger Current A branch of the North Atlantic water that flows around Iceland. See *Atlantic Ocean, currents.*

Irminger Sea A sea south of the Iceland-Greenland Ridge and north of 60° North Latitude. The water is present in a series of layers, with distinct regions of warm and relatively saline Atlantic water, cold Arctic water, and a mixture of the two. The name Irminger Sea is not recognized as a distinct entity by all geographers. See *Atlantic Ocean, currents.*

iron An element present in sea water as a hydrated oxide to the extent of about 0.01 mg/liter. Iron is also present in the form of consolidated deposits in so-called manganese nodules and in the remnants of meteorites. Some unconsolidated deposits have been commercially utilized, and iron deposits in nearshore locations have been dredged. The crucial issue is the relative cost of the ore and the fairly expensive process of dredging to obtain it. See *element, manganese.*

island A relatively small body of land surrounded by water. Islands may be either continental or oceanic in origin. A continental island is one that is part of a large, nearby continental mass of basement rock. Thus, Greenland is related to the rest of North America, and New Guinea is part of the Australian continent. The islands off the coast of Maine are hilltops of the submerged continental shelf.

Oceanic islands are those that rise from the ocean floor as volcanoes, or more properly as a series of volcanoes. The organisms living on any specific island depend on a number of factors, including the closeness of the island to other land, the temperature on and around the island, and the location of the island in terms of drift or wind patterns.

Island arcs are strings of volcanoes at

the edge of a crustal plate. See *archipelago, atoll, coral, ocean floor, seamount, volcano.*

island arc   A number of islands, most often volcanic, arranged in a curved chain. The concave side of the arc is usually close to a continent, while the convex or ocean side is most often bounded by a deep trench. See *archipelago, trench, volcano.*

isobar   A line on a weather map that joins points of equal pressure. The closer together the isobar lines appear, the greater the velocity of the wind parallel to the isobars. A series of isobars that are very close together is an indication of a great change in barometric pressure in a realtively small area. The greater the change in pressure, the more noticeable the change in weather. See *storm.*

Isopoda   An order of arthropods in the class *Crustacea.* There are about 4,000 species, ranging from terrestrial to freshwater and marine organisms. The sow bug is a terrestrial isopod and a good example of these almost flat, ovoid organisms.

Members of the Isopoda are generally segmented, with many similar segments, the abdominal segments looking very much like the thoracic ones. The number of legs varies from 4 to 16. Most specimens are small, ranging from 0.7 to 3.5 cm (0.25 to 1.5 inches) long. One stunning exception is *Bathynomus giganteus* of the Caribbean, which is 35 cm (14 inches) long. Many of the marine species are parasites that live on fish or shrimp. See *Crustacea.*

isostasy   The state of equilibrium of the constituents of the crust and mantle of the Earth. Isostasy explains the concept of continental flotation. The continents float in the lithosphere, the rocky layer of the Earth, as ice floats in water. See *continental drift, crust, lithosphere, mantle.*

isotherm   A line on a weather map connecting points of equal temperature. Isotherms are sometimes plotted against months of the year to show climatic variation over the entire year at a specific location. See *temperature, weather.*

isotope   One of two or more variations of atoms of a chemical element that differ from each other in their atomic mass but not in their chemical properties. Atoms are composed of three principal particles, protons and neutrons in their nuclei, and electrons orbiting the nucleus. Istopes of the same element have identical numbers of protons and electrons, but differ in the number of neutrons they contain.

The most common form, or isotope, of the element hydrogen has one proton, one electron and no neutrons. The isotope deuterium contains one neutron in its nucleus, in addition to a proton, and a third isotope, tritium, has two neutrons. Of these three isotopes, only tritium is radioactive.

The decay of radioactive isotopes, or radionuclides, is used as a means of dating rocks and fossils. If radionuclides are present in water, the water mass can also be dated. See *element, radioactivity.*

J

jaeger A skua. A large, dark-colored northern seabird, of the family Stercorariidae, that forces weaker birds to give up their food by making them drop or disgorge it.

Jan Mayen Island and Fracture Zone An island in the Greenland Sea, about 500 km (300 miles) east of Greenland, that belongs to Norway. It is the peak of the volcanic rise also called Jan Mayen. The last big blast of the volcano occurred in 1732.

The Fracture Zone runs northwest-southeast across the mid-ocean ridge north of the Island. See *Mid-Atlantic Ridge, mid-ocean ridges.*

Japan, Sea of A body of water that separates the Japanese islands from the rest of the Pacific. The Korea Strait, the Sea's southern boundary, has depths of more than 100 m (328 feet). The other connections to bodies of water are to the Sea of Okhotsk to the north and to the Pacific via shallow saddles. The average depth of the Sea of Japan is about 1350 m (4,400 feet), although depths of over 3,000 m (9,800 feet) are found off the island of Honshu. The Pacific Polar Front runs through the Sea at about 38° North Latitude. The water south of this line is warm and blue, while north of the line it is cold and green. The water of the Sea of Japan is also divided into deep, intermediate, and surface layers. The surface water shows considerable seasonal fluctuation in temperature and salinity. The oxygen content of the Sea is high, due in part to the extraordinarily high population of plankton.

Geologically, the Sea of Japan—particularly the northern half—seems to be an area of continental subsidence. The Korea Strait may also have been a land bridge in the Quarternary Period. This would account for the remains of elephants and mammoths that have been found on some of the islands. See *Kuroshio Current, Oyashio Current, Pacific Ocean, Tsushima Current.*

Japan, Trench A trench east of the Japanese islands and part of the almost continental trench line that is found in the western Pacific. The Kurile Trench is north of the Japan Trench; the Bonin and Marianas trenches are south of it. The deepest point of the Japan Trench is the Tuscarora Deep, at about 8,500 m (28,000 feet). See *Marianas Trench, trench.*

Java Head A headland at the western end of Indonesia at the entrance to the Sunda Strait. This headland is a famous landmark, and was used as such by clipper ships on the way to China. See *clipper.*

Java Sea A part of the western Pacific Ocean between Java and Borneo (Kalimantan). The entire bottom of the Java Sea is relatively flat and incised with well-defined river channels indicating the area's former state. This consisted of low-lying but dry land that was the route of the Asiatic faunal migration into the Indonesian islands. The sea is even now fairly shallow, with average depths of about 45 m (150 feet).

The current in the Sea flows west in winter—if one can call the colder season in a tropical region winter—and east from May to September due to the large inflow of river water. For a tropical sea, the Java Sea has low salt levels.

The oil deposits in the Java Sea are

continuations of the oil fields on the island of Java, and are presently being exploited. See *Flores Sea, Sunda Shelf.*

Jeanette A 19th-century American exploration vessel. On a polar exploration the *Jeanette* became frozen in the ice in September of 1879. The captain, George Washington deLong, and his crew eventually died of cold and starvation on the Lena River Delta (Siberia). They were attempting to sail close to the Pole by plotting a course from the Bering Strait to Wrangell Island. Then they would go overland. The idea of a Pole under water had not yet been attacked. The *Jeanette* was stuck in the ice in early September (1879) and drifted north of Wrangell. Bits of wreckage eventually wound up near Greenland. The *Jeanette* sank near the New Siberian Islands after drifting for two years. Frijdtof Nansen, the Norwegian explorer who established the fact that the North Pole was not on a continental mass, used the tragedy of the *Jeanette* in his arguments for this fact. See *Fram; Nansen, Fridtjof.*

jellyfish See *Scyphozoa.*

jet stream A long, narrow current of high-speed winds in the troposphere (the lowest segment of the atmosphere). The jet stream blows from a westerly direction, and can reach speeds of 400 km/hour (250 mph) or higher.

jetty A manmade rock structure usually at a harbor or river mouth or other area of ship passage. Jetties are constructed primarily to direct currents and to inhibit the shoaling or silting up of navigable channels. See *coast, groin.*

jib A triangular sail. It is set on a stay (supporting rod) extending from the first or foremast of a ship to the bow. As a verb, "to jib" means to swing such a sail or boom around to change the course of the ship. This is also called tacking. See *boom, mast, sail.*

John Dory A food fish of worldwide range. The John Dory tends to live in deep water, to depths of about 200 m (650 feet). It is gray, with a characteristic black spot on the side behind the operculum, or gill cover, and has a large ugly head and spines on both its dorsal and ventral surfaces. Most specimens are about 10 cm (3 to 4 inches) long and are very thin laterally but deep-bodied. *Zeus faber* is the Atlantic and Mediterranean species. In the Pacific the most common species is *Z. japonicus.* See *fish.*

JOIDES Joint Oceanographic Institution for Deep Earth Sampling. This 1970s project was a combined effort of the Scripps, Woods Hole, and Lamont-Doherty, oceanographic institutions and several distinguished university oceanography departments, notably that of the University of Florida at Coral Gables. The object was to take core samples in various places and categorize the bottom. The ship designed to do this was the *Glomar Challenger.* See Glomar Challenger, *Lamont-Doherty Geological Observatory, Scripps Institution, Woods Hole Institution.*

Juan de Fuca Ridge A site of volcanic activity in the Pacific Ocean near Vancouver, Canada. The area has been extensively explored in the 1980s. Sulfurous water of about 300°C has been found issuing from vents along the ridge at depths of almost 2 km (over 1 mile). The colonies of organisms found near these vents are unusual, and were until very recently unknown. See *mid-ocean ridges, sulfur bacteria, vent communities.*

Juan de Fuca, Strait of A narrow international waterway between the Olympic Peninsula of the State of Washington and Vancouver Island, British Columbia. The average depth of the strait is less than 100 m (about 330 feet). It is named for a Greek seaman in the employ of Spain who was reputed to have explored it in 1592. For the next 200 years the strait appeared now and again on navigational charts as part of a possible route between the Atlantic and Pacific oceans. However, it is a complicated waterway to use, since strong rip currents run in it. See *explorers and explorations.*

Juan Fernandez Islands   A group of islands that are outcroppings of the Juan Fernandez Ridge in the southeastern Pacific Ocean near Chile. Alexander Selkirk, a sailor who was left on the islands in 1704 after a fight, is the accepted model for Robinson Crusoe, the fictional character. The islands are typical of exposed rocky outcrops. See *Dampier, William; island.*

Junk   A flat-bottomed sailing vessel used by traders from the Malay Peninsula to China. The junk has square bows, a squared-off, built-up stern, and is equipped with rudders. The sails are large and strutted with long horizontal rods of bamboo; they are carried on either two or three masts.

Because of their flat bottom and low length-to-width ratio (usually 3:1) junks are slow-moving.

The Chinese have used these vessels for centuries as merchantmen, warships, and vessels of exploration. Chinese mariners and traders of the 15th century brought the junk to Borneo, the Philippines, Sri Lanka, the Malabar Coast, and Zanzibar in east Africa. See *explorers and explorations, sail.*

Jurassic Period   The second oldest portion of the Mesozoic era. The name is based on the limestone of the Jura Mountains of France, which contains a splendid record of ammonite (fossilized cephalopod) evolution. Gibraltar is another Jurassic formation, as is much of Yosemite in California.

The Jurassic period lasted about 30 million years, and ended about 135 million years ago. The location of continents and oceans in this period of geologic history is a matter of some controversy. The pre-Cambrian shields in Canada, Scandinavia, Central France, and the Harz Mountains of Germany were dry land, as was some of Australia. Much of the rest of what is now the Earth's continental landmass was submerged, according to some geologists. Geologists do agree that mountain building was occurring in several locales, including North America, the Caucasus, and Southern Europe during the Jurassic.

The fauna of the Jurassic included some of the most spectacular dinosaurs: the Brontosaurs, Stegosaurs, Plesiosaurs, Ichthyosaurs, and the proto-bird Archaeopteryx. The most common fossil invertebrates were, in addition to the very characteristic ammonites, the nautiloids, belemnites, gastropods, pelecypods, and echinoids. Silaceous sponges have also left a fossil record of Jurassic sea life, as have numerous corals. See *ammonites, Mesozoic Era.*

K

Kamchatka Peninsula A peninsula between the Sea of Okhotsk and the North Pacific Ocean-Bering Sea area. It is 1,200 km (750 miles) long in a roughly north-to-south line, and about 500 km (300 miles) wide at its widest point. The Kamchatka Peninsula is volcanic; there are two mountain ranges lying along the long axis of the landmass. Like Iceland it has geysers and geothermal springs. Unlike Iceland, the climate is much colder and there is considerable snowfall. See *Bering Strait, East Cape, Sea of Okhotsk.*

Kara Sea A body of water marginal to the Arctic Ocean east of Novaya Zemlya and south of Franz Josef Land. The Siberian plain lies to the south and southeast of the Sea.

The average depth of the Kara Sea is about 125 m (400 feet). It was formed after the last ice age, making it the world's youngest sea, which accounts for its very shallow depths. Almost half of it is less than 50 m (160 feet) deep, and its maximum depth is 620 m (2,000 feet). It is dotted with sand bars and islands. It is fed by the Ob, Yenisey, and Kara rivers. The volume of fresh water brought by these rivers means that the surface waters of the sea are almost fresh, with salinity levels as low as 1.0% near the mouth of the Ob. The salt content is about 3.3% near Franz Josef Land.

Commercial fishing and hunting are major activities on and around the Kara Sea. Petroleum deposits have been found in the area. The commercial activity of the region is to a great extent dependent on the weather, which is arctic and known for frequent storms year-round. See *Arctic Ocean, Barents Sea.*

kelp A group of large brown algae, or brown seaweeds, of the orders Laminariales and Fucales. Many are of commercial interest both as primary foodstuffs and as industrial raw materials.

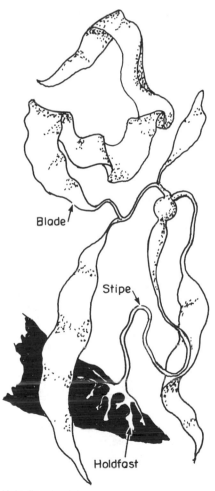

Blade

Stipe

Holdfast

Kelp *(Laminaria)*

The kelps resemble vascular plants upon superficial examination. They have a differentiated cell structure and distinct gamete forms. In the reproduction of Laminaria, called Devil's Apron, both sporophyte (having the full normal adult chromosome number) and gametophyte (haploid) generations exist. The large and visible kelp of the ocean is the sporophyte which in late summer produces zoospores. These zoospores constitute the haploid gametophyte generation, and may be either male or female (sperm or egg producing). The motile sperm are produced by antheridia and caught and retained by the oogonia, or egg-producing cells. The fertilized egg or zygote undergoes immediate growth into the recognizable kelp. This complex life cycle where the zygote and the adult plant are diploid (a pair of each chromosome) and the zoospores (sperm and eggs) are haploid is analogous to the reproductive cycle of ferns on land.

The kelps grow in large structures called macrocysts that may be as long as 60 m (200 feet), or as elaborately branched mats. These plants have structures that resemble the sieve tubes in land plants. The function of the tubes is the same: they are part of the plant's circulatory system and do what phloem does in vascular plants, transporting the products of photosynthesis to the various parts of the plant. See *algae, brown algae.*

Kepler, Johannes (1571–1630) A German astronomer and student of Tycho Brahe. Kepler's greatest contribution to astronomy was the mathematical treatment of the Copernican model of the solar system. His first important work was the *Rudolphine Tables of Planetary Motion* (the work Brahe began), named for Emperor Rudolph II, Kepler's patron. These tables were eventually published in 1627.

Kepler did careful observations of Mars for many years. These were finally published in 1609, along with observations of the nova of 1604. In his book *The New Astronomy,* he used the Copernican philosophy, his own excellent mathe-matics, and Brahe's careful measurements to plot the elliptical orbits of the planets. He also formulated three laws, known as Kepler's Laws, to describe the motion of the planets in the solar system. These laws later became the basis for Newton's theory of gravitation. All of this work was essential raw material for those who later devised accurate tables with which mariners determined position. See *Copernicus, Nicolaus; navigation.*

Kerguelen Islands A group of roughly 300 islands southwest of Australia which are volcanic outcroppings of the plateau that divides the South Indian Ocean Basin from the Atlantic-Indian Basin. The area is now a scientific base; in the 19th century the islands were known only to the occasional whaler and seal hunter. See *Indian Ocean, islands, plateau.*

Kermadec Trench Also known as the Tonga-Kermadec trough, this ocean deep constitutes a border of the southwest Pacific basin. It is associated with seismic activity. Like its Atlantic Ocean counterpart, the Puerto Rico Trench, the Kermadec Trench is part of a large slip-strike region. It is much deeper than the Puerto Rico Trench, with maximum depths of about 10,000 m (33,000 feet). See *Fiji Plate, trench.*

ketch A two-masted sailing vessel. Ketches were coastal fishing and trading ships that later were adapted to naval use as "bomb" vessels, with the bombs being the missiles fired by mortars—cannons with a very short barrel that fires in a very high trajectory. With the end of this military use, the square-sailed ketch was redesigned for recreational sailing. Ketches are now exclusively pleasure boats. See *sail.*

key A low or flat island or reef. Usually a coral-and-sand islet (e.g., the coral islets off the southern coast of Florida). See *coral.*

killer whale The largest of the dolphins; the males may exceed 9 m (28 feet) in length. It is an attractive black-and-white carnivorous animal that lives and hunts in packs. The large dorsal fin is a distinguishing characteristic. The range of *Orcinius orca* is worldwide. It has no natural enemies.

A great deal of writing about killer whales is highly exaggerated; in spite of their name, they do not attack everything that moves in the sea. Their usual prey are birds and small dolphins, although a large pack of killer whales may attack a weak baleen whale. There is little information available about the breeding behavior of *O. orca* other than that the young are black and orange. See *Cetacea, dolphin, whale.*

King William Island One of the large islands of the Canadian Arctic. It lies east of Victoria Island and was the site of the death of John Franklin, the famous explorer of the Arctic whose disappearance prompted many rescue expeditions. See *arctic; Franklin, John; Northwest Passage.*

kittiwake A type of North Atlantic gull. See *bird, gull.*

Kinorhyncha A class, and possibly a separate phylum, of microscropic worms. The Kinorhyncha have a spiny, segmented body but are not annelids. They feed on diatoms. See *Annelida, diatom.*

Korea Current A warm, north-moving surface current in the Sea of Japan. It turns east and becomes the Tsugaru and Soya currents. The former moves through the Tsugaru Strait into the Pacific Ocean. The Soya Current travels along the Japanese islands of Honshu and Hokkaido into the Sea of Okhotsk. See *currents.*

Korea Strait A body of water that separates Korea and Japan. The Tsushima islands are in the center of the Strait. The passage between the islands and Japan is also referred to as the Tsushima Strait. This waterway was the site of a major naval battle in 1905, when the Japanese destroyed the Russian fleet in one of the determining engagements of the Russo-Japanese War.

knot A measure of nautical speed. One knot = 1 nautical mile (6,080 feet) per hour. Originally, a log attached to a rope with evenly spaced knots on it was heaved overboard to measure the speed of a ship. By counting the number of knots let out in a specific time period, standardized at 28 seconds, the sailor could calculate the speed, hence the term "knots." The space between knots was standardized at 47 feet, 3 inches.

Krakatau, island and volcano Krakatau, or Krakatoa, is a small, uninhabited island in the Sunda Strait west of Indonesia, that erupted in a series of blasts culminating on August 27, 1883. Most of the island disappeared, and the resulting tsunamis drowned over 30,000 people on other islands. This explosion, while not as great as that of the volcano Tambora in 1815, was very well documented, probed, and photographed. Krakatau is still active as a volcano today.

It has been postulated that the surrounding islands and Krakatau itself were the remnants of a 17th century eruption that left only the fragments of a broken volcanic cone visible above sea level. By the 19th century a new volcanic cone had been created on one fragment, and this was the site of the famous eruption of 1883. A series of subsequent eruptions, with debris falling into the caldera (lava-filled crater), has produced a new island called Anak Krakatau or Child of Krakatau. The new cone emerged in 1928 and is still growing. See *tsunami, volcano.*

krill Small shrimp-like crustaceans of the genus *Euphausia.* They are distinguished from shrimp by their bristled tails. Krill are pelagic organisms ranging in size from 1 to 6 cm (about 0.5 to 3 inches). Most are transparent and many have light-producing organs, possibly as an aid in keeping the swarms of these animals together.

The enormous numbers of krill make

these organisms a significant food base that supports a large bird, fish, and whale population. Krill migrate vertically, and have been found at depths exceeding 2,000 m (6500 feet).

E. superba is a coldwater animal that swarms in the antarctic spring, feeding on plankton. In turn it is fed upon by baleen whales, particularly blue and finback whales. The krill population also seems to fluctuate for unknown reasons, with possibly very significant results throughout the food web of the polar regions. See Euphausiacea, whale.

Kurile Trench   A trench in the ocean floor that lies east of the Kamchatka Peninsula (U.S.S.R.) and the Kurile Islands north of Japan. The maximum depth is 10,542 m (34,587 feet). The Kurile trench is part of the broken line of deep water surrounding the Pacific. See Pacific Ocean, trench.

Kuroshio Current   Also known as the Japan Current, it is part of the North Pacific gyre. The Kuroshio Current moves northeast from the Philippines to the east coast of Japan. It has a high temperature of about 20°C, and a high salt content of 3.45%. The Current moves east of Taiwan and the Ryukyu islands chain, and in summer a branch, known as the Tsushima, goes through the Korea Strait.

At about 35° North Latitude the Kuroshio Current turns east, deflected by the cold Oyashio current. The turned stream moves east as the North Pacific Current, and disappears as a recognizable entity by the time it reaches Hawaii.

The waters of the Kuroshio are much darker blue than the water it moves through. Its rate of motion varies from 50 to 300 cm/second (1 to 6 mph). The Kuroshio has been called the "Pacific Gulf Stream," but it is not as well-defined as the true Gulf Stream, nor does it transport as much warm water, and it does not travel as far as the Gulf Stream does. It is, however, the major west-to-east current in the north Pacific Ocean. See currents.

L

Labrador Current A cold, low-salt (3.0 to 3.4%), surface flow moving south from the Davis Strait between Greenland and Baffin Island, and along the Canadian coast. The rapid introduction of very cold water into the warmer water coming up from the south creates the well-known fog banks off Labrador and Newfoundland. There is an area of upwelling off the Greenland coast that creates a source of nutrient materials, and these are the base for the high population density of arctic animal life. See *Atlantic Ocean, currents.*

Labrador Sea A body of water bounded by the coast of Labrador on the west, Greenland to the east, and Baffin Island to the northwest. At the Davis Strait, the Sea is only about 800 m (2,500 feet) deep, whereas most of the Labrador Basin drops to deeps of 3,500 m (10,000 feet). The bottom sediment is composed largely of *Globigerina* ooze.

The climate of the Labrador Sea is polar (i.e. very cold, dry, and windy). The water on the shelf is 2 to 3.5°C, which is colder than the water in the depths of the Sea. The oxygen content of Labrador Sea water is high and the salinity, particularly on the shelf, is low. In the Labrador Basin the salinity is a fairly constant 3.49%. Ice floes are a common feature of the shelf area, while the eastern part of the Sea is generally clear of floating ice. See *Arctic Ocean, Globigerina.*

lagoon A warm, shallow, quiet waterway separated from open sea either by sandbars, barrier islands, coral reefs, or some combination of the three. If the separating land is a bar or barrier island, the lagoon is long and narrow; if the lagoon is surrounded by a coral atoll, it is circular. Coastal lagoons can become marshes if deposited sediment cuts off their connection with the sea. See *atoll, coral, reef.*

Lamarck, Jean-Baptiste Pierre Antoine de Monet, chevalier de (1744–1829) A scientist, taxonomist, and developer of an evolutionary theory, Lamark was born to a family of the minor French nobility with a long military tradition.

Lamarck was destined for the Church as a career, but on his father's death escaped into the army, where he served for six years. After trying banking as an occupation, he turned to medicine, then a common first step toward a scientific career, but found science more to his liking than medicine.

Lamarck worked on physical principles following some already outmoded chemical theories, and finally, in 1778, published a major work, *Flore Française*, which he had begun while in the army. This established his reputation, and he became a member of the French Royal Academy of Science.

In the Revolutionary decade, Lamarck worked in the Jardin du Roi (which became the Jardin des Plantes in 1793–1794) and the Royal Cabinet, which re-emerged as La Musée d'Histoire Naturelle. While there, his major work was the reclassification of animals, which the Linnean classification scheme had put into broad categories. Lamarck worked on systemization and orderly grouping, and published this work in *Système des animaux sans vertèbres* (1801), *Philosophie zoologique* (1809), and the multivolume *Histoire naturelle des animaux sans vertèbres,* which appeared over the period from 1815 to 1822.

In these works Lamarck expounded on his philosophy of evolution by way of the inheritance of acquired characteristics, but his explanation of evolution was incorrect—characteristics acquired by an individual during its lifetime do not lead to genetic change, as Lamarck believed. Nevertheless, the concept that genetic change had occurred in the history of the Earth, and of the biota on it, was a revolutionary one and Lamarck gave this concept exposure long before Darwin did. See *Darwin, Charles; evolution; taxonomy.*

Lamont-Doherty Geological Observatory   An institution founded in 1949 by Columbia University, in New York City, to study earthquakes, submarine geology, marine biophysics, oceanography, chemistry, physics, and the body of work called polar studies. The Observatory and Laboratory maintain a research vessel, house a piston core collection of samples carefully removed from the Earth's crust, and maintain an index of organisms and fossils. The Lamont-Doherty Observatory pioneered the use of undersea photography, and its collection of photographs in this field is definitive. It also houses the reports of the JOIDES work, as well as the Atlantic, Antarctic, and Indian Ocean Surveys—total-ocean surveys done in the 1960s and 1970s. See *Ewing, William Maurice; explorers and explorations; Heezen, Bruce; JOIDES.*

lamprey   A primitive, jawless, cartilaginous vertebrate of the order Cyclostomata in the class Agnatha. These jawless fish are generally found in coastal, temperate waters on every continent except Africa. Most are burrow dwellers 15 to 100 cm (6 to 40 inches) long. They start life in fresh water, and with the exception of a few species migrate to the sea as adults. Most adult lampreys are parasites on fish. Lampreys have been partially responsible for a serious decline in the commercial fisheries of the Great Lakes. See *Cyclostomata, hagfish, parasite.*

Lampridiformes   Order of bony fishes that dates from the Cretaceous Period 135 million years ago. This group includes the ribbonfish, oarfish, and opah. There are about 21 species. The Lampridiformes are mesopelagic animals. They have distinctive trailing and retractible maxillae (ventral streamers attached behind the gills), but their outstanding characteristic is the absence of distinct fins. They have a dorsal ray that extends the entire length of the body, and a spiny construction above the head that looks like a crown. Their scales are cycloid but small, or absent entirely. There is a ductless swim bladder.

These fish are colorful and frequently very large. They may be 6 to 7 m (20 to 24 feet) long. They may have contributed to myths of sea monsters or sea serpents. See *sea monsters.*

lancet   A long, thin, deepwater fish that has a long, spiny, dorsal fin. The adults are 1.5 m (60 inches) and more in length. Lancets also have sharp and prominent teeth, feeding on other fish and invertebrates. The most common Atlantic species is the longnose lancet, *Alepisaurus ferox.* The Pacific species are *Alepisaurus richardsori* and *Alepisaurus borealis.*

lantern fish   Mesopelagic fish that have light organs around their disproportionately large heads and on their undersides. The pattern of distribution of the light organs is species specific, and may therefore be a mating recognition factor for the more than 150 species of lantern fish. The average size of the fish is about 2.5 to 15 cm (1 to 6 inches). They are usually found at depths of about 500 m (1600 feet) during daylight hours; at night they migrate vertically and feed at the surface. See *light organs.*

La Pérouse, (Jean-François Galoup), comte de (1741–1788?)   A French explorer and naval captain. After his service with the French Navy in the American Revolution, he was given command of a voyage of exploration to the far East and sailed with two ships, the *Boussole* and the *Astrolabe.* The explorers hoped to find the western approach to the Northwest Passage through the North American mainland.

La Pérouse touched the Asian mainland near what is now Vladivostok, and sailed through the strait between Sakhalin island and Japan. He landed on Kamchatka, in what is now Alaska, before turning south. The captain of the *Astrolabe* and some of its crew were killed by Samoans. La Pérouse continued on to Botany Bay, Australia and disappeared some time after February 7, 1788, his last dated correspondence. Subsequent French expeditions were sent to find La Pérouse, and one of these, led by Dumont D'Urville, succeeded in finding the remains of the two ships but no trace of La Pérouse. See *Dumont D'Urville, Jules Sebastien César; Northwest Passage.*

Laptev Sea Formerly known as the Siberian Sea, this body of water lies east of the Kara Sea and is fed by several rivers. The largest volume of water entering the sea comes from the Lena River. The Laptev is largely a shallow plain, and more than 50% of it is less than 50 m (160 feet) deep. The entire area is still recovering from glaciation; there is "relic" ice in its eastern region.

A thick layer of fresh water overlays the Laptev Sea. In the southeast the salinity ranges from 2 to 2.5%, falling below 1.0% in summer with the rapid inflow of river water. Very severe weather is the norm: the seashore has permafrost year-round, and the Sea itself is covered with ice for most of the year. The water temperature is below 0°C for most of its depth. Deep pockets of water from the Atlantic have been traced to depths of about 250 m (800 feet). These waters are significantly warmer than the very low-salinity water of the Laptev. See *Arctic Ocean, Kara Sea, Lena River.*

larva The pre-adult stage of many organisms, which does not structurally resemble the adult. The larva usually exploit a different food source from that used by the adult.

Larvacea A class of tiny, pelagic organisms in the subphylum *Tunicata,* that feed on nanoplankton These animals create a shell and move within it. *Larvacea* are found in all of the world's oceans down to depths of about 100 m (325 feet), but some have been found deeper. See *Tunicata.*

lateen A triangular sail. The word is a mangled version of "latin sail," because these sails were used on Mediterranean vessels. See *sail, square-rigged.*

lateral line A sensory organ along the head and sides of a fish. Some amphibians also have lateral lines. The "line," which shows on the side of a fish as a distinct streak of differing pigmentation, is made up of individual units called neuromes. These are groups of hairlike structures that move in response to a change in the surrounding water, caused by the movement of another animal or by a change in current flow. See *fish.*

latitude The angular distance north or south from the Earth's equator, measured in degrees, minutes, and seconds of arc. Points on the Earth's surface are located as being at a particular latitude by measuring the angle of incident light from the sun or a specific star above the horizon at a specific time. In the Northern Hemisphere the star used as a locator is the Pole Star, while in the Southern Hemisphere it is the dominant star in the constellation of the Southern Cross. Lines on a globe or map connecting points of identical latitude are called parallels of latitude. One degree (1°) of latitude = 111.1 km or 60 nautical miles; 1 nautical mile = 6,076 feet or 1852 m. See *astrolabe, chart, longitude, map.*

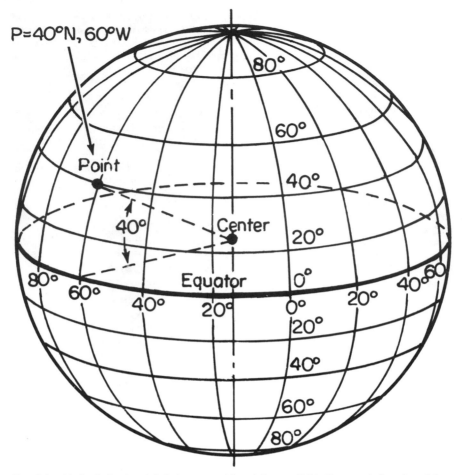

P=40°N, 60°W

Parallels of latitude (horizontal circles around the globe parallel to the equator) and meridians of longitude (vertical circles passing through the poles)

Laurasia   An archaic northern continent that resulted from the breakup of the still-older supercontinent Pangaea. Laurasia eventually broke up and formed North America and Eurasia. See *continental drift; Gondwanaland; plate tectonics; Wegener, Alfred.*

lava   Magma, or molten rock which pours out of a volcano or any other fissure in the Earth's crust. Lava is basaltic or rhyolitic, meaning that it has either a low or high content, respectively, of siliceous material. The former flows more readily, and when it cools it forms rock that is characterized as smooth (sometimes glassy, if it cools very quickly), rippled, or braided. The last type of rock is known by the Hawaiian word pahoehoe. Undersea lava flows are almost always of the pahoehoe type. This flows outward through tubes or natural vents or pipes in the crust. Clinker lava is a rough, pebbly type of lava and fed by open lava streams. See *basalt, magma, ocean floor, volcano.*

laver   A genus of red algae, *Porphyra,* which grow in sheets in water that is rich in nitrogen. Masses of laver are collected at the highwater marks of temperate seas. The laver is the "seaweed" most commonly used as food for humans both in Europe and the Orient. It is called *nori* in Japan. See *algae.*

Law of the Sea Conference   A series of United Nations conferences, the first of which met in Geneva in 1958. The initial agenda was the standardization of national limits to territorial waters. Chile was the first nation to claim a 200-mile strip of territorial water along its shore, and other nations soon followed this lead. Historically, the limit to territoriality had been the three-mile strip, because this was the furthest reach of naval artillery in the 17th century.

Although Chile's claim was not resolved, the Conference attempted to arbitrate claims to territorial waters and to look into the question of national rights to coastal mineral deposits. The 1960 meeting again tried and failed to achieve conformity.

The third meeting began in 1973 and continued for over 10 years. The great issue of this conference was the possibility of settling the various nations' territorial claims by establishing a 12-nautical-mile limit that would be surrounded by another 12-nautical-mile-wide "protected zone." Clearly, this left unresolved the problem of areas of overlap between adjoining nations. Furthermore, the proposed arrangement would have permitted six nations to claim over half the desirable coastal and continental shelf areas of the entire Earth. As a result of much adverse publicity, the treaty was never ratified.

leathery turtle   See *sea turtle.*

leeward   In the direction of the moving wind or a wave. See *windward.*

Lena River   A river in Siberia, the second longest (3,000 m; 9,900 feet) in the U.S.S.R. The Lena, flowing from mountains west of Lake Baikal, is one of the world's great rivers. it forms a huge delta in the Laptev Sea. Most of the water in the Lena is very cold because the water has its origins in melting snow. See *Arctic Ocean, Laptev Sea.*

leopard shark   A small shark (90 to 150 cm or 3 to 4.5 feet) long whose habitat is in shallow waters of the eastern Pacific Ocean. It is considered harmless, and feeds on fish and invertebrates. The gray-and-black barring pattern of the shark gives it its name. See *shark.*

Lesser Antilles   See *Antilles, Caribbean Sea, West Indies.*

light absorption and scattering   Seawater absorbs some of the light passing through it. It is a medium that is transparent to short wavelengths (the blue through ultra-violet end of the spectrum), and which absorbs longer wavelengths in the red and infrared regions of the spectrum.

The quantity of light reflected from the surface of water depends on the angle of the incident light and smoothness of the surface. If the Sun is directly overhead (90°), the reflectance is less than 5% and the remaining light is absorbed. As the angle of the Sun drops, the reflectance increases. At an incident angle of 30° the reflectance is 6%, and at 20° it is 12%; at 5° it is greater than 40%.

Light scattering within seawater is a function of the temperature and composition of the water. Also, if the sea surface is disturbed, it scatters more light than it would if it were smooth. Dissolved and suspended material (e.g., algae or krill) increases scattering, and conversely, seawater that scatters very little light (i.e., that is clear) contains less suspended material and is therefore less likely to support biological activity. See *insolation, krill.*

light organs   Luminous or light-emitting structures on an animal. Deepwater fish are examples of animals that have these structures. The structures consist of a light-emitting body, a color filter, and a reflector; the organ is rather like a headlight. Light organs form from dermal tissue; the usual sources of the light are pigments known as luciferins, which are species-specific. Fireflies are examples of terrestrial producers of luciferins. When luciferins are oxidized by an enzyme known as luciferase, light is produced as part of the chemical reaction. This light is "cold light" (i.e., there is no heat produced, such as when a light bulb burns).

Marine organisms ranging from bacteria through dinoflagellates and from crustacea to fish—both teleosts with bony skeletons and cartilaginous fish—have been known to produce light; however, there is no other class of vertebrate animals that does this.

Some hadal organisms that emit light seem to do so by a different mechanism: They contract and expand the pigment cells, or melanophores, in their bodies. When the melanophores contract, light is emitted; when they expand, the emission of light is blocked. See *benthos, color, hadal zone, photic zone.*

lighthouse   A structure on land that is a mariner's aid, producing a light as a danger signal and navigational aid. The precursors of lighthouses were shoreline bonfires maintained to warn sailors of treacherous rocks or currents. The most famous ancient lighthouse was the Pharos of Alexandria, built in the reign of Ptolemy II (283–247 B.C.). The Romans built lighthouses at Ostia, Ravenna, Messina, and in the outposts of the Empire at Dover and Boulogne.

The light produced by ancient lighthouses was firelight; only in the 18th century did oil lamps and parabolic mirrors first become used. Modern lighthouses use electric light. The electric lamps that are used are either incandescent or sodium or mercury vapor lamps similar to the ones used on modern highways. The optics are very similar to those in use hundreds of years ago. A catoptric (reflected) light uses a light source and mirrors to produce a light beam. The mirrors may be plane, parabolic, or spherical. Some lights use a prism to concentrate the beam of light and increase its carrying power. This is a dioptric light. A combination prism-and-mirror light is called a catodioptric light.

The first structures that bore lights were simple wooden towers. The characteristic tower that most people associate with lighthouses—inhabited by the keeper of the light—was first designed for England's Eddystone Light by John Smeaton in 1759.

While the necessity of warnings for navigational hazards has not diminished with the passage of time, the use of lighthouses has. Lightships, offshore platforms, and buoys perform the same sort of function, and lighthouses are now automated structures that produce powerful beams of electric light, and are designed to work in the worst of weather conditions with a minimum of human attention. The romance is gone. The increasing use and sophistication of onboard navigational equipment such as LORAN RADAR and SONAR has also somewhat decreased the need for warning lights. See *buoy, LORAN, navigation, SONAR.*

Ligurian Sea   The northern end of the Tyrrhenian Sea, which is in turn an arm of the Mediterranean. The Ligurian is surrounded by the province of Liguria, Italy, to the north, Tuscany to the north and east, and Corsica to the south. The Tuscan archipelago separates the Ligurian from the Tyrrhenian seas. The maximum depth of the Ligurian Sea is found near Corsica, and is greater than 2,800 m (9,300 feet). See *Mediterranean.*

limestone   One of the most plentiful rocks on Earth, made up principally of calcite ($CaCO_3$) and dolomite [$CaMg(CO_3)_2$]. About 20% of all sedimentary rock on Earth is either limestone, dolomite, or both. The two types of rock form the typical dome structures that are associated with petroleum deposits.

Limestone is a sedimentary rock, and

there exist three types which have different origins: Clastic limestone, which is derived from other calcareous minerals; organic limestone, which consists of the accumulated calcareous remains of once-living organisms; and precipitated limestone, which is the end-product of chemical reactions occurring in seawater.

Since the deposition of limestone has gone on continuously through much of the Earth's history, organic limestone beds are excellent fossil records. Fossilized blue-green algae, called stromatolites, can be traced from the pre-Cambrian to the Recent Period in such limestone beds. Most of the limestone was deposited in the pre-Cambrian and Lower Paleozoic eras, but this process is ongoing. Stromatolites are still forming in the Bahamas, Florida, and western Australia. Reef structures are other examples of organic limestone.

Dolomite is most often deposited in areas of high salt content such as the Persian Gulf and the Caribbean Sea. Since temperature is a controlling factor in the deposition of all carbonate rock, the kinds of rock and their location are a means of examining the climate of various places during the Earth's history. See *Cambrian, Devonian, geologic time, Paleozoic.*

limpet   A small, flattened snail found on littoral rocks. These animals are found worldwide in temperate zones. They all share, to varying degrees, the ability to trap water beneath their shell, and can therefore survive exposure on bare rock from one high tide to another. See *littoral, Mollusca, snail.*

Limulus   See *horseshoe crab.*

Linnaeus, Carolus (1707–1778) [Carl von Linné]   A Swedish naturalist and the founder of the presently used taxonomy—the naming of living things. His father was a serious botanist, and Linnaeus pursued his own botanical research while at the University in Lund and then in Uppsala, Sweden. His trips of exploration, undertaken while he was an improverished young professor, were to Lapland, in 1732, and to parts of

Sweden unknown to him, where he collected plants and minerals. Linnaeus finished his medical degree in Leiden, and stayed on in Holland until 1738, writing voluminously. His *Systema naturae (System of Nature)* (1735) and *General planterium* (1737) were his major works during this period. Linnaeus then returned to Sweden and gained a chair at Uppsala. He spent the rest of his life there, teaching, studying nature, writing and encouraging his students, who loved him.

Linnaeus knew that his designations of living things were arbitrary. He used several criteria to categorize flora and fauna: sexual organs and "cortex versus bark" in plants were but two. His use of Latin names came early, but in the last edition of the *Systema naturae* he introduced the binomial nomenclature for genus and species that made him and his classification scheme unique. He also used common names where possible. Linnaeus' treatment of animals is not nearly as good as that for plants, and his works on minerals are even less useful to a modern scientist. Not all scientists were convinced of the efficacy of the binomial nomenclature, particularly the influential comte de Buffon. See *Buffon, Georges-Louis le Clerc; taxonomy.*

Linschoten, Jan Huygen van (1563–1611)   A Dutch explorer. Linschoten was a merchant in Spain, and in connection with his business went to India in 1583. He did not return to Europe until 1590, having been shipwrecked on his return voyage. In 1594 he accompanied Willem Barents on a voyage to find the Northeast Passage to the Far East, but they reached only the Kara Sea. The next year saw Linschoten on another attempt at discovering the Northeast Passage. But the seven ship fleet turned back again, because they encountered impassable ice. This was Linschoten's last attempted long voyage.

Like Richard Hakluyt, Linschoten collected accounts of voyages. As a chronicler and early explorer he was an inspiration to other Dutch navigators. See *Barents, Willem; Hakluyt, Richard.*

lipid Any of a large, heterogeneous group of molecules including fats and waxes that are involved in a number of cellular functions and which, together with proteins and carbohydrates, are the principal components of living cells. The simplest lipids are long-chain fatty acids. Fats and oils are all lipids, but are only two types of substance within this group. Biological membranes such as cell membranes are constructed of complex lipid molecules. See *hydrocarbon, marine oil.*

lithium A soft, silvery white chemical element, and the lightest metal known. Lithium is present in seawater (0.017%) as the lithium ion, Li^+. The incorporation of lithium into rock is very slow: the turnover time of atomic lithium into the lithium ion is estimated at 20 million years. See *element.*

lithosphere The Earth's crust and upper mantle. The lithosphere is about 200 km (125 miles), thick, and is the layer thought to be broken up into the various crustal plates whose movement is referred to as plate tectonics. See *continental drift, plate tectonics, individual plates.*

littoral A term first used in the mid-19th century to denote the intertidal zone, or that part of the shore that is under water at high tide and exposed when the tide is low. This region has different biotas, depending on the type of bottom and temperature. Rocky bottoms, sand beaches, protected bays, salt marshes, mangrove swamps and coral reefs all are littoral environments. Each has a different spectrum of plants and animals. Some littoral zones, such as the Galapagos islands, are so specialized that the animals and plants found on them exist nowhere else. See *beach, coast, coral, estuary.*

littoral currents See *longshore currents.*

lobster one of a number of families of the class Crustacea. True lobsters are *Homaridae,* spiny lobsters *Palinuridae,* slipper or Spanish lobsters *Schyllaridae,* and deep sea lobsters *Polychelidae.* All are decapods with stalked eyes, and have chitinous, segmented exoskeletons that cover a three-part body. All members of these families also have five pairs of walking legs, of which one pair is the chelae, or claws. Usually, one of these claws is larger than the other.

Lobsters are most abundant in deep water of about 300 m (1,000 feet) or more. They live on carrion or live fish when they can get them. They are long-lived, maturing at about the age of five years, and fifty-year old specimens have been found. If they survive the predation of starfish, cod or rays, their only enemy is man.

Lobsters have been of interest to mariculturists for years. Attempts to "grow" them commercially have not been successful. While the animals have been studied by both pure and applied biologists, some aspects of their behavior are not well understood. The North Atlantic lobster, for example, has been observed moving in columns on the ocean floor. The reason for these marches is unexplained, as is the seemingly random cyclic rise and fall in this animal's population abundance. See *Crustacea, Decapoda.*

loggerhead turtle See *sea turtle.*

Lomonsov Ridge An 1,800-km (1,100 mile-long) ridge line that divides the Arctic Ocean into the Canadian and Siberian basins. The crests of this mountain range rise to heights in excess of 2,000 m (6,600 feet) above the ocean floor and come to within about 1,000 m (3,300 feet) or less of the surface. Unlike the Mid-Atlantic Ridge, the Lomonsov Ridge is not associated with volcanic activity. Nor is it part of the mid-ocean ridge system as a site of new rock formation. See *Arctic Ocean, aseismic ridge.*

longitude The angular distance east or west from a reference circle, or meridian, drawn from pole to pole around the Earth and passing through Greenwich, England—the so-called Zero Meridian, or Prime Meridian. Longitude is used, along with latitude, as a measure of location on the Earth's surface. Locations on the Earth's surface are described as degrees, minutes, or seconds of arc east or west of the Zero Prime Meridian. The International Date Line is based on the 180° (E or W) line. It is distorted considerably, as are all other meridian-based time zones, to avoid bisecting part of the same country or island group.

The concept of longitude was understood long before it could be accurately measured. That development had to await the standardization of accurate clocks that would work reliably on ships. See *chart, map, Ptolemy.*

long-line fishing A method of fishing used by commercial fishermen. The drifting of long lines is a practice used in the Far East, mainly in the tuna industry. The lines are 150 to 400 m long in any one section, and are dropped in several interconnected sections. The number of sections varies with the operator, the size of the boat, the number of crewmen, and the fish being taken. There may be as few as 10 or as many as hundreds of sections. Hooks are attached to the branch lines, and the number of hooks is also very variable.

The line, with its floats, is shot out of the boat at the start of the day's fishing. The crew then spends hours reeling in the line and the catch as it is marked by the floats. As the catch increases, the floats become submerged. Bottom long lines—set for cod, haddock, and flatfish, most often at depths of about 80 m—are attached to floats in the same manner as are the floating lines.

Long-line fishing may have contributed to the overfished state of some areas of the world's oceans. The tonnage of fish taken in this manner is decreasing. See *fishing.*

longship A galley used by the Vikings. This was their all-purpose coastal raiding, trading, and emigrant ship. The heyday of the longships was about 600 to 1100 A.D. They ranged in size from about 40 to 80 oars, and also carried a single, square sail, but in fjords, on rivers, and in narrow channels depended solely on their oarsmen.

Longships were not used in winter, when they were pulled up on the coast. They were essentially open boats with no cover for their oarsmen, who were also assault troops. They were not very well designed, but the large crew made them workable. See *galley.*

longshore currents Currents produced by waves landing on a shore at some angle with the shore other than zero. The direction and velocity of the rebound wave deflected by the shore is largely a function of the local topography. Longshore currents can move coastal sediment, and are responsible for a good deal of the natural "drift" or erosion of some sandy beaches. See *beach, currents.*

lophophore A fold in the body wall of an invertebrate that encloses the mouth. The lophophore is a very complex external structure, consisting of a crown of ciliated tentacles that are used to entrap food. It is characteristic of several primitive coelomate phyla, including the Bryozoa, Entoprocta, and Phoronida.

LORAN Long Range Navigation, developed in the United States in the 1940s. LORAN consists of pulsed transmissions from two pairs of radio stations. Measurements of the time interval between the arrival of each of these signals makes it possible to take a position line and plot it on a chart, which covers about 1140 km (700 miles) of sea. The technique is also used in charting, in which the ionosphere is a "bounce-back" or rebound medium. If one is aiming at the sky, the range of LORAN is 2,280 km (or 1,400 miles). See *echo location, navigation, SONAR, triangulation.*

Lord Howe Ridge A feature of the ocean floor in the South Pacific, also known as the New Zealand Rise. It rises to 1,000 to 2,000 m (0.6–1.2 miles) above the seafloor. This elevation runs from about 40° South Latitude to Lord Howe Island to the northwest, and then north into the Coral Sea. See *Coral Sea, Pacific Ocean.*

low water   The lowest level of the tide before it turns and starts to rise again. See *high tide, tide.*

lugworm   A marine annelid, or polychaete worm of genus *Arenicola,* that lives in sandy, intertidal waters. It is a bottom dweller and like earthworms, can grow quite large. Common specimens are over 20 cm (8 inches) long. See *Annelida.*

luminescence   See *bioluminescence, light organs.*

lunar day   The time from one lunar transit over the meridian to the next. This time interval is 24 hours and 50 minutes.

lunar tide   That part of the tide that is due to the gravitational attraction of the Moon. The path of the Moon's motion is fairly complex, which explains the variation in the tide at any given location from one day to the next.

The moon takes a lunar month—28.5 days—to revolve around the Earth. During this circuit it reaches a point where it is closest to the Earth—its perigee—and a point at which it is farthest away, its apogee. The change in the distance between the Earth and the Moon during the Moon's revolution is reflected in the tidal range at a given location. The closer the Moon is to the Earth, the greater its gravitational attraction.

The plane of the Moon's orbit intercepts the ecliptic (the plane of the Earth's orbit) at an angle of 5°. The plane of the Moon's orbit also rotates (precesses) in an 18.6-year cycle. This means that a complete tidal cycle is 18.6 years. See *solar tide, tide.*

lungfish   A bony fish of the subclass Dipnoi. Lungfish are freshwater species related to the coelacanth, and can survive for long periods in deoxygenated water or in almost dry streams. In earlier times, there were nine families of these fish, and some were marine. The coelacanth, the sole surviving marine relative of the lungfish, is, however, thought to have originated as a freshwater animal. The bony mouthparts of some fossil lungfish suggest a diet of shelled invertebrates, which these fish crushed in their mouths. They are important in the evolutionary history of the Earth because they may be ancestral to terrestrial animals.

These ungainly. thick-bodied animals move sluggishly, and are air breathers. Most have a single lung that is dorsal to the viscera. They breathe by gulping air. However, the juvenile forms of lungfishes have gills. The African and South American species estivate: in summer, when their rivers become very shallow, they bury themselves in mud capsules that trap both air and moisture. This enables them to survive the hot, dry weather.

Lyell, Charles (1797–1875)   A British geologist born in Scotland. His most influential work was his *Principles of Geology,* considered by his contemporaries to be a basic work that explained the uniformitarian theory of geology. Lyell, like other uniformitarians, believed that the geologic formations of the Earth took millions of years to develop. He also believed in a much greater antiquity of man than would be possible from the biblical account. See *Darwin, Charles; evolution; geology; Hutton, James; Murchison, Roderick.*

M

Mackenzie River   A river in northwest Canada. The Mackenzie and its tributary, the Finlay, form the second largest river system in North America. From its headwaters near Great Slave Lake, the Mackenzie moves northwest, keeping to the east of the Rocky Mountains along its 1,700 km (1,000 mile) run to Mackenzie Bay on the Beaufort Sea. The Bay is east of the boundary between Alaska and the Yukon Territory of Canada. See *Beaufort Sea*.

mackerel   Commercially important torpedo-shaped, carnivorous fish of the North Atlantic and also other waters. They are related to tuna and are distinctive because of the row of small fins posterior to their symmetric anal and dorsal fins. The common mackerel *(Scomber scombrus)* is about 30 cm (1 foot) long with wavy black or dark blue bars on a sea-green background. The underside is silvery white. Large schools of mackerel are exploited commercially, but the fish is also sought as a sport fish. Related species are the chub *(S. choos)*, Pacific mackerel *(S. japonicus)*, Spanish mackerel *(S. commerson)*, and king mackerel *(S. cavalla)*. The more tropical fish tend to be larger and heavier. See *fish, fishing*.

mackerel shark   A large, fierce, pelagic shark of the genus *Lamna*. These shark are fished for food. They are colored various shades of gray, are about 3 m (10 feet) long, are ovoviviparous, and feed on fish of the temperate zone such as herring and mackerel. The name mackerel shark comes from the patterning on this shark, which resembles that of the mackerel.
   Other sharks in the family *Isuridae* are the white and mako sharks. The mako is

fished for food. The porbeagle *(L. nasus)*, a small viviparous shark, is notorious for voracious eating and the destruciton of commercial fishing nets. See *ovoviviparity, shark, viviparity*.

macrobenthos   Large organisms that live on or near the ocean bottom. Macrobenthic organisms are further classified by size. The organisms that pass through larger mesh screens in an array of screens of graduated size openings go into a larger size classification. The classification is as follows:

Megabenthos	large bottom dwellers: lobsters and crabs.
Macrobenthos	those animals that are larger than a mesh opening of 0.5 mm: clams, oysters, annelid worms, barnacles, and amphipods.
Meiobenthos	organisms that are small enough to pass through a screen of mesh size 0.5 mm, but large enough to be retained by a sieve of mesh size 0.05 mm. These include flatworms, roundworms, polychaetes, and tiny crustacea.
Microbenthos	benthic organisms that are small enough to pass through a sieve of 0.05 mm mesh opening. The bacteria, protists, and algae are in this group.

See *benthos*.

macroclimate   The study of meteorology and climatology on either a worldwide scale, over long periods of time, or both. See *climate, weather.*

Macruridae   Deep-sea fishes related to the cods, the grenadier or rat-tail fish. This group of about 300 species are bottom dwellers in deep ocean waters. They are 30 to 60 cm (1 to 2 feet) long and have huge heads that taper into small, slender bodies. Some grenadiers have light organs, others grunt. See *benthos, fish, light organs.*

Magellan, Straits of [Estrecho de Magallânes]   A sea channel, discovered by Ferdinand Magellan, through southernmost Chile and Argentina, connecting the Atlantic and Pacific oceans. The passage is a 560 km (350-mile)-long, rocky, 3.5 to 30 km (2 to 20-mile)-wide, circuitous course through islands, bars, and sometimes shallow channels at the southernmost reaches of South America. The average latitude is 54° South Latitude. The weather of the region is dominated by almost constant cold and fog. See *explorers and explorations; Magellan, Ferdinand.*

Magellan, Ferdinand ([Magalhães, Fernão] (1480?–1521)   A Portuguese navigator who discovered the Straits (named for him) linking the Atlantic and Pacific oceans. While Magellan's later voyages were made for Spain, his first voyages, to India in 1505, and again in 1511, when he was a participant in the capture of Molucca, were for his native Portugal.

A papal bull divided the world's unknown regions between Spain and Portugal, with all the territory east of the Cape Verde Islands reserved for Portugal and the rest for Spain. Magellan believed that the Moluccas were close to the Americas and approachable through Spanish lands. He therefore applied to Spain to finance an expedition that would look for a way through the American landmass to the Moluccas, or Spice Islands. Like Columbus, he hoped to find a passage in temperate waters. Magellan's ships left Spain in 1520 and followed the South American coast into cold, inhospitable waters through the autumn. In October Magellan found the straits between the South American mainland and the Tierra del Fuego archipelago, and sailed through them in 38 days. The flotilla almost starved on the long run across the Pacific to Guam and the Philippines. They landed there in March, 1521, and were cordially received. Magellan, however, intervened in a local war and was one of the casualties; he was killed April 27, 1521. The surviving men returned to Spain later that year and were thus the first men to have circumnavigated the globe. See *explorers and explorations.*

magma   Fluid rock. Magma is the liquid material that is produced at the mid-ocean rift valleys. When it solidifies it forms igneous rocks. See *mid-ocean ridges, volcano.*

magnesium   (Chemical symbol, Mg)   A light, malleable ductile metal used industrially. It is commercially extracted from seawater because magnesium is the second most plentiful metallic element in seawater. Its concentration in the sea is about 1350 milligrams per liter of water. Some magnesium is also found in sediments, and some is present in the manganese nodules found on the seafloor.

The chlorophyll molecule incorporates magnesium as its central atom. The importance of this metal therefore cannot be overestimated, since with the exception of the vent communities, all oceanic life depends on chlorophyll and the process of photosynthesis. See *chlorophyll, elements, manganese nodules, minerals, water.*

magnetic anomaly   The change of the Earth's magnetic field. This phenomenon was discovered by magnetic measurements of the ocean floor. Strips parallel to the mid-ocean ridges were found to have stronger or weaker magnetism, depending on the magnetic iron content of the ocean-floor rocks. This was explained in the 1960s (1963) by Drummond Matthews of Cambridge University, and Fred Vine, as being the result of basalt formation. As new rock is extruded in ridge areas, it is affected by the Earth's magnetic field. This is now positive, or normal, although in some previous geologic times the Earth's magnetic field has been negative or weaker than normal. The alignment of iron-containing crystals in the rocks points one way in normal periods, and is reversed in negative periods. This alternation of iron-crystal formation has been correlated with the formation of new rock at mid-ocean ridges. Since alignment occurs wherever new rock is formed, the same patterning also occurs with volcanic creation of new rock on land. See *mid-ocean ridges.*

magnetic field   The region over which a magnetic body exerts its effects. The axis of rotation can be thought of as a bar magnet with the northern end arbitrarily called the positive end, the southern one is then the negative end. Compasses or magnetic (iron-containing) minerals point to the positive end of magnets. This included the earth as magnet. The geographical position of the magnetic poles varies irregularly. Also the magnetism of Earth changes in intensity—the N. Magnetic Pole attracts compasses less strongly and goes to almost no magnetic declination and gradually reverses polarity. A magnetic field produces detectable magnetic waves. The Earth is such a body. See *geomagnetism, magnetic anomaly.*

magnetic poles   The two regions, in the Northern and Southern hemispheres, respectively, where the Earth's magnetic field is strongest. The magnetic poles are close to the geographic poles.

The North Magnetic Pole, the place toward which all compass needles point, was first found by Sir James Ross in 1831 on the Boothia Peninsula. While it is close to the geographic North Pole, the two are not coincident, and the position of the magnetic pole is not fixed, nor is its variation constant. The location of the

pole and the magnetic field of the Earth are constantly changing. This variation is called secular variation. It is the end-product of the movement of the liquid ferrous core of the Earth. Since the variation is erratic, regular surveying is necessary (using a ship with no iron on it) to ascertain the location of the magnetic pole.

In 1947 the North Magnetic Pole was 320 km (200 miles) further north than it had been when Ross found it. Twelve years later, it was another 320 km north.

In addition to wandering, the magnetic field of the Earth has at times in the past reversed polarity. The North Magnetic Pole has, in the Earth's history, sometimes even been in the southern hemisphere. At present, the strength of the Earth's magnetic field is varying and also diminishing. If this continues, the magnetic field may weaken to the point at which compasses cease to work and the aurora disappears. See *aurora; Flinders, Matthew; magnetic anomaly; Ross, James Clark.*

mako shark   *Isurus oxyrhinchus,* a mainly tropical mackerel shark. It often swims just below the surface of the water, with only its characteristic dorsal fin protruding. The mako does not approach shore, living only in deep waters. These sharks are fairly large, usually being longer than 3.5 m (11 feet). The mako is sometimes used as a food fish. See *fish, shark.*

Malacostraca   A subclass of the Crustacea, in the phylum Arthropoda, comprising heavily armored crustaceans. Among the Malacrostraca are the crabs and lobsters, as well as mantis shrimp, euphausiids, and all of the other decapods. See *Arthropoda, Crustacea, Decapoda, Euphausiacea, Stomatopoda.*

Maldives   A string of coral atolls southwest of the southern tip of India. The Maldives are the overwater part of the Chagos-Laccadive Ridge. There are fifteen principal islands and hundreds of uninhabited islets in the group. See *Indian Ocean, mid-ocean ridges.*

mammal An air-breathing endo-thermic (warm-blooded) animal whose young are born alive and fed on milk. There are over 4,000 species of mammals. The aquatic mammals have lost all or the greater part of their hair, but have otherwise retained the mammalian characteristics: of a four-chambered heart, biconcave red blood cells, a diaphragm breathing muscle, and both the hard and soft palates of the oral cavity.

Sea-living mammals have special adaptations to their marine existence, such as powerful tails; lowered metabolic rates, permitting the use of less oxygen; a great fat layer for insulation; and milk that is very high in both fat and protein content.

There is no evidence of mammals before the Triassic Period. They appeared first as land animals; modern marine mammals occurred later. Seals and walruses (the order *Carnivora*) date from the Paleocene Epoch, toothed whales *(Odontoceti)* and baleen whales *(Mysticeti)* from the Upper Eocene. The *Sirenia* (sea cows) are earlier, probably of mid-Eocene origin. See *Cetacea, pinnipeds, porpoise, seal, Sirenia, whale.*

manatee A large, herbivorous, tropical mammal, of the genus *Trichechus,* of coastal waters, estuaries, and slow rivers. *T. manatis* is found in Florida and the Caribbean, *T. inunguis* in the Amazon and Orinoco systems, and *T. Senegalensis* in west Africa, where it lives in small herds. These slow-moving animals are from 2.5 to 4.5 m (8 to 15 feet) long, and weigh about 700 to 800 kg (1,500 to 1,700 pounds). Manatees have poor senses of sight and smell, and therefore are likely to be hunted or killed in accidents. Attempts are being made, notably in Florida, to preserve the manatees. See *mammal.*

manganese Manganese is an essential trace element. It is found in plant and animal tissue and is classified as a micronutrient. Its metabolic function is unknown. As a metal it is hard and brittle and resembles iron but isn't magnetic.

Manganese is also found in trace amounts dissolved in seawater (0.00 mg/l).

manganese nodules Nodules found on the ocean floor, in which manganese is the dominant element. These nodules, which are greater than 3 cm (1.5 inches) in diameter and contain at least 26 other elements, were first discovered by the *Glomar Challenger* expedition. Manganese also occurs in the form of sand or rocks, as coatings on rocks, and as a filler in coral reefs.

The nodules seem to be built up in layers, although the mechanism of their formation is still only a matter of conjecture. It is possible that when metallic elements are added to seawater saturated with these elements, the elements, or salts of other compounds containing them, precipitate. As these precipitated substances descend to the ocean floor, they then attract other, similar materials that have reached the saturation point. At the ocean bottom, the precipitated material crystallizes and sometimes incorporates biogenic materal nodules.

Manganese nodules are formed in areas of little or mild bottom current, and since nodules of specific composition form in specific areas, they represent signatures of the areas where they are formed. The rate of nodule formation has been estimated as about 6 million tons (metric) per year, and manganese nodules therefore represent a potentially significant source of minerals.

mangrove A tropical to semitropical plant (genus Rhizophora) that grows in salt marshes and tidal estuaries. Mangroves are found in Florida, in the Caribbean, along the eastern coast of South America, in West Africa, in Southeast Asia, and in Australia. They grow as trees, with their roots in the water, to heights of 10 to 12 m (30 to 40 feet), and have edible fruit. The roots produce offshoots which project several feet away from the parent and produce a new tree. The mangrove also grows "knees"—root structures that prop up the tree. Since

map   167

the mangrove tree stands in water-soaked ground, the tree needs such props to keep its heavy trunk upright. The "knees" are also fairly porous, and provide an air supply for the submerged root system. The bark of the red mangrove contains tannins and is used in the manufacture of leather. See *estuary, salt marsh.*

mantle   A term that has two meanings in the context of marine science. Geologically it refers to the layer of the lithosphere (the rocky part of the Earth) that lies between the two discontinuity layers: the Gutenberg discontinuity below and the Mohorovicic, or Moho, discontinuity above. The mantle consists of peridotite, which is largely olivine, a dark green, dense rock composed of magnesium and iron silicates.

Biologically, the mantle or pallium is the membrane in a mollusk that surrounds the soft body parts. The entire body, other than the foot, is within the space called the mantle cavity. The mantle in the adult organism is lobed, with the lobes secreting the upper and lower halves of the shell (calcareous valves) in species that have a shell. The mantle isthmus connects the lobes of the mantle and secretes the ligaments that hold the valves in place.

The molluscan mantle is folded at its margin, which is the site of growth of the organism. The outermost fold secretes the shell (if there is one); while the middle fold is the sensory perception system. The middle fold also carries any tentacles that the animal may have, and in the scallop, bears the animal's *eyes.* The inner fold, which is closest to the gut, is muscular and controls the flow of water into and out of the animal. The cephalopods, which lack shells, have a very thick, strong mantle covering. See *clam, Cephalopoda, oyster, Pelecypoda, scallop, snail.*

map   A rendition of the Earth or part of it, usually on a flat surface. Mariners have almost always made and used maps. Maps have even been constructed by primitive peoples lacking a written language, such as the Polynesian islanders of the South Pacific, whose stick charts were accurate representations of island groups and showed island landfalls at the intersections of the sticks.

Ptolemy's map of the ancient world (2nd century A.D.) is a reasonably accurate depiction of the world he knew. The decline of Western Europe after the disintegration of the Roman Empire and the early Middle Ages brought the mapmakers' art to a standstill. Scientific cartography revived with the Age of Exploration. Maps made by Mercator in 1569 showed the congruence of the American and African coasts. The concept of longitude appeared in 1690, and was a function of the development of good chronometry. Sailors' maps, or portolans, gave accurate representations of the coastlines by the 13th century, and showed ocean depths by the 18th century.

Devices for measuring distance and speed, and for determining direction, were attempted in several societies. The Chinese had a compass by the 1st century A.D. Astrolabes of varying complexity and functionality were used by Arab sailors to calculate latitude. The oldest known European seachart (1184) was shown to Louis IX as he prepared for his departure on the Crusade of 1270.

To measure the speed of a vessel at sea, a log attached to an evenly knotted line was heaved overboard. By counting the number of knots run out per hour—using an hourglass—the Renaissance sailor could calculate the distance his vessel traveled and the time required for the journey.

Maps are made in several styles. They depend on the emphasis the maker puts on some specific feature. Since it is not possible to accurately depict all of the features of a curved Earth on a flat surface, some sacrifice of one feature is necessary. Cylindrical projections, of which the Mercator projection is the best known, have rectilinear parallels of both latitude and longitude. This produces great distortions at the poles but is accurate in the depiction of coastal features. Conic projections begin from some focal position: if the position is in space, the projection is said to be orthographic; if the position is on the surface of the Earth, the projection is said to be stereographic. A projection from the center of the Earth is a gnomonic projection.

The international authority for nautical maps is the International Hydrographic Bureau, with headquarters in Monaco. See *chart; latitude; longitude; Mercator, Gerardus; Ptolemy.*

marginal seas   Large areas on the boundaries of oceans that have effects on the entire ocean. The Caribbean, including the Gulf of Mexico, and the entire Mediterranean are marginal Atlantic seas. The South China Sea and the Sea of Japan are marginal seas of the Pacific, and the Arafura and Timor are marginal seas of the Indian Ocean. See *ocean.*

Marianas Islands   A group of islands in the Pacific Ocean, east of the Philippines. The Marianas are small, tropical, typical coral-reef and volcanic islands. They were found by the Magellan expedition (1521) and called Los Ladrones (the thieves), but were renamed in 1668 in honor of the widow of Philip IV of Spain, and opened for colonization. Their existence was largely unnoticed by the rest of the world until World War II. Saipan, the site of a major battle, and Guam, are two of the principal islands. See *coral, Pacific Ocean.*

Marianas Trench   A deep ocean trench east of the Marianas Islands. It was first sounded in 1899. Ever-increasing depths have been recorded in the Trench, the latest figure being 11,708 m (38,635 feet). See *Pacific Ocean, trench.*

mariculture   The cultivation, under appropriate environmental conditions, of marine plants and animals, in analogy to agriculture. For example, marine algae are raised by mariculture in Korea, Japan, and other locales where they are important dietary items. The Chinese have been raising carp in ponds for centuries.

The formation of lagoons for trapping fish is an ancient, marginally maricultural technique. It is, however, not very successful, since the creation of a lagoon changes the living conditions of the entrapped organisms. The establishment of appropriate living conditions is a difficult problem, but the success or failure of any mariculture project depends on it.

The Romans raised oysters in baskets almost identical to the baskets used today, using an identical technique in which larval oysters were gathered, placed in the baskets to provide mechanical protection, and left in their natural environment to grow. This technique is successfully used today in Brittany, the Mediterranean, and in Japan. In Japan, oysters are raised for both food and the pearl industry. The same methodology also works for mussel cultivation, and some clams and scallops are also cultivated under this type of system. Direct use of the natural environment is simpler than trying to duplicate this environment, since different larval stages of an animal require differing growing conditions and a different food supply. Another approach to mariculture involves the release of tank-raised juveniles into their natural habitat, followed by recapture of the adults, a technique that is more like ranching than cultivation. This method is used routinely to increase the yield of salmon, trout, and striped bass stocks.

Currently, the Japanese are experimenting with the raising of squid and octopus by mariculture. While the cultivation of lobsters is still eluding experimenters, shrimp can be raised in semi-wild surroundings. The larva are fed on plankton in indoor tanks, and then released to protected waters. These attempts have, however, not yet been commercially viable.

Nevertheless, it is estimated that at least 20% of the fish and shellfish taken commercially can be raised in tanks or other enclosures. French efforts are concentrated on the turbot, sole, bass, sea-dace, shrimp, scallop, and lobster. Carnivorous species are more difficult to maintain in captivity because it is necessary to maintain a captive population as foodstuffs and to cultivate also the necessary bacteria and plankton as foodstuffs for the desired crop or the intermediate feeder population. See *fishing*.

marine archeology   The systematic study of layers of material on the ocean bottom, with the careful mapping, siting, and sifting of artifacts. Some marine archaeological expeditions have historical and cultural objectives, others focus on sunken treasure.

The first large-scale, successful hunt for sunken treasure was made by William Phips, who in 1687 brought a cargo of more than 31,000 kg (68,000 pounds) of gold and silver to the Royal Dock near Greenwich, England. Phips had searched for and found, at a depth of about 12 m (40 feet), the wrecked *Nuestra Senora de la Concepcion,* a Spanish galleon that sank in 1641 north of the island of Hispaniola, in the Caribbean Sea. It had been a part of the annual treasure convoy from the Spanish colonies back to Spain. Phips' salvage was the result of a forty-day diving session using native divers, with no equipment other than the "Bermuda bell," a primitive diving bell made of a weighted wine barrel.

The first modern marine archeological exploration was initiated in 1958, when Peter Throckmorton, an archeologist who learned to use SCUBA equipment, used the reports of sponge divers to locate a Bronze Age wreck that had gone down around 1200 B.C. near the island of Rhodes in the Mediterranean. An expedition led by George Bass of the University of Pennsylvania extensively explored the site 280 m (90 to 95 feet) down. An archeological grid was established and photographed. The calcareous accretion on the wreck was removed and new maps and photographs taken. This process was repeated as each artifact emerged. The remaining wooden parts of the ship were raised to the surface, cleaned, and preserved.

In 1961–1962 the same two archeol-ogists and a team of fifteen others worked on a 7th Century ship, a Byzantine wreck located 36.6 m (120 feet) below the surface near Bodrum (Turkey). A three-dimensional grid was laid over the cargo ship, and each section was carefully examined. The cargo of amphorae and sarcophagi was cleaned in situ by carefully breaking away calcareous accretions and then vacuuming them away, examining each artifact repeatedly, and sifting the removed material. Cameras were used extensively. The structure of the ship was preserved down to its makeshift repairs. It was then raised and immediately immersed in a solution of polyethylene glycol to fill in the cellular structure of the ship's wood and keep it from crumbling once it was out of the water and dried.

The techniques developed by Bass and Throckmorton were used again on other ships. A Roman cargo vessel found on the bottom in the Gulf of Taranto (between the "heel" and "sole" of the Italian boot) was carefully examined in this way. The Swedes have spent years working on the warship *Vasa,* the pride of Gustavus Adolphus' navy, which sank upon launching on August 10, 1628. It was rediscovered in Stockholm harbor in 1956 and brought to the surface in 1961. The intact ship has been completely preserved and studied, and is now in a museum that has been built around it near its original launch site. The hull of Henry VIII's flagship, the *Mary Rose—* which sank in Portsmouth harbor in 1545 with its full complement of 400 men— was found in 1968 and finally raised, and is being totally preserved.

And in 1985, after years of searching, Mel Fisher, who had been surveying the Florida waters for downed galleons, finally found the silver store of the *Neustra Señora de Atocha* which had sunk near Florida in 1622. Fisher was certainly more careful not to damage his site than William Phips was, even though his objective, like Phips', was to recover treasure rather than practice archeology.

The efforts of the last several years have contributed enormously to the understanding of ancient shipbuilding techniques. Before the careful examination of ancient wrecks, the only information available was based on contemporary drawings and coins, from which it was not certain just how much artistic creation and how much adherence to the model had entered into each rendition. The raising of wrecks has also opened up an entirely new area of social history, the extent and variety of ancient maritime commerce, which formerly could only have been speculative.

marine oils   Fish oils, characterized by a high fatty acid content. The fatty acids that constitute these oils are carbon compounds whose molecules consist of chains of carbon atoms to which hydrogen and oxygen atoms are attached by chemical bonds.

At one or more points these chains are linked through atoms of oxygen, to a molecule of some alcohol, usually glycerin (glycerol). Whether a fat is a liquid or a solid depends on its length and structure, and on the temperature at which it exists. Animal fat of either terrestrial or marine origin is composed of fatty acids that are 12, 14, 16 (the predominant number) or 18 carbon atoms long, with a small quantity of acids of shorter or longer carbon length. The significant difference in marine oils is the degree of unsaturation, or number of carbon atoms per molecule that are doubly rather than singly bonded to neighboring carbon atoms. This is much higher (there are more double bonds) in marine oils than in most other fats.

The natural marine oils have a variety of uses. They are of commercial interest as edible oils, as drying agents for the paint and coating industry, and in the tanning of leather. Marine oils must, however, be actively sought, as opposed to the fat of terrestrial animals, which is largely a byproduct of the meat-packing industry. The fish most widely sought as oil-producers are the California sardine (pilchard), menhaden, and herring. Whale oil is becoming increasingly rare due to overhunting.

Fish liver oil, which is rich in vitamin A, is stored in the livers of cod and some sharks. Generally speaking, fish rich in liver oil do not have very oily flesh, whereas oily fish do not store great quantities of oil in their livers. See *fish oil.*

marlin   A group of large, tropical food and sport fish of the order Perciformes of the family *Istiophoridae.* The several species of marlin are characterized by a long body with a long nose that terminates in a spear. Marlin are carnivorous, deep-sea animals that are sought by both sport and commercial fishermen. The most common species are listed below.

marsh   A brackish-water area adjoining salt water. The ecosystem of a marsh is highly specific for climate, tidal height and the amount of salt in the water. Typical tropical marshes are dominated by mangroves. More temperate marshes have grasses as the pre-eminent plant species. The most usual halophyte flora (salt-tolerant plants) are cordgrass *(Spartina)* and its relatives. The regions near tidal creeks have the least iron, sulfur, or both in the soil, and the greatest plant growth. In turn, they support large animal populations.

The ecological relationship of tidal

Most Common Marlin Species

Blue marlin *(Makaira nigricans)* (Atlantic Ocean)	400+ kg (900 pounds)	Blue with a silvery white underside
Black marlin *(M. indica)* (Indian Ocean)	700+ kg (1500 pounds)	Blue-gray to black, with stiff pectoral fins
Indian marlin *(M. audax)*	120 kg (260 pounds)	Barred upper surface
White marlin *(M. albida)* (Atlantic Ocean)	40–45 kg (90–100 pounds)	Greenish-white

See *fish.*

marshes and open sea are only now beginning to be understood. The marsh supports a large aquatic population of plankton, which is eaten by crustaceans and mollusks, which in turn either supply food for or feed on the insect population. The insects, in turn, are prey for egrets, gulls, rails, and ducks. The amphibians, snakes, and turtles of the marsh exist on this base, as do the rabbits, mice, and deer, and the carnivores that prey on them. The use of marshes as a breeding ground for many species that are oceanic as adults is slowly making the study of marshes and their life-forms important to commercial interests. Until very recently, the average approach of the marine engineer to a marsh was to drain it. See *duck, estuary, heron, littoral.*

Marshall Islands   An island group east of the Caroline and Marianas islands; the most easterly part of Micronesia. The Marshall Islands form two parallel chains lying in a northeasterly to southwesterly direction. They are typical volcanic and coral oceanic islands. The islands were first noted by European sailors in 1529 when Alvaro Saavedra of Spain claimed them for his king, Charles I (Emperor Charles V). Kwajalein is one of the largest islands; others in the group include Bikini and Eniwetok—both of which were used as sites for early above-ground atomic tests. See *atoll, coral, island, Pacific Ocean.*

Mastigophora   The class of flagellated protists, consisting of unicellular, tailed organisms. There are over 5,000 species. See *dinoflagellates, flagella.*

Maury, Matthew Fontaine (1806–1873)   An American oceanographer known for establishing the first "sea lanes." As a midshipman he circumnavigated the globe in the expedition of 1826–1830. He was promoted to lieutenant, but an accident prevented his going to sea, and he was given shore duty at the Depot of Charts and Instruments, an office that later became the U.S. Naval Observatory and Hydrographic Office.

Maury issued form-logbooks to captains, and used their notes to produce wind charts, which were published in 1848. These charts made possible shorter passages using areas of the ocean that had the most favorable winds—the first establishment of "sea lanes." Maury was the American representative at the first meeting of the International Hydrographic Bureau. He attempted soundings of the Atlantic in preparation for the work involved in laying the trans-Atlantic cable. His classic work, *Physical Geography of the Sea*, first appeared in 1855. It was to be a standard in the field, and went through many editions. Although it contained errors, it was the best compendium of information available, and was written in a very readable, informative style by someone who obviously liked what he did. Maury's recommendation for different travel lanes for eastbound and westbound steamers created internationally honored sea lanes designed to avoid collisions. The vastly increasing use of the ocean for travel and freight shipment made this kind of regulation necessary.

During the Civil War, Maury worked on the electric torpedo, which the Confederate Navy was trying to develop, and went to England as a special envoy. He served with Emperor Maximilian in the attempt to establish a European empire in Mexico. He returned to the United States in 1868 and taught meteorology at the Virginia Military Institute in Lexington, Virginia until his death.

Mediterranean   An extension of the Atlantic Ocean between Europe and Africa. The term Mediterranean is also used to describe a type of landlocked sea: an arm of an ocean surrounded by land. The Caribbean is an example of a Mediterranean-type sea.

The area of the Mediterranean is about 2.88 million km^2, or 1.15 million square miles. The tides are semi-diurnal and usually very shallow; the average tide is no more than 1 m (3.3 feet) high. The predominant surface current is an inflow of cold, oxygenated water from the Atlantic Ocean through the Strait of Gibraltar. The water then circulates along the North African coast to the eastern end of the Mediterranean, where some breaks off to make the circuit of the Ligurian Sea. The return flow is along the European coast back to Gibraltar, where the warmed-up, oxygen-poor water leaves the Mediterranean below the inflow of fresh cold Atlantic water. The Mediterranean loses vast quantities of water through evaporation, and is therefore considerably more salty than is the Atlantic. The salinity at the eastern end is about 3.6%.

The present Mediterranean is a much smaller version of the primordial ocean that encompassed almost half the Earth in the Jurassic Period. The Aral, Caspian, and Black seas were once integral parts of the Mediterranean rather than totally landlocked lakes. The present Sea may be thought of in terms of two distinct basins, an eastern and a western one, which may be subdivided into distinct regions. The western Mediterranean may be subdivided into the Alboran Basin, between Gibraltar and the Gulf of Lions; the Balearic Basin to the east of the Alboran and ending at Corsica and Sardinia; and the Tyrrhenian Basin east of Corsica and Sardinia and west of Italy. The eastern end of the western Mediterranean is the Strait of Sicily. The eastern Mediterranean may be subdivided into the Adriatic Basin, east of Italy; the Ionian Basin, from the Strait of Sicily to Greece and Crete; the Aegean Basin, north of Crete; the Sea of Marmara to the east of the Dardanelles; and the Levantine Basin east of Greece.

The average depth of the Mediterranean is about 1,500 m (4,600 feet), although there are considerable deeps. Those of the French coast near Marseille (the Calanques) have been well explored. The greatest depths, of more than 5,000 m (16,000 feet) are in the Hellenic Trough, which makes a wide arc around the Greek peninsula and Crete. The Balearic Abyssal Plain covers a triangular area encompassing most of the bottom of the Balearic Basin. The base of the triangle lies along the Algerian coast, and

depths of more than 2,700 m (8,000 feet) are reported. Most of the rest of the Balearic bottom is the delta cone of the Rhône river. This is one of the few western rivers that delivers any appreciable volume of water into the Mediterranean.

The Mediterranean is relatively low in nitrates, phosphates, and other nutrients, since most of its water source is the Atlantic, not land runoff. The eastern basin is more impoverished in terms of dissolved minerals than the western, and is also saltier.

The dominant winds in the Mediterranean are the mistral, a northwesterly cool wind that blows into the sea from the Alps and brings very clear skies; the bora, a cold, dry winter wind; and the khamsin (humid) and sirocco (dry) hot winds that move from Africa into the Sea and into Europe and, depending on their velocity, may transport large quantities of sand.

The dominant bottom feature of the eastern Mediterranean basin is the delta cone of the Nile. The Nile delta is not as imposing as that of the Rhône. Nile silt has not built up as much area as would otherwise be expected, because of the active subsidence in the region, which is near the edge of a crustal plate. The quantity of Nile sediment has not significantly increased since the construction of the Aswan Dam in 1970.

There are many islands in the Mediterranean, which range in size and historical importance from tiny ones in the Greek archipelago to Mallorca, Crete, Corsica, Cyprus, Sardinia, and Sicily— the largest islands. The general climate of the whole region is warm and relatively arid. See currents, delta, islands, seas, Tethys Sea.

medusa   One of the two forms of all adult cnidarians. Jellyfish have the umbrella or bell-shape of the medusa with their mouth at the bottom center of this structure. The medusa moves by expelling water. See Cnidaria.

megabenthos   See macrobenthos.

megaloplankton See *plankton.*

Meinesz, Vening (1887–1966) A Dutch geodesist known for undersea gravity measurements. His full name was Felix Andries Vening Meinesz.

Geodetic measurements are concerned with the shape and size of the Earth, with the exact position of points on the Earth's surface, and with variations of the Earth's gravity field. Meinesz began his career by doing a gravimetric survey of the Netherlands in 1911. He designed a dual pendulum apparatus for measuring the effect of gravity, but surface waves created excessive perturbations. He then used submarines, since waves are damped exponentially by depth. This technique of submerging his pendulums made very exact mapping of the ocean bottom possible.

Between 1923 and 1939 Meinesz used eleven different ships and made over 800 observations of the seafloor. His apparatus was used by all explorers until the 1950s. After that, spring gravimeters were used on ships equipped with shock-absorbing platforms.

The Earth's gravity field was extensively explored by Meinesz. He plotted gravity anomaly belts of island arcs that were thousands of kilometers long. They lay along trenches and volcanic islands, and clustered around earthquake foci. Meinesz thought that gravity anomalies were a result of compression of the Earth's rigid crust. He did not think in terms of plate tectonics. See *ocean bottom, sounding.*

meiobenthos See *macrobenthos.*

Mekong River A major southeast Asian river. The Mekong rises in the Tibetan highlands, flows south through Tibet and the province of Yunnan, China, and then forms the boundaries between Burma and Thailand, and then between Thailand and Laos, and eventually flows through Cambodia and the southern end of Vietnam to the South China Sea.

For much of its roughly 4,200-km (2,600-mile) course, the Mekong is a tur-

bulent stream with falls and white-water regions. The delta is vast, and is certainly Vietnam's greatest asset: It is the best-producing rice-growing area in the world.

Melanesia The collective name of the group of Pacific islands northeast of Australia. Fiji, the Bismarck archipelago, the New Hebrides, and the Solomon Islands comprise this group. See *atoll, island, Pacific Ocean.*

Mendocino Escarpment A sharp drop in the seafloor of the North Pacific Ocean. This "break" extends from the California coast toward a point north of Midway Island. See *crust, fracture zone, subsidence.*

menhaden See *herring, marine oil.*

Mercator, Gerardus (1512–1594) [Gerhard Kremer] A Flemish cartographer who produced a global map in 1538. Very much influenced by the global map of Ptolemy, Mercator made a determined effort to produce a map that was as good as his 2nd century inspiration. He followed the production of the map by a globe of the Earth in 1541 and a celestial globe in 1551.

After his appointment as professor of cosmography at the University of Duisberg, Mercator changed his approach. He redesigned his map on the basis of parallels of latitude perpendicular to the meridians of longitude. His new maps were published in 1569, and he planned an atlas, the first installment of which appeared in 1585; it was never finished.

The increasing use of Mercator's rectilinear maps by French explorers led to their general adoption. These maps show coastlines very distinctly in spite of their distortions of shape and size. The error or distortion on mercator maps is the result of the shape of the map. The Earth is spherical, not rectilinear. Thus, on a map on which latitude and longitude lines form right angles, the north-south distances become more exaggerated than do the east-west distances as latitude increases. The best example is the appearance of Greenland. On a mercator map Greenland appears to be larger than South America, which is certainly not the case. See *maps, navigation.*

mercury   A chemical element which in its usual state is a reactive, silvery-white, liquid metal. Mercury has been used extensively in the chemical industry since ancient times in spite of its known toxic effects. While the toxicity of the compounds of mercury has been long established, it was not until the mid-1960s and 1970s that the potential of mercury for environmental contamination was fully understood. The manufacture of chlorine, sodium carbonate, and certain other chemicals are among the industrial processes that use mercury. The effluent of plants manufacturing these products eventually contaminated aquatic as well as terrestrial environments. Minamata Bay in Japan became famous because of the tragic effects of mercury poisoning observed in many residents of this heavily industrialized area. This was the first documented example of the entrance of mercury metal into the ecosystem of a region. Mercury effluent entered the Bay from industrial runoff. Compounds of mercury—which are much more soluble than is the metal itself—were formed in the water by anaerobic or sulfur metabolizing bacteria, and when they were incorporated and concentrated in the fish population, they eventually poisoned all of the consumers of the fish, including humans. See *element, pollution.*

meridian   A circle of longitude passing through both the North and South poles of the Earth.

meroplankton   See *plankton.*

mesoderm   The central tissue layer of an embryo. In vertebrates it gives rise to the tissues that form the skeleton, circulatory system, and muscles.

mesopelagic   A term referring to the oceanic water column from depths of 200 to 1,000 m (700 to 3,300 feet). This is the aphotic zone: no sunlight penetrates to this depth. See *pelagic environment, photic zone.*

mesozoa   Very simple animals that usually contain from 10 to 20 cells. These cells are usually undifferentiated. The mesozoan group is an intermediate one between unicellular and colonial organisms. There are about 50 species. The phylum was created to encompass those few organisms that did not seem to fit into the taxonomic classificaiton of either Monera or Protista. See *Protista, colonial animals, Monera.*

Mesozoic Era   The time in the Earth's history following the Paleozoic and followed by the Cenozoic Era. The Mesozoic lasted about 130 million years, (from 230 million to 100 million years ago). The name was suggested by British geologist John Phillips in 1840, to commemorate intermediate life forms. The Mesozoic is divided into the Triassic, Jurassic, and Cretaceous periods.

Landforms of the Triassic were eroded in the Jurassic, and then overlain by warm seas in the Cretaceous. The greatest botanic variations were in the development of seed-bearing plants. The gymnosperms (coniferous plants) developed first, and were then followed by the angiosperms (flowering plants). The Mesozoic is the Age of the Vertebrates. On land, the spectacular diversification of reptilian forms led to the development of the dinosaurs. Most of these highly specialized reptiles lived in the Jurassic and Cretaceous periods.

In the sea, Paleozoic corals were replaced by modern types during the Jurassic Period. See *Cretaceous, Jurassic Period, Triassic Period.*

Messina, Strait of   A waterway separating Italy and Sicily; Homer's much-feared Scylla and Charybdis. The main current runs south for six hours and then, after a short period of slack water, reverses. The north-to-south current is less strong. When, however, the current is running northward and is augmented by a sirocco, it is possible for it to be a navigational risk even today.

During the Pleistocene epoch the strait did not exist. The water level in the entire area around Sicily was lower, and Sicily was not an island but an extension of the Italian peninsula. See *Mediterranean, sirocco, strait.*

metamorphic rock Igneous or sedimentary rock which on subjection to heat, pressure, or both recrystallizes or metamorphoses.

Meteor Research Expedition An oceanic exploration voyage funded by Germany, which took place from 1925 to 1927. This was one of the most significant ocean voyages of exploration, and its many traversals of the Atlantic resulted in carefully mapped bottoms of both the North and South Atlantic basins. The expedition's chemical analysis of seawater established the circulation patterns of all the oceans, and the resultant effect on nutrient dispersal and in turn on plankton growth. See *Antarctic Water; Atlantic Ocean; Ekman, Vagn Walfrid; mid-atlantic ridges; Wüst, George.*

meteorology The study of the atmosphere and its phenomena: weather and climate. Weather and climate depend on the interactions between the atmosphere and the oceans. These interactions are extremely complicated, and meteorologists tend to consider the atmosphere and the oceans as a single system.

Such ocean currents are wind-driven. In summer the ocean absorbs most of the radiant heat from the Sun, and this heat flows back into the air in winter, particularly where warm water meets cold air and cold water. The interface between cold North Atlantic water and the warm Gulf Stream is an example. Another is the Kuroshio Current off the east coast of Asia. The addition of warm, moist air to a region of dry, cold air produces major weather instabilities, such as the Polar Fronts, which are more or less permanent features of the high latitudes.

Atmospheric pressure, or the pressure of a column of air, is about 1 kg/cm² (14.7 pounds/square inch) at sea level. If the air is cold and dry, that figure is correct, but if the air is warm or wet it is less dense and the pressure is lower. Therefore, on a stationary Earth, air flow would be horizontal from high pressure areas to those of low pressure. However, the Earth's rotation produces the dominant effect on wind, and with increasing altitude, the Westerlies increase in velocity and reach a maximum at about 32° North Latitude, where they are known as the Jet Stream.

Clouds are formed as warm air and water vapor rise from the ocean and the mixture expands because the atmospheric pressure decreases. If large quantities of water rise rapidly, they condense as ice crystals in the high troposphere forming small wispy cirrus clouds. Cumulus clouds are the result of very local, rapidly (6,000–12,000 m; 20,000 to 40,000 feet) rising warm, moist air and these large, elongated clouds can grow into thunderclouds.

Wind movement in the atmosphere is of two types: either continuous, (i.e., of the jet stream type), or seasonal, as are the monsoons. Local manifestations, in addition to cloud formation, are cyclonic winds.

Horizontal movements of air are called frontal advances. If cold air displaces warm, the front is a cold front; if warm air moves a cold body of air aside, it is a warm front. In some areas the heat-flow pattern is such that fronts are stationary. The polar convergence is such a zone.

Meteorological instruments are used for the study of weather patterns and the prediction of weather phenomena. These instruments include the two oldest ones: the thermometer and barometer, used for the measurement of temperature and air pressure, respectively. The 20th century has added radar for tracking storms, infrared heat sensors, a variety of recording and photographic devices, and sophisticated methods, such as balloons, kites, rockets and satellites, for putting these devices in the air and maintaining them in place. See *coriolis effect, cy-*

clone, jet stream, monsoon, weather, wind.

Mexico, Gulf of   A body of water encompassed by the eastern coast of Mexico and the southwestern United States, with outlets to the Carribean at the Yucatan Channel and to the Atlantic Ocean through the Florida Straits. The Gulf is roughly 1.6 million km^2 (620,000 square miles), and relatively shallow; the Sigsbee Abyssal Plain has an average depth of about 3,600 m (11,500 feet).

There are wide continental shelves south of Texas and Louisiana, north of the Yucatan (the Campeche Shelf), and east of Florida. The delta of the Mississippi River is built of silty sand and extends almost to the edge of the continental shelf. The Sigsbee Knolls, in the center of the Sigsbee Plain, are thought to be the tops of salt domes similar to those on the Texas-Louisiana shelf. The domes, like those on the continent, are an indication of oil deposits. The knolls are composed of foraminiferal deposits. The silty deposits of the Mississippi and other rivers emptying into the Gulf are spread by bottom currents and surround the Knolls but do not cover them.

The major water inlet to the gulf is through the Yucatan Channel, where the sill depth is between 1,500 and 1,900 m (4500 to 5700 feet). Cold South Atlantic water comes into the gulf through the passages between the Windward and Leeward Island chains. The average salinity of the gulf is high, usually above 3.5%, although areas near shore may be brackish due to heavy river outflow. The temperature on the surface varies for the most part with latitude; the water temperature is fairly constant below 1,000 m (3,300 feet).

The tides in the gulf are fairly gentle. Storm surges due to hurricanes are the exception. Tides in many places are diurnal, with only one high and one low tide each day. See *Caribbean Sea, Florida Current, Gulf Stream, Yucatan Current.*

Mexico Trench   A cleft in the seafloor that marks the eastern edge of the Pacific crustal plate where it abuts the small Cocos Plate. This depression runs north-south, into the Gulf of California. See *crust, Pacific Ocean, plate tectonics, trench.*

microbenthos   See *macrobenthos.*

Micronesia   A group of western Pacific islands east of the Philippines; it includes the Marianas, Carolines, Marshalls, Gilberts, and some smaller islands. See *atoll, Pacific Ocean.*

Mid-Atlantic Ridge   The undersea mountains of the Atlantic Ocean.

mid-ocean ridges   An interlocking series of continuous undersea mountain ranges. Together they constitute the longest series of mountains on Earth, extending over 56,000 km (35,000 miles). Each mountain range has its own name. The north-to-south ranges are the Mid-Atlantic Ridge in the Atlantic Ocean, the Reykjanes in the North Atlantic and Arctic, and the East Pacific and Mid-Indian Ocean rises in the Pacific and Indian Oceans, respectively. The Carlsberg Ridge runs southeast into the Mid-Indian Rise from the Arabian Plate. The Southwest Indian Ocean Ridge and the Southeast Indian Ocean Ridge separate the Antarctic Plate from the South American, African, and Nazca plates. The Southeast Indian Ridge rises between Antarctica and the Indian and Australian plates.

Seismic sounding has shown that the thickness of the Earth's crust under a ridge is not greater than that elsewhere on the ocean floor. Thus, the current theory holds that molten rock rises under the ridge crest, is ejected as lava into the water, and then flows onto the valley floors on either side of the ridge, solidifying as it cools and gradually widening the oceans as it pushes continental blocks apart. The slopes of the ridges eventually form all of the suboceanic crust. There is constant creation of new rock at the crest of a ridge and disappearance of old rock at the edge of the slope under a continental plate.

The ridge itself is a mountain complex rising from the abyssal floor in a sharp ascent. If the Mid-Atlantic Ridge is taken as the model ridge system, a ridge bisects the ocean between continents, with the ridge crests at the midpoints. The crests are the high rift mountains; the rift valleys between them, and the plateaus from one crest to another, are scarred by deep fracture zones called slip faults. The crests drop down to a series of lower mountains and plateaus that form the flanks of the ridge.

Rift valleys, which are depressions that lie parallel to the sharp peaks of the ridge center, are the sites of shallow-focus seismic activity. The line of epicenters of such seismic activity follows the rift valleys, including the East African Rift Valley, which is terrestrial. The existence of a rift valley and its accompanying ridge was predicted near Easter Island because of the earthquake activity in the region. Other characteristics of rift valleys are high heat flow and magnetic and gravitational anomalies.

The Pacific Ridge system has a more rounded crestline than does its Atlantic counterpart. The ridge system of the Indian Ocean is extremely complex since there are intersections in it. The junctions of ridge systems are areas of great interest: for example Macquarie Island, south of the southern island of New Zealand, has literally been pushed to the surface by an abutment of crustal plates.

The *Challenger* voyage "discovered" the Earth's undersea ridges, which had been unknown in all the millenia that men had sailed on the ocean's surface. The first evidence that the sea bottom was not featureless did not come until the 1850s, when Matthew Maury recorded what he called the Dolphin Rise. (A ridge and rise system are roughly the same. The difference is one of intensity: a rise is gentler and does not have the axial rift valleys. It is the result of a greater volume of lava that spreads more rapidly.) This was the first indication of the existence of the Mid-Atlantic mountain system. See *individual plates; Maury, Matthew Fontaine; Mid-Atlantic Ridge; oceanic crust; plate tectonics*.

migration A term used to describe regular animal journeys along well-defined routes, particularly those involving a return to breeding grounds.

There are three types of migrant fish: (1) *Oceanodromous fish* (always in the sea), such as the herring. Herring all belong to one species, but there are non-mixing local populations that live in one area and spawn in another and at different times. Thus there are several groups of herring living in the North Sea. The Cod is also an oceanodromous migrant. It travels, as does the tuna, greater distances than does the herring. 2) *Anadromous* fish (which spend most of their lifespan in the sea). These are the salmon, which return to their native rivers to spawn, and with some species, to die. 3) *Catadromous* fish are essentially freshwater species that spawn in the sea, such as the eels.

Other animals also migrate. Sea turtles spend their entire lives at sea, usually in deep water. Some have been tagged and were found to travel over 2,000 km (1,400 miles) to their native beaches to lay eggs. The most familiar migrants are birds. Coastal birds stay on the continental shelves, pelagic ones have small nesting areas and great flying ranges. Some petrels fly almost from pole to pole. The prime example is the Arctic tern. Its breeding areas are northeastern Europe, Asia, or North America but it lives in Antarctica.

Humpback whales are good examples of migrating mammals. They winter in warmer water and move toward the poles for the greater food supply that follows the planktonic bloom. Not all individuals move in such a migration, and not all have specific breeding areas. Seals and walruses do, however. The Pribilof Islands are vast seal nurseries.

Copepods are among the best-traveled planktonic migrators. Their movement is a vertical one. Many nektoplankton migrate to or near the surface at night and then drop down from 100 to about 1,000 m (330 to 3,300 feet) in daylight. For a small animal, this is an enormous distance. The predators, fish and pelagic birds, follow these ups and downs.

Some polychaetes move from inshore rocks to the open sea to reproduce. The Palolo worm rises vertically from coral reefs to the surface to cast off its tail segments, which contain the eggs or sperm. Crabs mate close to shore and lay eggs in brackish water.

A great deal of research has been devoted to unraveling the question of how animals "know" when to migrate, and what clues lead them back to some communal or ancestral spot. There is evidence that the development of gonadal tissue is stimulated by changes in light intensity, and thus that migratory animals respond to a change in season before they respond to a weather change. They do use the Sun's position to navigate. There are theories of the use of "radar" and the magnetic field of the Earth by flying flocks of birds. See *anadromous, catadromous, eel, herring, polychaetes, salmon.*

Milne Seamount A submarine mountain of volcanic origin in the North Atlantic, to the west of the Mid-Atlantic Ridge. It is east of the mid-ocean canyon that separates the Ridge from the Grand Banks off the coast of Newfoundland. See *canyon, Grand Banks, seamount.*

minerals Naturally occurring chemical elements or compounds, usually solid and crystalline, obtained from the ground or from the sea. The sea has been used as a source of minerals for centuries: ancient coastal civilizations made salt. The winnowing of sodium chloride from the sea has largely been abandoned, but some other alkali salts are now obtained by evaporation of sea water. The commercial production of different minerals from the sea is recent. The sea is divided into regions, each of which yields different products that must be extracted using different methods.

Seawater contains almost every element; the criterion in deciding which to extract is cost. For sodium, magnesium, and bromine, the removal from seawater is cost-effective.

Beaches yield building materials, such as the sand used for concrete, and limestone (calcium carbonate). A variety of useful minerals is found either on shore or just off it. Diamonds are mined near the coast of Southwest Africa, for example, iron ore off the Japanese coast, and gold near Nome, Alaska. The drowned river valleys of Indonesia and Malaysia are a source of tin. The continental shelves are mined for sulfur and explored for petroleum. Some of these probes have yielded good results, notably in the Gulf of Mexico and the North Sea.

The sea bottom is harder to get to commercially, and will therefore be the last region to be exploited for minerals. However, the research vessel *Glomar Challenger* is part of a project to develop deep-sea petroleum on a commercial scale. Also on the bottom are manganese nodules, which contain copper, nickel, and cobalt in commercially attractive quantities. Since there are large reserves of the nodules and new ones accumulate, they are a renewable resource.

The red clay of the ocean bottom contains some rare metals in amounts of less than 1%. This is enough to make such a find exciting, but recovery would be expensive, and awaits some cost-effective technique for bringing the clay from the bottom to the surface.

The methods of retrieving minerals from the ocean bottom vary with the mineral in question and the depth at which it is found. Oil is pumped up onto offshore platforms by the same mechanism used to pump it out of the ground. The platforms are large and heavy enough to supply support for the pumping machinery. They are most often fixed in place, but need not be. Sulfur is also obtained by pumping. For removing tin ores, ocean-going dredges and draglines are used.

The effort put into a sea-mining installation is a function of the value of the material retrieved. For rare metals, more costly methods, such as the construction of a tunnel from nearby land, or the use of submersibles or work crews that spend long periods (a week or more) underwater might be considered. In Antarctica there exist mineral deposits of major international importance; however, the difficulties of a sustained effort in that very hostile environment require a multinational project. Such a project is currently under discussion. See *diving, manganese nodules, seawater.*

Miocene Epoch   The fourth epoch of the Tertiary Period of the Cenozoic Era. The other epochs of this era have been the Paleocene, Eocene, Oligocene, and Pliocene—the last being the last epoch of the Tertiary Period. The name Miocene was suggested by the geologist Charles Lyell in 1833. He meant to imply a smaller number and diversity of fossils than had been dated from the Pliocene. The epoch lasted more than 10 million years, during which the climate was generally mild, with warm to temperate conditions experienced almost as far into the polar regions as the poles themselves. There was considerable volcanic activity at this time, and much mountain building. The Caucasus, Himalayas, Appennines, and Pacific Coast Range all grew during this time, and during the Miocene the European Alps became folded and deformed. The ocean overlay considerable parts of the globe, and when it receded, left great banks of sediment thousands of meters thick.

On land the Miocene was the time of the dawn redwoods. The great forests receded and left large grassland areas, the savannas. The horse and the camel, typical grassland animals, evolved in this time, while the oredonts—carnivorous doglike mammals—disappeared.

To a great extent, the marine fossils of the Miocene resemble the present marine biota. There was a luxuriant expansion of gastropod families in this epoch. Oysters and sea urchins were also abundant and distributed worldwide. The seagoing mammals, sea cows, dugongs, and whales adapted further to the marine environment. See *Oligocene, Pliocene, Tertiary.*

Mississippi River   A 4,120 km (2,470-mile) long river flowing south through the United States from northern Minnesota to the Gulf of Mexico. The Mississippi and its largest tributary, the Missouri, form the largest river system in North America, in terms of both length and volume. While the annual output of the Mississippi-Missouri river system varies enormously from one year to the next, it is the great river system of the continent, draining about 12% of North America.

The source of the Mississippi is Lake Itasca in northern Minnesota. The lake is at an elevation of over 440 m (1,463 feet) in a region glaciated during the Pleistocene. The river moves over a series of falls in the mountain ranges between its source and the Minneapolis-St. Paul region. It is navigable from that point to the Gulf of Mexico.

The upper Mississippi is fed by Minnesotan snowfall, but south of its confluence with the Ohio, which drains the well-watered northeast, over 50% of the volume of water in the combined river is of eastern origin. The Missouri River component is less than 20% of the total, since the sources of the Missouri's water are the Rocky Mountains and the more arid prairie.

South of the junction of the Ohio with the Mississippi, the river is about 1,360 m (4,500 feet) wide. It meanders considerably, changing its channel as it moves over the floodplain. The floodplain is the part of the river valley which can flood in periods of high river volume, notably in spring. A floodplain can be miles wide. It has changed its course many times in the past, and much of the center of North America bears scars and marks of the ancient river's lateral movement.

The junction of the Mississippi with the Missouri near St. Louis is roughly 1000 km (600 mi) from the sea, but with its meandering the river's true path is almost three times that length.

The Mississippi Delta begins in Louisiana at the junction with the Red River. The lowest reaches of the river break up into a network of small, shifting channels called bayous. See *delta.*

mizzen   The mast behind the main mast on a sailing vessel. The term also refers to those sails carried on the mizzenmast. See *sail.*

MODE   Mid-Ocean Dynamics Experiment, a study of short-term, small-scale weather phenomena in the North Atlantic. The United States and Great Britain cooperated on this venture in 1973. The objective was to identify oceanic-atmospheric interactions so as to establish what were called "weather" patterns in both air and water. This makes the analogy between the two and equates oceanic eddies with cyclonic storms, such as hurricanes. See *ocean-atmosphere interactions, storms, weather.*

Moho   An abbreviation for Mohorovicic discontinuity, a region within the Earth in which the density changes suddenly because there is a change in the composition of rocks from those of the Earth's crust (granites and basalts) to those of the mantle below (olivines and pyroxenes). The crust of the planet is from 4 to 60 km (2.5 to 40 miles) thick, varying from one location to another, but thinnest beneath the sea. The discontinuity was named for its discoverer, Andrija Mohorovicic. See *discontinuity layers, mantle, plate tectonics.*

molecule   The smallest unit of a chemical compound that retains the characteristic properties of that compound. A molecule is a group of atoms held together by strong chemical bonds. Molecules vary in size and composition from the very small hydrogen (H_2) molecule, composed of two atoms of hydrogen, to proteins and starches, which have

thousands of atoms in a single molecule. See *atom, protein, salt.*

Mollusca   A very large phylum of invertebrates. Next to the insects, this group has the greatest number of living species and individuals. They inhabit both terrestrial and aquatic ecoystems. The individual members of the phylum are usually encased in calcareous shells produced by their body covering, the mantle. The mantle is the distinguishing feature of the mollusks. The marine mollusks are found on rocky or sandy bottoms, from dry land to the littoral region to deep sea.

The taxonomic classification of the mollusks is being revised. There is agreement that these animals have a long geologic history. Their fossil record dates from the early Cambrian. The mollusks include the well-known groups of single-shelled animals (snails), the two-shelled (oysters, clams, scallops), the plated animals (chitons), and unshelled animals (the squid and octopus).

Most mollusks have bilaterial symmetry with the right side being a mirror image of the left. In the gastropods (snails), however, one side is highly developed as the foot, the large muscle that the animal uses for locomotion. Many mollusks move slowly. Some, like the snails, slide along on a mucus layer; cephalopods, the swiftest group, move by jet propulsion; scallops flap their shells in a half jet-propelling, half hopping movement; oysters are sessile (stationary) as adults, as are mussels; clams burrow.

The internal structure of the mollusks is as varied as their exterior. Basically, all have a combination head-foot organ that consists mainly of the foot, and mantle-covered viscera (internal organs). They have a mouth, salivary gland, esophagus, and stomach, that digests their food, which among the shelled members of the group usually consists of detritus. Snails and their relatives have a specialized mouthpart, the rasping radula. The mollusks respire through paired, ciliated gills, which are also in the mantle cavity (the space within the mantle). The rate of respiration is controlled by sensory organs, also part of the mantle, which regulate the incurrent and excurrent siphon, the means by which the animal brings in water and expels the products of digestion, carbon dioxide, and nitrogenous waste in the form of ammonia or urea. The molluscan heart lies dorsal to the viscera. One of its prime functions is to maintain turgor (muscular rigidity) in the foot. The blood, which has some red pigments, like the blood of the annelids, also controls the tentacles, siphons, and other appendages. There is a nerve net in all the members of the phylum, with that of the cephalopods being quite complex and specialized.

Reproduction among the mollusks is quite varied. Chitons, some snails, scaphopods, most bivalves, and the cephalopods have separate sexes (dimorphism). Most gastropods are hermaphrodites, but fertilization usually requires two individuals, although each animal has male and female gonads. In many organisms, particularly the bivalves, fertilization is external. Most molluscan larvae are free-swimming.

The bivalve mollusks are commercially fished, and have been part of the human diet for thousands of years. They have also been raised commercially by mariculture. Some gastropods (e.g., the abalone) are eaten, as are the cephalopods. Oyster drills and teredos are pests. Some members of the phylum Mollusca are highly toxic, such as the cone shells; others prey on each other. Many are hosts to parasites, such as the foraminiferans that cause red tides, and most support large predator populations. See *abalone, chiton, Cephalopoda, clam, oyster, pearl, scallop, sea hare, Sepia, shell, snail.*

mollusk   An animal belonging to the phylum Mollusca. All members of the phylum have soft bodies that are covered by a mantle, a thick membrane. In many mollusks the mantle secretes a calcareous shell. Clams, snails, and scallops are all examples of shelled mollusks. Octopi and squid are mollusks that have mantles but no shells. See *Cephalopoda, clam, gastropod, mantle, Mollusca, octopus.*

molt   Also called ecdysis. The shedding of an exterior. On land the familiar animals that molt are birds and reptiles. In a marine context, the significant animals that molt are the arthropods. In order to grow larger, an arthropod splits its chitinous (hard exterior) shell at its fuse lines and walks away from it. A new shell is secreted to replace the older, smaller one. The soft-shelled crab is merely the ordinary blue crab caught soon after molting. The hermit crab doesn't molt: it moves out of a smaller shell and looks for a larger one. See *Arthropoda, chitin, crab.*

Monera   A kingdom of very simple organisms that lack distinct cell nuclei. Bacteria and blue-green algae are examples. The genetic material of these organisms is interspersed with the rest of the protoplasm of their cells, rather than being contained in a nucleus.

Monoplacophora   Very ancient animals belonging to the phylum Mollusca. The monoplacophorans are "living fossils," along with the horseshoe crabs and coelacanths. They have been found only in the Pacific, in warm temperate-to-tropical areas of Central and South America. They were differentiated from Paleozoic limpets in the 1940s, but were known only as fossils until recently. A live specimen of the species *Neopilnia* was found in mud off the Pacific coast of Central America at a depth of about 2900 m (11,700 feet).

These single-shelled organisms look like limpets. Their relatively flat shell is about 4 cm (1.6 inches) in diameter, and covers a very un-molluscan body. The digestive, excretory, and muscular systems of the animal do not resemble those of the limpet.

The diet of the monoplacophorans may be detritus. See *fossil, limpet, mollusk.*

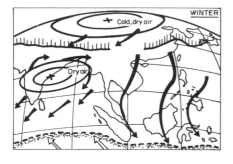

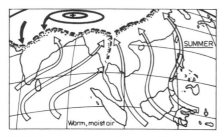

Typical monsoon weather conditions

monsoon A wind that changes direction with the season, most often associated with Africa and South Asia. Winter or dry monsoons tend to blow from northeast to southwest. In summer the wind direction is reversed and the monsoon brings rain.

Monsoons have their source in the shapes of land and water masses, the general circulation of winds about the Earth, and the different heat-retaining properties of land and water.

The Indian monsoon produces warmth and rain as it moves north until July, after which the entire subcontinent is hot and wet. By September or October, colder air moving from Tibet reverses this pattern, and the prevailing weather is hot and dry. Since larger land areas produce greater monsoonal effects, the monsoon in Indonesia is not as obvious as the Indian monsoon in terms of temperature change and rainfall accumulation, but the seasonal change in wind direction is marked.

In Africa the Hamattan is a hot, dry wind from the Sahara. It moves from northeast to Southwest to the Intertropical Convergence, where it meets hot, moist air from the Bight of Benin. This line of convergence moves from the equator in winter to about 15° North Latitude in July.

In the Americas, monsoonal developments in the Gulf of Mexico are weak and barely distinguishable from normal seasonal variations. See *wind*.

Moon The Earth's sole natural satellite. It has a diameter of 3,476 km (2,160 miles), and its mean distance from the Earth is 384,000 km (238,857 miles). The Moon rotates on its axis in $27\frac{1}{3}$ days, and revolves about the Earth in $29\frac{1}{2}$ days. It shines by reflection of the Sun's light, and during a complete revolution about the Earth, four phases are observed: the new Moon, first quarter (half Moon), full Moon, and last quarter (half Moon). The Moon exerts an attractive force on the Earth. This gravitational force is most obvious in its effect on that part of the Earth's surface that moves easily, the water. Since the Earth is not perfectly spherical, the gravitational force exerted on different parts of Earth varies and gives rise to different tide heights in different places.

There have been several theories of the genesis of the Moon; one that was reasonably popular around the turn of the century was that the Moon was "ripped out" of the Earth's crust and that the Pacific Basin is the result of the Moon's departure. Currently it is believed that the Earth and the Moon formed simultaneously. See *lunar day, lunar tide, tide*.

mosasaur A Cretaceous aquatic lizard of worldwide distribution. Mosasaurs were contemporary with the ichthyosaurs and plesiosaurs. While most were about the size of the modern porpoise, some were much larger, at 9 m (30 feet) long. Mosasaurs had paddlelike limbs and elongated heads and bodies. Their jaw, like that of their modern relatives the monitor lizards, had a hinged mandible (lower jaw). This enables the modern lizard to consume large prey quickly and presumably functioned in the same manner for the ancient lizard. The mosasaur preyed on fish and cuttlefish. See *Cretaceous Period, ichthyosaur, plesiosaur.*

Mozambique Current   That part of the Indian Ocean gyre that flows between Mozambique on the African mainland and the Malagasy Republic (Madagascar). It is a south-flowing stream along the southern coast of Africa, and most of it eventually doubles back and flows north along Madagascar. It feeds into the Agulhas Current. See *Agulhas Current, currents, gyre, Indian Ocean.*

mud   Sediment, consisting either of rock fragments or organic detritus, in which the particles are smaller than 0.06 millimeters. See *mud flat.*

mud flat   A stretch of level ground, behind the backshore, that on its sea side merges gradually into the slope of the beach. The characteristic plants of mud flats are eel grass (*Zostera*) and turtle grass (*Thalassia*). The best known animals of the region are the fiddler crabs (*Uca*). See *backshore, beach, estuary, fiddler crab, Zostera.*

Mullet   A worldwide, abundant food fish of the family Mugilidae; there are about 100 species known. Their usual habitat is tropical to temperate waters. Mullet can tolerate low salt conditions well, and are frequently found in brackish water. The usual adult fish is a silvery, large-scaled individual about 30 to 90 cm (1 to 3 feet) long.

Mullets are omnivores, and do well on a largely plant diet. The common species, *Mugil cephalus*, has been successfully cultivated. See *fish.*

Murchison, Roderick Impey (1792–1871)   A Scottish geologist who defined the earlier periods of the Paleozoic Era (395 million to 570 million years ago). He joined the Geological Society of London in 1825, and then explored extensively in France and in the Alps;

first with Adam Sedgwick and then with Charles Lyell.

Murchison's major work centered on the Silurian strata. He published this in 1835 and followed it with a long investigation of Old Red Sandstone.

Profound disagreement with Sedgwick over their joint work in Devon led Murchison to a long controversy with his former partner in the assignment of dates to the Early Silurian and Late Cambrian systems. The disputed layer was named Ordovician by Charles Lapworth well after the deaths of the disputants.

Murchison was commissioned to form a party with the prominent geologists, Philippe de Verneuil and Alexander Keyserling to survey European Russia. This group explored and named the Permian deposits. Murchison returned to Britain, where he served as director of the Geological Survey of the British Isles from 1855 until his death. See *Cambrian Period; Devonian Period; Lyell, Charles; Sedgwick, Adam; Silurian Period.*

murex   Any of a group of widely distributed, tropical-to-temperate marine snails of the family *Muricidae* that live in rocky, shallow water. They have a characteristic attractive, heavy, spined or frilled shell. The snails feed on other mollusks by boring a hole in the shell of these prey and sucking out the contents. *Urosalpinx cinerea*, known as the oyster drill, is a murex species of commercial importance since it destroys a valuable cash crop.

Many murex species produce a yellow fluid that turns to red or purple in sunlight. *M. branoris* is a Mediterranean species that is the source of Tyrian purple—the royal purple of the ancient European and Near-Eastern world. *M. pecten* is the Venus comb, an Indian Ocean shell much prized by collectors. See *Gastropoda.*

Murray,   John   (1841–1914)   A Scottish oceanographer born in Canada. He went to Scotland at the age of 17 to study medicine at the University of Edinburgh, but left without a degree in 1872. By that time he had already been on one major oceanographic collecting voyage, and was therefore a reasonable choice as an assistant to Charles Wyville Thomson, professor of natural history at the University of Edinburgh and organizer of the H.M.S. *Challenger* expedition sponsored by the Royal Society. Murray spent three and one-half years on board this ocean-going research vessel.

He carefully established that the already-known single-celled protozoans of the genus *Globigerina* were ocean surface dwellers whose skeletons eventually sank. To do this, Murray had extremely fine nets constructed that were rigged to lie on and just below the surface, and that could be towed without fouling. He collected and named radiolarians, diatoms, and pteropods. His work established the terrestrial origin of the red clay of the ocean bottom, and the amazing slowness of its sedimentation rate.

When Thomson died in 1882, Murray had spent years sorting the hundreds of H.M.S. *Challenger* specimens, and was named editor of the forthcoming report of the voyage. He also wrote the summary of this report and the section on deep-sea deposits with Alphonse Renard, Belgian geologist. Murray's deduction that the antarctic ice had a continent beneath it reinvigorated exploration of the South Pole.

In addition to the H.M.S. *Challenger* report—a massive, multi-volume work—Murray and Hjort produced in 1912 a famous textbook on oceanography, *The Depths of the Ocean,* which became a standard in the field. See *H.M.S. Challenger; Thomson, Charles Wyville.*

Mysiidacea   An order of tiny, shrimp-like, free-swimming crustaceans. The average adult is about 18 to 20 mm (0.25 to 0.80 inches). These animals are found in temperate waters, and range from the intertidal zones to hadal regions. The deep-dwellers are usually larger. There are more than 600 species of *Mysiidacea.* See *benthos, Crustacea.*

mussel   A bivalve mollusk well-represented in temperate to tropical waters worldwide. *Mytilus edulis* is the edible mussel; it is a blue-black, slender bivalve, 5 to 15 cm (2 to 6 inches) long, that attaches itself to pilings or rocks by byssal threads. The California species *(M. californianus)* is brown. An Atlantic species, the horse mussel, is *Modiolus capex.*

Strands of mussels can be seen in the water of intertidal zones. Unlike the oyster, the mussel is not permanently fixed in place. Mussels are raised commercially. Besides being consumed by humans, they are also eaten by birds, such as gulls and ducks, and by starfish and flounder. See *byssal threads, mantle, Mollusca.*

Myzostomaria   A class of flattened, almost disk-shaped annelids. They are tiny and are parasites on echinoderms, particularly crinoids. Fossil sea lilies bear the scars of these exterior parasites. See *Crinoidea.*

N

Nansen, Fridtjof (1861–1930) A Norwegian oceanographer, polar explorer and humanitarian. He was trained as a zoologist and was a professor of zoology at the University of Christiania (now Oslo), Norway. Nansen's interest in oceanography as a separate science led to his directorship in 1901 of ICES—the International Council for the Exploration of the Seas. In 1908 he became professor of oceanography at the renamed University of Oslo.

Prior to his directorship of ICES, Nansen had undertaken extensive explorations in the arctic. He tried to reach the North Pole, and in 1895 achieved 86° 14′ North Latitude, the farthest north anyone had previously reached. His work in the arctic included charting parts of the Kara Sea and doing careful sampling work while aboard the *Fram*, a vessel built to withstand ice pressure—an ever-present danger in arctic exploration. After observations of the movement of pack ice from the unanchored ship, Nansen realized the relationship of wind to current, but could not adequately explain this relationship without mathematical expression. He suggested the problem to his colleague Ekman, who developed equations to explain what is known as the Ekman spiral—current moving at a 45° angle to the wind.

Nansen was also known for his humanitarian efforts on behalf of World War I refugees, and in 1922 he was awarded the Nobel Peace Prize. See *Ekman, Vagn Walfred; Fram; Kara Sea; Nansen bottle.*

Nansen bottle A device designed by Fridtjof Nansen and Otto Pettersson for collecting deep-water samples. It contains a series of concentric insulated rings with a thermometer mounted in the assembly. The bottle sinks, mouth down and closed, and is then tripped open at a particular depth. It thus collects water at a given depth, then closes to isolate the sample, and maintains the sample at the temperature it had when collected. The thermometer gives a reading of that temperature. This device has undergone much redesign in the years since its introduction. See *Nansen, Fridtjof; oceanography.*

Nares, George Strong (1829–1915) A British admiral and surveyor, and the ship's captain for most of the voyage of the H.M.S. *Challenger* from 1872–1876. Nares had gone to sea as a boy, and served on many scientific and technological voyages. He was in the Arctic from 1852–1854, then in the waters north of Australia, surveying that continent's coast. Nares was then assigned to the Mediterranean, where he did surveying that was the first step to laying telegraph cables.

Prior to his service on the *Challenger*, Nares was involved in the project of charting the currents off Gibraltar. He left the *Challenger* in 1875 for an Arctic expedition (1875–1876). His son, John Dodd Nares (1878–1957), was also an admiral and a hydrographer. See *Challenger, H.M.S.*

narwhal An arctic porpoise related to the beluga whale. The narwhal is a large mammal, about 5m (16.5 feet) long at maturity. The young narwhal is usually quite dark—almost black—but turns gray with a light underside as it grows. The adults are almost white.

The narwhal's incredible characteristic is its visible tusk: males exhibit a long, spiral-grooved left tusk that may be 2.5 m (8 feet) long, while their other tusk remains buried in the gum; females have both tusks in the gum. The tusks are probably sexual attributes and are sometimes broken or scarred in fights between rivals. The tusks are thought by some to have originated the unicorn legends, or at least to have contributed to them. Narwhals live in herds and feed on fish, usually cod and halibut, and crustacea. See *mammal.*

nautilus   A cephalopod mollusk most frequently found in the South Pacific and the Indian Ocean. The true nautilus, also called the pearly or chambered nautilus, is differentiated from the paper nautilus (Argonauta), which is more octopoid. The pearly nautilus is the only genus of the Nautiloid order present in today's oceans. Paleontologically, the true nautiluses are an important group that once had many genera and species, now largely extinct; however, the animals alive today are not living fossils, but rather are different from the individuals that lived in ancient seas.

Present-day adult nautiluses have coiled shells that are about 25 cm (10 inches) in diameter. The shell is usually brown with white bands, and is composed of 35 to 40 compartments. The animal always lives in the largest, last-constructed compartment, and uses the empty ones as a flotation device by controlling gas flow within the shell. The nautilus is striking in appearance: it has a lensless eye, can move fairly quickly, and has upwards of 90 tentacles. It lives in relatively deep waters, 100 to 400 m (330–1,300 feet) below sea level, on slopes of fringing or on barrier reefs. It migrates up at night and feeds on carrion, to which it is probably attracted by smell. Until fairly recently there were few living nautilus specimens known and therefore much of the life-style of the nautilus was conjectural. See *Cephalopoda, Mollusca, paper nautilus.*

Nautilus   The name of several famous ships. The first was a reasonably functional submarine designed by Robert Ful-

ton, who is remembered as the designer of the first steam-powered ferry. The fictional Captain Nemo's submarine was named *Nautilus,* as was the first U.S. nuclear-powered submarine. See *Fulton, Robert.*

navigation   The application of observation to the problem of plotting a course for a ship or airplane moving from one place to another. Bearings are taken with respect to the position of the true North Pole (from the point to the Pole), or to the magnetic North Pole, which is the place on Earth that a compass points to.

While the properties of magnetic iron were known in antiquity, the relationship between iron and the Earth as a magnetic object was unknown. Compasses were introduced possibly as early as the 12th century. Without them, ocean voyages were very adventuresome and much more dangerous than they are today.

Early sailors did their navigation by simply moving from one known point on the shore to another. The difference between such nagivating and piloting is almost nonexistent. The only aids available to the early coastal sailors were sounding rods or lead lines to determine depths, and windvanes.

Travel across the open ocean was much more difficult: a sailor would fix on an established point to determine his position, and then take aim on his destination by dead reckoning (i.e., taking the straight line or course from the starting point to the destination). Since the surface of the Earth is curved, the straight line may be a great circle, and without any fixed object to sight on, holding a course was very difficult.

Long voyages that took ships out of the sight of all land were nevertheless undertaken. The Phoenicians were very experienced in navigating the Mediterranean, and, late in their history, they also moved out of sight of land to Cornwall, England, and its offshore islands. The Norsemen and the Polynesians—at roughly the same time—also moved out of sight of land. All three seafaring cul-

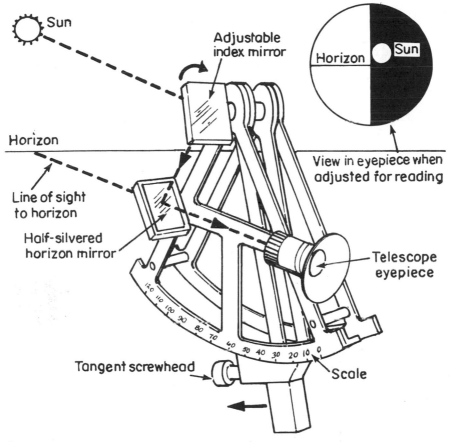

Sun

Adjustable
index mirror

Horizon   Sun

Horizon

View in eyepiece when
adjusted for reading

Line of sight
to horizon

Half-silvered
horizon mirror

Telescope
eyepiece

Tangent screwhead

Scale

Sextant

tures knew and used celestial navigation.

Celestial navigation, which was a great advance for scientific navigation in general, involves the position of a ship relative to the position of the Sun, or the Moon, or a star on the celestial sphere— an imaginary, hollow construction surrounding the Earth and, like the Earth, making one rotation about its axis every day.

The rapid development of navigational aids and instruments in the 15th, 16th, and 17th centuries displaced the astrolabe with a succession of other instruments such as the backstaff, the quadrant and its variant the octant, and the sextant. With these instruments a good estimation of latitude was possible. Using a sextant, the user measures the altitude of a celestial body from the horizon as the reference point. The Sun (or Moon, or some particular star such as the North Star, Polaris) appears at a point some degrees above the horizon at a particular time of a particular day. Tables have been constructed which, when compared with the data read from the sextant, translate these observations into the number of degrees that some position is above the horizon and north (or south) of the equator. At the equator, each of the 360 degrees of longitude of the circumference of the Earth corresponds to about 110 km ($1° = 110$ km) or 70 statute miles. As the meridians of longitude converge with increasing latitude, the degrees of longitude (and the time zones) shrink. Good clocks and extensive tables for the correction of instrument readings have resulted in the accurate location of ship positions with respect to the vessels' longitude.

The use of clocks to calculate longitude from a prime or first meridian was recommended in 1530, but it took another 250 years before the Harrison chronometer finally made it possible to measure longitude in this simple manner. Trigonometric tables (epherimids) are needed to solve these spherical trigonometric problems, and the first were published by the decree of King Alfonso X of Castile in 1252. The standard text on spherical trigonometry, a reworking of Greek mathematics by John Holywood (Johannes Sacrobosco), a famous 13th-century cleric and mathematician, appeared in Europe at roughly the same time as the tables.

The idea of a prime or 0° meridian of longitude predated Ptolemy, but became internationalized only in the 17th century. An observatory was created at Greenwich, England by Charles II in 1674, and John Flamsteed, the Royal Astronomer, was assigned the task of updating and revising the lunar tables, which were finally published as the Nautical Almanac in 1765.

The 20th century was another period of rapid technological development in the field of navigational instruments. The 20th century also brought radio and several other techniques that are based on radio waves. Radio has made dead reckoning much simpler on the sea, and it is a vital tool for the flight navigator. Radio waves, long wavelength radiation bands, are used in hyperbolic navigation, a technique based on the observation that radio waves move around the Earth's surface in great circles. When they intersect, they indicate position. RADAR can be used as a distance measure in addition to its locating capability. LORAN is a radio-based technique that operates by triangulation (i.e., establishing a position relative to that of two known signal broadcasting points,

and it is used for both surface navigation and flight.

Once a position on the surface of the Earth is determined in degrees of latitude and longitude, this must be translated into distances. Distances on all standard international charts are measured in kilometers or miles (statute or nautical). A statute mile as measured on land is 5,280 feet, the nautical mile is 6,076 feet or 1,852 meters (1 kilometer = 1,000 meters; 1 meter = 3.3 feet). Velocity is given in kilometers per hour (km/hr) or in knots (nautical miles per hour).

Charts are prime navigational aids. They show latitude, longitude, landmarks, and navigational sightings. These may be lights, buoys, water depths, and markers on the sea bottom. Charts for airplanes also show airports, control points, and elevations. See astrolabe; Bowditch, Nathaniel; charts; Davis, John; latitude; longitude; maps; Maury, Matthew Fontaine; sextant; stick charts.

Nazca Plate and Ridge   The Nazca Plate is one of the Pacific Ocean segments of the Earth's crust. It abuts the South American Plate to the east. The point of contact is at the Peru-Chile Trench. The smaller Cocos Plate is north of the Nazca.

The Nazca Ridge is a line of seismic activity lying perpendicular to the South American coast. See crust, evolution of ocean, Pacific Ocean, plate tectonics.

neap tide   A tide with the smallest range, which occurs during the first and last quarter of the lunar month. See low tide, tide.

nearshore   That part of the beach between the shoreline and the line of the breaking waves. See backshore, beach, coast.

Necho II   The Pharaoh who, according to the Greek historian Herodotus, commissioned a circumnavigation of Africa. Necho is documented as having sent, in about 600 B.C., an expedition south from the Red Sea. The sailors kept close to the coast and continued along it to Gibraltar, which they reached three years later. The lengthy voyage was interrupted by at least one farming season when the crew stopped to raise some crops to replenish their food supply.

Necho is also the pharaoh who is supposed to have attempted and then abandoned the construction of a canal across the Sinai Peninsula from the Mediterranean to the Red Sea—more or less where the present Suez Canal is now.

nektobenthos   Self-propelling benthic organisms. They live close to, but not on or in, the ocean bottom. See *benthos, nekton.*

nekton   All free-swimming oceanic animals that move in the water independently of wave and current action. This includes a vast array of crustaceans, mollusks, fish, reptiles, and mammals. Most of the nekton are carnivores, and some are also scavenging organisms. Many

species are commercially exploited by fishing, and some are cultivated. See *fishing, individual animals, mariculture, plankton.*

nematocyst   The stinging cell, one of many such thread-containing capsular structures on the surface or tentacle of a cnidarian (e.g., anemone, jellyfish). Upon stimulation, which may be chemical or tactile, the nematocyst ejects a coiled, hollow, often barbed thread that delivers a toxic substance to stun or kill prey or enemies. The thread is not retracted; instead, a new one is regenerated by the nematocyst. See *Cnidaria, cnidoblast, Portuguese man of war, toxin.*

Nematoda   A phylum of round-bodied, unsegmented worms found in all oceanic habitats from intertidal zones to hadal ones. It is the third-largest invertebrate phylum from the standpoint of number of species (about 15,000), the largest being *Mollusca,* and the second largest *Arthropoda.* Nematodes are frequently the most abundant multicellular animals in bottom sediments. While most marine species are free-living, parasitic nematodes may be found in fish, mammals, birds, and other invertebrates. See *parasite, worm.*

Nematophora   A phylum related to the *Nematoda.* This group of worms consists of free-living adults that have parasitic juveniles. Nematophorans live on

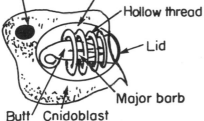

(1) Before discharge of the nematocyst

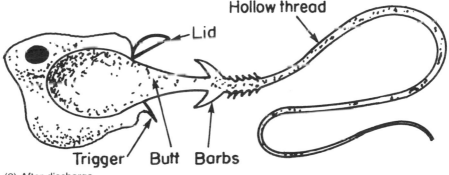

(2) After discharge

arthropods, including shrimp and crabs. See *Nematoda*.

Nemertea   The phylum of ribbon worms. There are over 800 species. See *ribbon worm*.

Neridae   A family of worms of the class Polychaeta, of the phylum Annelida. The most common genus is *Nereis*. These worms are usually bottom dwellers that live on other worms and zooplankton. They range in size from about 2 to 90 cm (0.75 to 33 inches), and are found in the temperate waters of the North Atlantic or the Pacific Ocean. They are usually brightly colored, may be red, brown, or green, and they are known for their strong jaws. Their respiratory systems are developed, they have gills. Like many other polychaetes, the majority of species of Neridae are dioecious, with distinct male or female individuals; some other species are hermaphroditic. See *hermaphrodism, Polychaeta*.

nets   The use of nets to catch fish is older than written records. Nets range in size and type depending on locale, type of fish sought, and the level of industrialization of a culture. Nets of the surrounding type—seines and purse seines—are used to trap pelagic fish such as herring and tuna, which are found near the surface and in open water. The demersal fish, bottom dwellers such as cod and haddock, are trapped in Danish seines, which also surround their catch. These are bag-shaped trawls that are pulled along the bottom. Gill and drift nets are fish traps. Lift nets are used beneath a school of fish that is attracted upwards by a light. See *fishing*.

neuston   Organisms that use the surface film of water for support. Neuston includes snails, protozoa, and cnidaria. These animals attach themselves to the water-air boundary by adhering to the more cohesive layer of water molecules at the surface. See *surface tension*.

New Hebrides Basin   A part of the Coral Sea located east of Australia, northeast of the Bellona undersea plateau, and west of the New Hebrides island chain. The islands in the basin are volcanic. While some parts of the islands have been dated to origins in the Tertiary, the still-active volcanoes are younger. See *Pacific Ocean, Tertiary Period*.

New Hebrides Trench   The point of contact between the Australia Plate and the Pacific Plate. The trench runs down the western edge of the New Hebrides islands chain. The deepest point, 7661 m (25,281 feet), is the maximum depth in the Coral Sea. The existence of the trench was deduced because of the volcanic activity in the area. Its location was finally established in 1910. See *New Hebrides, trench, volcano*.

nickel   A relatively rare, silvery-white metallic element with commercial importance in alloys and as a catalyst. It is found dissolved in seawater in concentrations of about 0.002 mg/liter. It is also a constituent in manganese nodules, in which the average content of nickel is about 2% (ranging from 0.01 to 2.35%). Manganese nodules are nevertheless a good source of nickel, since their recovery may eventually become cost-effective. See *element, manganese nodule, minerals*.

Niger   A West African river that empties into the Gulf of Guinea in the South Atlantic Ocean. The Niger is about 4,200 km (2,600 miles) long, and is the great river of West Africa. The amount of water carried by it is enormous. The resulting delta has an area of about 24,000 km² (about 14,000 square miles) of built-up mudflats. The River is channelled into a network of waterways that are obstructed by shifting sandbars, making most of the waterways unnavigable. See *delta*.

Nile   An African river flowing from Lake Victoria in Uganda north into the Mediterranean in Egypt. The Nile is the most significant North African river; it is the longest river in the world with a length of about 6,650 km (4,130 miles). It runs almost directly north into an area of the Mediterranean that was once a great gulf. This area has since been filled by the Nile with silt from the Ethiopian highlands. The delta of the Nile is 16 to 25 m (50 to 75 feet) above sea level and about 160 km by 250 km (100 by 155 miles) in area, and consists of extremely fertile soil. The northern periphery is broken by extensive salt marshes. The Nile cone on the floor of the Mediterranean is not as well-defined as is that of the Rhône.

The River is of enormous importance to the countries that it flows through. For both Egypt and the Sudan it is almost the only source of water. In antiquity the incredible grain harvest of the Nile valley fed most of the populations living in the eastern Mediterranean, and for a considerable portion of its very long history, Egypt was a grain exporter. The construction of the Aswan High Dam in 1970 brought much change to the valley and the configuration of the River: Huge areas behind the Dam were flooded permanently (Lake Nasser), while sites downriver of the dam are no longer inundated as they were for thousands of years. Seawater irrigation has brought significantly more salts into areas that did not have this problem before 1970. Since most of the mountain silt that comes down the upper Nile is now trapped behind the Dam, the delta has ceased to enlarge. Moreover, because the silt also acted as a fertilizing material, the agricultural production of the Nile valley and delta now depends on chemical fertilization, which increases the problem of runoff. See *delta, Mediterranean.*

nitrogen   A colorless, odorless gaseous element that constitutes about 78% of the total volume of the atmosphere. It is present in air as the inert diatomic N_2 molecule, and as a variety of nitrogen oxides, NO_x, where x is some number (0, 1, 2, 3) of oxygen atoms. Other nitrogen oxides are N_2O_3 and N_2O_5, both of which are products of combustion. Nitrogen is present in water largely as an inert constituent of dissolved air, but more importantly in the form of the dissolved nitrite (NO_2^-), nitrate (NO_3^-), or ammonium (NH_4^+) ions. The concentration of ammonia (NH_3), and to an extent also of NO_3^- in water, is a function of the decomposition of plankton. Thus, it is greatly increased in spring in shallow water, where most of the plankton population is found. Some oceanic water is richer in nitrogen than other waters; the water of the North Atlantic, for example, is relatively nitrate-poor.

Some free-living cyanobacteria are nitrogen-fixing organisms. They inhabit a variety of places, including tidal flats, marshes, coral reefs, coastal areas, and the open sea. While nitrogen fixing is more efficiently done on land, the volume of water in the ocean is so great that a considerable proportion of all nitrification occurs in the sea. The nitrogen-to-phosphorus ratio may be a factor in controlling plankton blooms, in addition to the temperature, trace metals, herbivores, and available light. See *element, nitrogen cycle, plankton.*

nitrogen cycle   The cycle of chemical processes that exchange nitrogen between the atmosphere and the Earth. Nitrogen is present in the atmosphere as a diatomic molecule, N_2. On land, nitrogen-fixing bacteria convert this relatively water-insoluble gas to a soluble ion, either NO_2^- or NO_3^-, which is taken up by autotrophic organisms and incorporated into protein. Cyanobacteria do this in the sea. Upon the death or degradation of a nitrogen-containing organism or material, free nitrogen is produced and may again become part of the hydrosphere. This then provides the nitrogen necessary to form the proteins in marine plants and in the animals that feed on them. Denitrifying bacteria return nitrogen to its elemental gaseous state. See *autotrophs, carbon cycle, hydrosphere, peptide formation.*

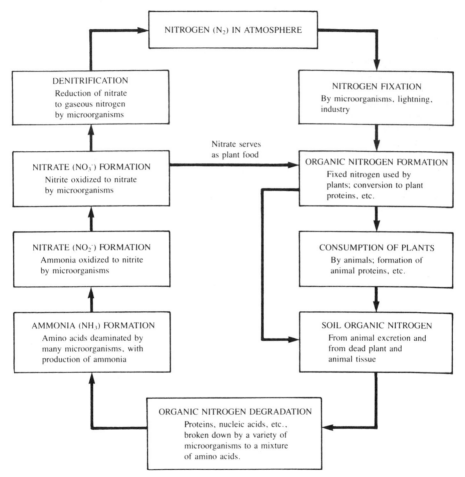

nitrogen fixation   The conversion of atmospheric nitrogen, N_2, into a form usable by a plant. The usable form is either the nitrite ion, NO_2^-, or the nitrate ion, NO_3^-. Nitrogen fixation is most often associated with land plants, such as clover and alfalfa. In the ocean, there are species of bacteria and algae that do this. See *nitrogen cycle, nutrients.*

nitrogen narcosis   A loss of critical thinking capacity, accompanied by giddiness and euphoria, which the French marine explorer Jacques Cousteau called "rapture of the deep." At depths of about 100 m the concentration of nitrogen in the blood acts a a narcotic, and a diver loses control of the capacity for critical judgment. The depth at which this phenomenon occurs varies with the diver and with individual dives. See *Cousteau, Jacques-Yves; diving.*

NOAA   The National Oceanic and Atmospheric Administration (NOAA), an agency established by the U.S. Department of Commerce to encompass a number of previously existing services and to give them all a common direction. These services are the:

National Ocean Survey (formerly the U.S. Coast and Geodetic Survey), which creates nautical and air charts, conducts oceanographic surveys, makes tide and current predictions, and prepares navigational charts for American waters.

National Weather Service (formerly the U.S. Weather Bureau), which forecasts the weather and conducts storm watches.

National Marine Fisheries Service, which does research on economically important species of fish and other marine organisms. This agency also oversees Marine Mammals Protection and the Endangered Species Act.

Environmental Data Service, which studies the effect of environment on the nation's food and energy supply. It operates the National Climatic Center, the National Oceanographic Data Center, and the National Geophysical and Solar Terrestrial Data Center.

National Environmental Satellite Service (NESS), which maintains and collects data from the NOAA satellites.

Environmental Research Laboratories, which study ocean-atmosphere processes and the effect of pollution on coastal ecosystems.

Office of Sea Grants, which gives funds to research institutions.

Office of Coastal Management, which exists to help coastal states in the development of management systems for their coasts. It also operates marine sanctuaries.

Noctiluca   A genus of marine dinoflagellates which when disturbed become luminescent. See *dinoflagellate, light organs.*

North American Basin   An Atlantic Ocean region of rather poor definition, west of the Mid-Atlantic Ridge and east of the continental rise to the Blake Plateau. The Bermuda Rise is almost central in the North American Basin. The Nares and Hatteras Abyssal Plains are encompassed by the Basin, as is the Puerto Rico Trench. The deepest point of the Atlantic is in this true, "island arc" trench. See *Atlantic Ocean.*

North American Plate   A continental plate that extends from the Mid-Atlantic Ridge to the west coast, where it abuts the Pacific Plate. The southern boundary areas of the North American Plate include the South American Plate east of the Antilles chain, and the Caribbean Plate. The southwestern boundary is the small Cocos Plate. There is seismic activity at the boundary edges of the Caribbean, Cocos, and Pacific plates. The famous San Andreas fault is the result of the slip of the Pacific Plate northwest relative to the North American Plate. See *Atlantic-type margin, crust, Pacific Plate, plate tectonics.*

North Atlantic Current   Part of the surface current of the North Atlantic. When the Gulf Stream moves into the Grand Banks region, it becomes a weak, divided flow; it is also known as the North Atlantic Drift at this point, and drifts eastward. The water is warmer than the surrounding water, which is of polar origin. The mixture results in the polar front, an area of permanent fog and low pressure that is relatively stationary north of the Current. See *current, Gulf Stream, gyre.*

North Atlantic Deep Water   Cold water of low salinity sinking either east or west of Greenland and then spreading southward. Not until the influx of Mediterranean water does the salinity reach a more normal level for oceanic water. The North Atlantic water is characterized by a high oxygen and relatively low nutrient content. The latter is in striking contrast to the waters of the antarctic, which support very large populations of living organisms. See *Antarctic Water.*

North Cape Current   A North Atlantic current that flows around the northern tip of the Scandinavian peninsula, bringing the warmer, saltier water of the Atlantic into the Barents Sea. It was discovered in the mid-19th century, when warm water was noted in the Barents Sea and the existence of this branch of the North Atlantic gyre was charted. See *Atlantic Ocean, Barents Sea, currents.*

North Pacific Current   The eastward flow of water in the Pacific Ocean in the northern hemisphere. It originates as the Kuroshio Current and moves from Japan toward North America. See *currents, gyre, Kuroshio Current, Pacific Ocean.*

North Sea   A fairly shallow arm of the North Atlantic between Great Britain and northern Europe. The southern end of the North Sea is the shallower part, averaging less than 40 m (150 feet) deep. At the Dogger Bank one finds average depths of 12 to 13 m (40 feet), with some spots as shallow as 5 m (about 15 feet). The area is an excellent fishing ground. The catch is herring, flatfish, mollusks, and crustaceans.

The bottom of the North Sea slopes downward north of the Dogger Bank to the Scandinavian shield and the shallow Viking Bank. A narrow, deep channel to the north encircles the southern end of Norway and runs into the Skaggerak—part of the North Sea between Norway and Denmark.

North Atlantic water enters the North Sea mainly from the north, not via the Straits of Dover, where a high sill acts as an effective barrier. The water is generally cold, from 2 to 7°C in winter and about 13 to 18°C in summer at the surface. The salinity, which is greatest in winter, is less than 3.5% in the north and west, and because of the relatively warm and salty Atlantic inclusions, the North Sea very rarely has a frozen surface.

The Sea is an epicontinental one; it covers an area that had been exposed in the Wurm glacial period. The bottom is continental and made up of the local material. Helgoland (or Heligoland), the rocky island off the coast of Schleswig-Holstein (Germany), is also a continental fragment. Peat bogs are submerged under the North Sea and are continuous with the peat deposits in England, Ireland, and on the European continent.

The early history of the North Sea area goes back to the late Permian Sea (280 to 225 million years ago), which was in a hot region. Intense evaporation produced the salt domes in North Germany and Denmark, and those in the North Sea that are currently being tapped for oil. The closing of the Tethys Sea in the Mesozoic (225 to 65 million years ago) produced the beginnings of the current shoreline. The land bridge between Dover and Calais was broken about 10,000 years ago by the rise in sea level. The northern Netherlands drowned at roughly the same time, and so did other parts of the North Sea coast such as the Godwin Sands off the Kentish coast of England—submerged sand dunes that caused many medieval shipwrecks.

Charming tales exist in many North Sea areas that abut these submerged coasts: the Germelshausen legends in Germany and the "drowned village" stories in Ireland, Jutland, and the Netherlands. They are all related to the same legend of supposed, ghostly church bells sounding under the water. See *Atlantic Ocean, Dogger Bank, Norwegian Sea, Permian Period.*

Northeast Passage   A sea route to India and the Spice Islands around the north coast of Eurasia. Norse sailors had found their way to the Barents and White seas by the 10th century, and trade between Russia and Norway using a northern sea route was established by the 14th century.

By 1525, Russians were theorizing the existence of a sea route from the White Sea to China, an idea that had also occurred to the English, who tried to reach the fabulous lands to the east while avoiding Spanish and Portuguese ships. Several English mariners were involved in the attempt to find a Northeast Passage, but none went further east than the Kara Sea.

Local Siberian sailors used the rivers to the Kara and the Laptev seas. In 1648, Dezhnev rounded the eastern tip of Asia and found the strait named for its rediscoverer, Vitus Bering. Russian attempts to map the far north were made in the 18th century, and established the existence of a Northeast Passage. However, it was blocked by ice and too difficult to pass. The whole of the Passage was finally traversed by N.A.E. Nordenskjold on the *Vega* (1878–1879). Others who used the passage were deLong in the *Jeannette*, Nansen using the *Fram*, and Toll in 1900. Although modern icebreaking equipment has made the passage more accessible, it is a shallow one, and the fact that there is little of commercial value in the area has left the passage largely unused. See *Bering, Vitus; Dezhnev, Semyon Ivanov; Nansen, Fridtjof; Wilboughby, Hugh.*

Northwest Passage   A northern sea route from Europe to Asia sought by European sailors trying to reach China

and the Spice Islands—Indonesia and Malaysia. Cabot's attempt in 1498, on his second voyage, cost him his life. Cartier, in a similar attempt, was first hopeful when he found the St. Lawrence River, only to be disappointed when the great gulf at the entrance to the river led to a progressively narrowing waterway.

The search for a sea route to the Orient intensified as the Turks gained total control of the eastern Mediterranean. After Pope Alexander VI (1493) divided the unknown world into future Spanish and Portuguese territories, as formalized by the Treaty of Tordesillas (Spain), the only known routes to the lucrative spice and silk trade were in the hands of the Spanish and the Portuguese. These were the routes through the Straits of Magellan at the tip of South America, and the sea route around the tip of Africa, the Cape of Good Hope. Both the English and the Dutch sought another way.

The attempts at a western passage were primarily but not solely made by English sailors in the 16th and 17th centuries. Gilbert's original petition for a charter called for the search for a Northwest Passage. He was followed by Frobisher, Davis, and Hudson. Luke Fox and Thomas James in 1631 proved that Hudson Bay was not a route to the Pacific. De Ulloa tried, as did Juan de Fuca, to find the "Straits of Anian" from the western edge of North America. After these unsuccessful attempts it was not until the 18th century that interest revived in finding the Northwest Passage. The North West Company, which had been formed to commercially exploit the northern regions of North America, revived the search as a public relations measure. One result was Mackenzie's discovery of the river named for him. James Cook on his third voyage charted the Alaskan coast, thus giving reliable information about it to the mapmakers of the day.

The British Admiralty established a prize, periodically amended, to reward explorations that approached the pole or particular meridians of longitude. It was the stimulus for much subsequent exploration. John Ross in 1818 sailed into the

Davis Strait to Lancaster Sound. He turned back at a line of hills that he called the Croker Mountains. The Secretary of the Admiralty, John Barrow, was skeptical of the existence of hills, and sent Ross' second in command, William Parry, to find the hills in 1819. They were gone. The hills were either icebergs or folded upthrust ice on floes. Parry wintered in the arctic, the first European to do so successfully. He later made two more arctic journeys. George Lyon and Frederick Beechey also sought the Passage by going through the Bering Strait. John Ross returned to the arctic in 1829 and spent several winters icebound. During the winter of 1831 his nephew, James Clark Ross, discovered the magnetic pole. This series of voyages ended with the death of John Franklin, with James Clark Ross subsequently leading the first of many expeditions that went in search of the missing Franklin. In 1850 Robert McClure discovered the existence of the sea lanes that made up the Passage. Leopold McClintock in 1857 found the remains of the Franklin party.

The much sought-for passage between the Atlantic and Pacific oceans was finally sailed through by Amundsen. He began in June 1903 at the Atlantic end, and arrived in Alaska in August of 1906. The Passage is ice-free about three to four months a year. It is actually a series of interconnected channels which skirt the islands and permanent ice of the Canadian Northwest Territories. The submarine U.S.S. *Nautilus* traversed the passage while submerged in 1958. The southernmost route from west-to-east is used to supply American radar stations in the area. Some Alaskan oil is moved by tanker out of the Point Barrow area and through the Northwest Passage when it is relatively ice-free. This was first done in 1969 by the tanker *Manhattan*. That journey was the first realization of the centuries-old dream of a free New World passage for commerce. See *Baffin Bay, Bering Sea, explorers and explorations, Hudson Bay, individual explorers*.

Norwegian Current   A continuation of the North Atlantic Current, which car-

ries warm, salty Atlantic water north along the coast of Norway to Svalbard (Spitsbergen), where it becomes the North Cape Current. See *current, gyre, North Cape Current*.

Norwegian Sea   A body of water that extends north-to-south from Svalbard (Spitsbergen) to the Faroe Islands. On an east-to-west axis it extends from Norway's coast to Jan Mayen Island and to Iceland. The deepest point in the Norwegian Sea is about 4,000 m (13,000 feet). The Sea is divided into basins and separated from the rest of the North Atlantic by a series of undersea ridges. The Mohns Ridge is the mid-ocean ridge and the Sea's western boundary. The bottom is volcanic, with foraminiferan and diatomaceous residues. An earthquake belt runs from Jan Mayen Island to Spitsbergen as a continuation of the Mid-Atlantic Ridge.

According to one version of the formation of the Norwegian Sea, the Scottish mountains, Norway, Bear Island, Spitsbergen, Greenland, and Newfoundland all constituted one continuous landmass which was then separated by faulting and oceanic spreading, a process that has gone on since the Permian Period 280 to 225 million years ago. See *Atlantic Ocean, evolution of oceans, Foraminifera*.

Norwegian Trench   The undersea perimeter of Southern Norway from the Skagerrak to the Norwegian Basin of the North Sea; it is about 250 to 600 m deep. The origin of the Trench is disputed. Some researchers believe that it is a glacial scour, others think it is a tectonic rift. See *glacier, trench*.

notochord   The single, rodlike dorsal support structure of chordate animals. The notochord develops after the infolding of the blastula—the hollow ball of embryonic cells. This phase of early embryonic development is called gastrulation. The notochord is almost always replaced. In vertebrates it is replaced by a spinal column. Other animals, such as the tunicates, absorb the notochord and

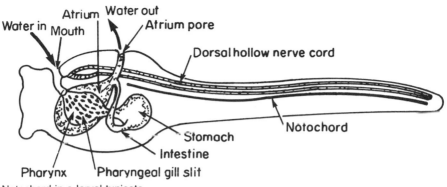

Notochord in a larval tunicate

develop into radially symmetric adults rather than bilaterally symmetric ones. Some primitive fish such as the hagfish and lamprey retain the notochord in adult life. See *hagfish, lamprey, vertebrate, Tunicata*.

nudibranch   An unshelled marine snail, also known as a sea slug. It is further characterized by having no gills or mantle cavity. These shallow-water animals have striking branched body outgrowths, called cerata, which are involved in respiration and function like gills. The cerata may be complex in structure; they are frequently brilliantly colored, making the nudibranchs the most ornately colored mollusks.

The nudibranchs feed largely on sea anemones. Some families swim instead of crawling about. While most nudibranchs are small, there are some larger species, especially those of Australia's Great Barrier Reef, which may attain the giant size of about 30 cm (12 inches). See *Gastropoda*.

Nummulites   A genus of foraminiferans. Thousands of species of these animals flourished in seas of the Tertiary Period. They were particularly successful in the Eocene Epoch, and their fossils form a significant proportion of Eocene limestones. See *Eocene Epoch, Tertiary Period*.

nurse shark   A shark, *Ginglymostoma cirrata*, common in the Gulf of Mexico; also an arctic shark, *Laemargus microcephalus*. Nurse sharks are also found in the Pacific Ocean. They are yellow-brown to gray-brown, usually longer than 4 m (14 feet), and are known to have attacked humans. See *shark*.

nutrients   Chemical substances necessary for the maintenance of life in the sea. They are generally the same materials needed for the growth of marine plants, consisting of nitrites, nitrates, phosphates, ammonia, and silicates.

The surface layer of the water is relatively poor in all nutrients. An intermediate layer with a concentration of all dissolved material that increases rapidly with depth overlies the layer of maximal concentration at about 500 to 1,500 m (1,550 to 5,000 feet).

The varying concentration of nutrient material in this layer is dependent on many factors, including the inflow of river water, light, temperature, insolation, currents, and upwellings. All nutrients are recycled. The death of organisms or excretion of waste by organisms releases material into the water, where it is exposed to oxygen and the nutrients are regenerated. The turnover rate depends on all of the factors listed above, and differs from one ocean to another. Thus, phosphate is cycled about seven or eight times a year. A complete turnover of nitrate ion takes twice as long. See *currents, minerals, nitrogen, phosphorus cycle*.

O

ocean-atmosphere interchange
The interchange of water and air between the oceans and the atmosphere,
and the complex of causal relationships
that affects the combined air-water system. On a large scale, the air currents or
winds generated by temperature differ-
entials move in specific patterns. Air is
also moved into water and water into air.
Heat, salt, and other gases than air are
also exchanged from one fluid to the
other.

Gross movements of large quantities
of fluid (water or air) are influenced by
the Earth's rotation and the shape of its
ocean basins. They in turn produce ocean
waves and currents. The ocean is the
recipient and storage unit for a surplus of
radiant solar energy. This is then released
to the atmosphere at differing rates
through the year. The consequence of
this heat exchange is worldwide weather.
See *currents, meteorology, weather.*

ocean basins Those parts of the
Earth's oceans that are neither continental nor part of the structures of the mid-
ocean ridges.

One theory of basin development,
which depends on continental plate tectonics, is that the continents are granitic
(siliceous or sialic) or acid igneous, and
float on the denser mantle rock or sima
(silica/magnesium) by displacing their own
weight. As the mid-ocean rift separates,
the heavier basic mantle rises, becomes
seafloor, and eventually dives under and
into a trench at a specific region known
as a subduction zone. The mantle is then
remelted, releasing granitic (or sialic) lighter
rock, which is deposited as lava or granitic intrusions. These accrete to form island arcs, which eventually coalesce to
form continents. The regions between

the continents are ocean basins. They
are, according to some researchers,
"crustless" (i.e., they are on the naked
plates as compared to crust-bearing or
continental plates). See *mid-ocean ridges,
plate tectonics, trench.*

ocean floor The substrate underlying
the water of the world's oceans. The
ocean floor is a three-layered structure
overlaying the denser, more viscous
mantle. The topmost layer extends down
to a depth of about 2 km (1.2 miles) and
is composed of unconsolidated sediment
that is almost entirely of terrestrial origin.
The middle layer is composed of lava of
two types. The upper level of this layer
is 0.5 km (0.3 miles) thick and consists
of what is known as pillow lava. It consists
of a series of rapidly cooled bubbles derived from the volcanic activity of the
deep sea. The deeper portion of the middle layer is about 2 km (1.2 miles) thick
and marked by feeder dikes or lava sheets
intruded into other rock. The lowest layer
of the ocean floor is approximately 5 km
(3 miles) thick and consists of gabbro—
coarse-grained igneous rock.

The major features of the ocean floor
are the ridges or rises, trenches, and fracture zones. Fracture zones are troughs
traversing the oceanic ridges. They are
seismically active regions, sites of plate
shifts. Frequently they are named; the
Mendocino Escarpment, west of the California coast is a sheer drop with clearly
defined magnetic anomalies on either side.

A new computer-imaging technique for
mapping the ocean floor is based on the
same technique used in creating maps
from satellite-transmitted information. It
is so sensitive that it has revealed tiny
bumps and valleys which are clues to
seamounts and trenches that had previ-

ously been undiscovered. See *crust, fracture zone, mantle, mid-ocean ridges, volcano.*

oceanic crust   The floor of the ocean. It may, but usually does not, extend under the continental areas. The ocean bottom is only about 5 km (3 miles) thick. It is largely basalt, and is the youngest rock on Earth. See *basalt.*

Oceanic Islands   Islands arising from the seafloor rather than from continental shelves. These islands are likely to be volcanic in origin. An undersea volcano will build a mound toward the surface as material moves up from the mantle to the plate surface, where it is subjected to erosion just as a terrestrial rock formation is subjected to erosion. The volcanic mount may be worn away as the volcano becomes extinct. If it is reinforced and augmented by coral accretion, or if there are several volcanoes in the same area, the aggregate structure may have a sufficient pile of material to emerge as an island. Hawaii is an example of this.

Lines of volcanoes of the above type exist in the Pacific. The best examples are the Marquesas and the Samoan and Hawaiian islands.

Oceanic rises may have several peaks that emerge as islands. These are near but not directly on the rise crest. Reunion, Ascension, and Easter islands are such islands. Other islands, such as the Aleutians, are associated with the edges of crustal plates. They are in unstable areas at the edges of trenches, and experience frequent earthquakes. See *atoll, coral, island, mid-ocean ridges, seamount.*

oceanic rises   See *mid-ocean ridges.*

oceanographic equipment   Equipment and instruments used in the study of the oceans and their flora and fauna. Since oceanography is itself a composite field of study, the equipment for any investigation depends on what is being investigated.

For the analysis of seawater, if it is the physical properties of the water that are of interest, the equipment might consist of sampling bottles of the Nansen type or others; thermometers, both recording and ordinary; and photometers for measuring water transparency. Nansen bottles collect water at a specific depth and record the ambient temperature at the time and place of its collection. To determine the chemical properties of the water, such as the salinity, nutrient content, pH (acidity), and dissolved gases, chromatographs, spectrophotometers, and pH meters are used. Physical oceanographers use depth-sounders, satellite maps, and devices to measure the electrical fields and pressure of the Earth under its water envelope.

The velocity of ocean currents can be measured electrically by using a predetermined reference point. Since seawater is an electrolyte (i.e., conducts an electric current), the amount of current carried will change. Subsurface currents can be tracked by radio waves, a technique that is accurate down to depths of abut 8 km (5 miles).

Sediment samples can be scooped up fairly easily where the ocean is not very deep. Small samples can also be vacuumed up. If a large sample is needed, however, this presents a problem, since dredging is the only available technique for this, and the weight of the dredge requires very heavy duty equipment, which is difficult to control from a moving ship. Similar difficulties are encountered in core sampling, another useful technique, which produces samples arranged in chronological order, with the oldest on the bottom.

Biological sampling is done by means of large nets of the type used by the fishing industry. These nets are intended to capture rapidly moving fish traveling below the surface. For plankton, nets are designed to have a specific mesh-size, so that a given volume of water can be examined for its marine biota.

Oceanographic laboratories range from permanent ones, such as those at Woods Hole, Massachusetts, and the Scripps Institution at La Jolla, California, to those created for a specific expedition. They include floating laboratories on ships such as the *Glomar Challenger,* or those on fixed platforms, and on unmanned floating panels. The Floating Instrument Platform, or FLIP, is a University of California creation that is towed horizontally and then righted. It is 108 m long and when upright, 17 m is underwater. It contains laboratories, cranes for manipulating sea-sampling equipment, and living areas for the crew.

oceanographic vessels Ships used for scientific exploration of the sea. In the past, such vessels have been fairly nonspecific in design. They were either converted warships, yachts, or merchant ships. H.M.S. *Challenger* was a corvette; the *Atlantis,* built in 1930 by the United States, was a ketch. Since World War II, the increased interest in oceanography has been reflected in a number of new ships designed to support a working team of scientists on board, and to act as support stations for submersibles. Since heavy equipment—including submersibles—is lifted by the mother-ship, the latter must have a sufficiently sturdy hoisting capability and stability. Coring equipment is also necessary, both for bottom sampling and probing for important minerals and petroleum.

Oceanographic platforms, which perform many of the same functions as research vessels, may be as simple as an unmanned buoy that houses recording devices, or as complicated as an offshore oil rig anchored to the ocean bottom. Like a ship, the platform functions as laboratory, workplace, and home for those based on it.

See *Atlantis; H.M.S.* Challenger; Fram; Albatross; *Albert I; Cousteau, Jacques-Yves.*

oceanography An umbrella term for the science of the sea. This encompasses the study of currents, winds, and tides; sea-air interactions; seawater; the geology of seafloor features; and the biota of the sea. Since man has been exploring this environment for thousands of years,

considerable information, particularly geographic and commercial (fishing), has been collected. However, the systemization of this body of information is less than 200 years old. The scientific societies devoted to the study of the sea are younger still: The first, the International Council for the Exploration of the Sea, began in Stockholm in 1899.

The voyage of H.M.S. *Challenger* (1873–1876) has been taken by oceanographers as the beginning of their science, because it was the first systematic government-sponsored expedition entirely dedicated to scientific exploration.

Earlier explorations frequently included naturalists: Captain James Cook travelled with Joseph Banks (among others), and Robert Fitzroy had Darwin with him. There were also captains who were amateur scientists, such as Alexander Scorsby, who contributed to information about the sea by bringing home materials and by recording observations for scientists to work on. An example of the latter activity was the careful recording of specific gravities of water samples by Chappe d'Auteroche, which resulted in Lavoisier's work on the salinities of diverse waters.

Preceding the voyage of H.M.S. *Challenger,* the establishment of the U.S. Coast Survey (today the National Ocean Survey) in 1807 was an early realization of the importance of the ocean to the young American nation. The original charge of the Coast Survey was to chart coastal waters, since commerce was vital. Later, the Survey's chief, Alexander Bache, a descendant of Benjamin Franklin, directed studies of the Gulf Stream. He believed that it was a study that was essential to the American nation as a whole, not just to the coastal states.

In 1830, the U.S. Navy's Depot of Charts and Instruments, now the Naval Oceanographic Office, was founded as a library for information about the sea. Its director from 1842 to 1861 was Commander Matthew F. Maury of the navy, who made it a center for information on marine weather conditions, sea lanes, and currents.

S. P. Baird, an assistant secretary of the Smithsonian Institution, initiated the U.S. Fish Commission (today the National Marine Fisheries Service) to help maximize fish catches. The research vessel *Albatross* and its expeditions came from this, as did the laboratory at Woods Hole, Massachusetts. These early beginnings in the United States were all made before 1900.

The expedition of the H.M.S. *Challenger,* led by Charles Wyville Thomson, himself a botanist, included a number of other professional scientists. Among them was John Murray, who finished the monumental task of producing the report of the voyage after Thomson's death in 1882. The fifty volumes of the report were finally completed in 1895.

It had by then become a matter of national pride to have an ongoing oceanographic research effort. The end of the 19th and the early 20th centuries saw the Germans mapping the oceans in the *Gazelle, Valdivia,* and *Deutschland.* The Norwegians explored the Arctic in the *Fram* and the *Michael Sars,* the French were in the Pacific in the *Travailleur* and *Talisman,* and the Russians explored using the *Vitiaz.* Prince Albert I of Monaco used several yachts named *Alice* or *Hirondelle* on a number of expeditions. These were not the only expeditions; many more were financed for short periods or limited studies.

The first combined effort at oceanography was funded by ICES—the International Council for the Exploration of the Seas. The members were Norway, Sweden, Denmark, Finland, Germany, the Netherlands, Russia, and Great Britain. The United States returned to the scene with the foundation of the Scripps Laboratory in California in 1925. The Woods Hole Oceanographic Institution on Cape Cod was created in 1930.

There were significant expeditions by several nations between the World Wars and directly after World War II. They concentrated on a number of subjects, such as the biological aspects of oceanography, the problems of mapping, currents, and weather observations. The combined efforts of many nations have been more the norm in the post-World War II era. The 18-month long IGY (International Geophysical Year) in 1957–1959 approached many problems. The ICITA (International Cooperative Investigations of the Tropical Atlantic) and IIOE (International Indian Ocean Expedition) were both done in the early 1960s. See *Albert I; Agassiz, Alexander;* H.M.S. *Challenger; IGY; Maury, Matthew Fontaine.*

oceans   The bodies of salt water that cover more than 70% of the Earth's surface. This vast area is divided into a number of oceans and seas. Their names depend on both the reference used and the convention followed. Attempts at standardizing the definitions and boundaries of the oceans is the charge of the International Hydrographic Bureau, based in Monaco. The criterion for calling a body of water a specific ocean or sea is that it have some definable and independent land contour, bottom, topography, and wind and water circulation in both the horizontal and vertical planes.

The evolution of oceans has been a topic of continuing interest for geologists. The many, often conflicting theories about this were clarified once the idea of continental drift was accepted. Many bits of evidence concerning the present size and shape of the Earth's ocean basins could then be unified. Thus, despite continual continental sedimentation, the ocean floor is about ten times younger than are the continents—200 million years old versus about 2 to 3 billion years old. The ocean floor is also somewhat denser than the continental rock.

One theory explaining the bottom topography of the ocean is that hot lava rising at the crests of mid-ocean ridges pushes older rock aside and drives the continental plates further apart. Another idea is that the cold, dense slabs of sea floor sink in zones of seismic activity (e.g., the Aleutian Islands and Peru-Chile Trench), and that this pulls the rest of a plate toward the sunken zone. According to the latter theory, seafloor spreading is not inexorably unidirectional, but can be changed by major events. One such event was the collision of India with the Asian landmass, which altered the movement of all Indian and Pacific plates. The rate at which plates move is 2 to 20 cm/year, which is enough to account for the present size of the ocean basins.

The hydrosphere (the water component of the Earth's surface) may have come from the degassing of the primordial Earth, which occurred about 500 million years ago. The early, acidic ocean of the young Earth must have dissolved some silicates and carbonates due to its high hydrochloric acid level, and also dissolved sodium in rocks to produce a salinity level close to that of the present oceans.

The mineral content of the oceans has been in a steady state for about 100 million years. Thus, as materials are added by runoff from continental areas, new sedimentary material is deposited, which is subsequently recycled as new land areas or reintroduced as gases to the atmosphere. The concept of the ocean as a complex solution in equilibrium has replaced the earlier one of the ocean as a large collecting body. See *continental drift, evolution of oceans, ocean floor.*

octopus   Any eight-armed cephalopod mollusk of the order *Octapoda,* genus *Octopus.* These shell-less mollusks are bottom dwellers in warm to temperate shallow water, and some also occur at abyssal depths. Their range in size is from about 5 cm to 5.5 m (2 inches to 22 feet). The bag-like body conceals a sharp beak, and their muscular arms are equipped with rows of suckers. The largest ones can have an armtip to armtip length of up to 9 m (30 ft). The octopus is a predator. *O. vulgaris,* the common octopus, feeds on crustaceans. Some species are plankton feeders.

The octopus is a solitary animal that lives among rocks or in holes. Its major sense is keen eyesight; like other cephalopods, the octopus is deaf. It can alter its color to match the surroundings, and does so within minutes. If startled, it pro-duces ink as a camouflage, and moves rapidly backward by water-jet propulsion. See *Cephalopoda, mantle, paper nautilus.*

Odontoceti   The family of toothed whales. See *baleen, individual species, whale.*

offshore zone   A relatively flat zone extending from the breaker line on the beach to the edge of the continental shelf. See *coast, continental shelf, petroleum.*

oil   A term often used interchangeably with fat. Generally oils are liquids at room temperature, while fats are solids. Whether they are called oils or fats, these substances are composed chemically of triglycerides—compounds consisting of glycerol combined chemically with long-chain fatty acids. If, in the fatty acid chain, one or more of the methylene ($-CH_2-$) groups that—linked to one another—make up the "backbone" of the chain, are replaced by a doubly-bonded carbon (-CH-) group, the oil is called unsaturated. When there are several -CH- groups in the molecule, the oil is said to be polyunsaturated.

Marine oils are commercially important products. Whales are hunted for their oil, which is rendered from the blubber or body fat on factory ships—huge, mechanized versions of the 19th century whalers. Sperm whale oil is quite special because it contains alcohol esters called spermaceti, a material used in cosmetics or pharmaceuticals. Fish oils are generally extracted from the livers of sharks, cod, menhaden, sardines, tuna, and mackerel by steam distillation, a process that extracts the oil without heating it to temperatures at which decomposition might occur.

Fish oils are used extensively in the manufacture of lubricants, soaps, and paints. The medicinal use of cod liver oil is based on its high vitamin A and D concentration. Certain fish oils are also being investigated for their possibly beneficial effect in preventing or retarding atherosclerosis.

Okhotsk, Sea of An arm of the northern Pacific Ocean, separated from it by the Kurile Islands and the Kamchatka Peninsula. The Kurile Straits are the outlet to the Pacific. The Nevrlesky (Tatar) and LaPérouse (Soya) straits lead to the Sea of Japan. The Okhotsk Sea is surrounded by a mountainous shoreline. The average depth of the water is about 775 m (2,350 feet), with a maximum depth of over 3,370 m (10,000 feet). The bottom is in part continental shelf and the remainder is of island origin. The southern end is the most ancient and also the deepest part.

There is considerable precipitation in the Sea of Okhotsk, since it is an area in which cold continental air meets warmer maritime air. There is an ice cover for about six months of the year. The waves are noticeably higher than in the Pacific in general. The water temperature is relatively warm. The salinity, particularly at the surface, is low.

There is a warm deep current in the Sea, that is also highly oxygenated. This anomalous water property supports a huge phytoplankton population. This, along with the resident algae, support a large mollusk and echinoderm population that is in turn a food supply for the commercially important salmon, seal, and crabs. See *Dezhnev, Semyon Ivanov; Pacific Ocean.*

Oligocene Epoch That part of the Tertiary Period that followed the Eocene Epoch and preceded the Miocene. Very few terrestrial Oligocene fossils are closely related to modern biota; however, most marine Oligocene fossils are very similar if not identical to modern forms. The Foraminifera were extremely numerous in both species and numbers of individuals in this epoch.

Geologically, a major event of this span of time, which began about 38–40 million years ago and lasted about 12 million years, was the upthrust of the Tethys Sea bottom, resulting eventually in the formation of the European Alps.

Warm climates prevailed as the sea encroached on land areas and retreated again in several cycles of advance and retreat. These cycles are each named, usually for a locale. At some point in the Oligocene, South America was isolated from North and Central America, allowing divergent faunas to develop in these different areas. See *Tertiary Period.*

Oligochaeta A class of worms of the phylum Annelida. The common earthworm is an oligochaete. The class is widely distributed in water (both in fresh and marine environments) and in soil. The individual members of the class range in size from those that are only millimeters long to the great giants of Australia that are 3 m (10 feet) long and more.

Setae or bristle-like projections and a clitellum (the thickened band around the body, which is closer to the "head" than to the tail) are characteristic of the Oligochaeta. In marine specimens, the clitellum is much less conspicuous than in terrestrial ones. On the earthworm this organ produces the gelatinous material in which the eggs mature and hatch; however, this semiliquid nest is not needed in a water environment. Oligochaetes are hermaphrodites. See *Annelida.*

Omnivore An animal that ingests both plant and animal tissue. See *heterotroph.*

ooze Soft sediment on the ocean floor, composed largely of the remains of minute organisms. Oozes are characterized by the organism that contributes the major part of their composition: they may be foraminiferan, diatom, or radiolarian. See *diatom, Foraminifera, ocean floor, radiolarian.*

opholite A basaltic fragment of old marine origin. It is the result of the horizontal spread of what had been a vertical slice of ridge crest. The Trovdos opholite on Cyprus is typical. Other opholite sites are in the Shetland Islands, the Bay of Islands in Newfoundland, the Ural Mountains, and numerous sites along the India-Asia suture line. See *basalt, crust, evolution of oceans, Pangaea.*

orca See *Cetacea, killer whale.*

Ordovician Period Part of the Paleozoic Era, preceded by the Cambrian and followed by the Silurian Period. The name became established in 1873, and is derived from a tribe of pre-Roman Britons. The epochs encompassed by the Ordovician are variously named in different locations, and these local names are used to describe the rocks found there. The entire period lasted about 70 million years, and ended about 430 million years ago. During the Ordovician Period sedimentation on the continental shelves of Gondwanaland built up landmass and closed the Iapetus Ocean, which has been called the proto-Atlantic. Cincinnatian seas covered most of mid-North America, while in South America Ordovician formations occurred in Venezuela and Colombia, and south to Argentina. Well-defined fossils dating from this time have also been found in Australia.

The fossil record of the Ordovician is extensive: brachiopods are the most characteristic fossils, graptolites are generally found, and cephalopods, trilobites, and corals are known, as are some crinoids, gastropods, pelecypods, and ostracods. Primitive fish also proliferated in the Ordovician. See *Cambrian Period, Paleozoic Era, Silurian Period.*

Orinoco A 2,900 km (1,800-mile)-long river flowing through Venezuela from the Brazilian border to the Colombian border and east into the Atlantic Ocean. It is often compared to the Niger River in content and configuration. Its delta is quite large, built by its own sedimentation in a relatively tranquil area, and augmented by deposits from the Amazon River brought to the northeast by offshore currents. See *Amazon, Atlantic Ocean, Humboldt, Friedrich Heinrich Alexander Von.*

Ortelius, Abraham (1527–1598) A Belgian (Flemish) geographer who was a friend and contemporary of Mercator, the famous cartographer; he was almost as influential as his friend. Ortelius had extensive, first-hand knowledge of geography, having traversed a considerable part of the known world. At Mercator's urging, he compiled the *Theatrum Orbis Terrarum,* an atlas of 53 maps, published in Antwerp in 1570.

Because the atlas collected information useful for sea voyages, and appeared at a most propitious moment in European history, it was a huge success in spite of its errors which were the result of using locally obtained, and often inaccurate, recordings for depths and distances.

In 1575, Ortelius was invited by Philip IV of Spain to be the King's personal geographer; he accepted and continued his work by adding more maps to the atlas; eventually seventeen more were included. Ortelius spent the remainder of his life collecting maps and attempting to codify nautical notation.

osmosis The movement of water across the barrier of a semipermeable membrane, such as a cell membrane. The water flows from a region with a greater concentration of water to one of lesser concentration. This means that water moves in the direction of a greater salt content. If water outside a cell has a greater salt content than water in the cell, water will move out of the cell, dehydrating it.

Animals in the sea have chloride levels in their bodies that mirror the chloride content of the seawater (i.e., they are isotonic with the seawater). Some animals can adjust to changes in the salt content of the water, not becoming bloated when it drops, nor becoming dehydrated if it increases. The hagfish is such an animal. In most bony fishes the kidneys and gills are the barriers to the entry of salt.

The mechanism by which migrating animals adjust to changes in the salt concentration of water is not completely understood. Fish such as the salmon may go only one way—such as from fresh- to salt water—without need for an adjustment in the mechanism to deal with an increase in the salinity of the ambient environment. Other animals live to return again to an environment of different salt (especially chloride) content. They can therefore change the permeability of the membranes of their gills, digestive systems, and kidneys.

Aquatic mammals have the lowest blood salt levels of all aquatic animals. It is, however, higher than that found in terrestrial mammals. See *chlorine, minerals in seawater.*

osprey   Also known as the fish hawk. A large, fish-eating hawk known on all continents. It has a wingspan that ranges from about 60 to 65 cm (2 to 2.25 feet), is a dark brown bird with a white underside and some white head feathers. The osprey fishes by hovering over and then seizing its meal. It breeds on rocky islands, cliffs, and in some trees. The osprey population, which had been declining since about 1900 because of inroads on its habitat, was severely crippled by the introduction of chlorinated insecticides. As with other predatory birds, the effect of DDT accumulations in the adult birds disturbed the mechanism by which the eggshells are formed, resulting in thinner then normal shells. This meant that adult ospreys, in attempting to cover their eggs in order to hatch them, crushed them. The osprey has been making a comeback since the ban on the use of DDT. See *fish, pollution, predator-prey relationship.*

Ostracoda   Tiny arthropods known as mussel or seed shrimp. The name is derived from their appearance. They have seemingly lost their segments, and have a large carapace. They are small (1 mm or 0.04 inches) bottom dwellers that feed on plankton or detritus. Some are parasitic. The 2,000-odd species that exist now include some freshwater and terrestrial forms. The 10,000 extinct species date from the Upper Cambrian to the Pleistocene. Some genera can be used as index fossils (characteristic reference points) in the exploration for mineral deposits, especially for petroleum. See *Cambrian Period, fossil, petroleum, shrimp.*

otter   A semiaquatic mammal of the weasel family; usually a riverine animal. The sea otter *(Enhydra lutris)* of the North American West Coast was hunted to near extinction in the early 20th century. Protection in 1910 led to a gradual increase in their numbers, and now they are claimed to be a threat to abalone beds.

The otter is a strong swimmer. It feeds on fish and also on mollusks (usually abalone), which it opens by smashing them on a stone. Otters are playful, large (75 to 120 cm or 30 to 50 inches long), thickset (15 to 40 kg or 32 to 70 pounds) animals. They have thick red-brown to dark brown lustrous fur. Their heads and faces are usually gray. They float on their backs in groups (between gorgings), and generally seem to be enjoying themselves. See *abalone, predator-prey relationship.*

overfishing   See *fishing industry, whale.*

ovoviviparity   A property of some invertebrates, fish, and reptiles involving the incubation of growing young within the body of the female. The eggs are fertilized internally and the developing embryos are nourished by large eggs rather than placental tissue. The young are hatched within or right after extrusion from the parent. Several species of shark exhibit this characteristic. See *fish, shark.*

oxygen   A chemical element that as a gas (O_2) forms about 21% of the Earth's atmosphere, and in combined form, about 50% by weight of the Earth's crust. It is also a key constituent of organic compounds and all living matter. Dissolved in seawater (H_2O) as a gaseous molecule (O_2), the concentration of oxygen varies considerably from 4 to about 6.5 ml/liter (milliliter of oxygen per liter of water). The concentrations of oxygen in the water covering the continental shelves are higher—in the 5 to 8 ml/liter range. The concentrations of oxygen in tropical surface waters are generally low.

The oxidation of organic material and the metabolism of animals depletes the

Earth's oxygen supply, while the activity of algae in the photic zone of the oceans replenishes it. The zone of minimum oxygen concentration is that at intermediate depths (800 to 1,500 m).

Tiny amounts of the allotropic form of oxygen, O_3—known as ozone, are present in seawater. This form of the element is much more reactive than is O_2. The greatest concentration of O_3 on Earth—although the quantities are still very small—is in the stratosphere, where it is produced by a photochemical reaction.

The oceanic dumping of waste materials of all sorts—industrial, sewage, agricultural, and radioactive—removes oxygen from the water. A crude measure of the quantity of oxygen needed to react with waste material and incorporate it harmlessly into the environment is the biological oxygen demand, or BOD. Since the supply of oxygen is relatively constant, areas with a high BOD are those with a diminished flora and fauna. See *nitrogen, pollution.*

Oyashio Current   A northern Pacific Ocean flow, also known as the Kurile Current. It is a cold, polar current whose counterpart in eastern North America is the Labrador Current. The Oyashio flows southwest along the Kamchatka peninsula after having moved west across the Bering Sea. It meets the Kuroshio at about 35 to 37° North Latitude, and some of the cold water sinks below the warm current. All of the current is traceable because of its low temperature (4 to 5°C) and low salinity, (3.37 to 3.4%) which is in contrast to the warm (about 12 to 18°), depending on the season, and more salty (3.45 to 3.50%) Kuroshio.

The Oyashio carries along a much greater plankton and nutrient load than does the Kuroshio. Hence, the Japanese have given it the name it has, which

means parent. See *currents, gyre, Kuroshio Current.*

oyster   A gray, bivalve mollusk with one convex and one flat shell, of either the Ostreidae (true oyster) or Aviculidae (pearl oyster) family. Edible oysters are native to almost all waters of the Northern Hemisphere. They have been a foodstuff sought after by humans for millennia, and a cultivated foodstuff certainly as long ago as Roman times. Neolithic middens (garbage dumps) contain oyster shells.

Oysters range in size from 7 cm (2.5 inches) to the 30-cm (12-inch) diameter of *Crassostrea gigas* of Japan. The young spat is a ciliated, planktonic swimmer (the larva is called a veliger) for a few days after hatching. It then attaches itself to a rock and filters water for food. Oysters are harvested commercially when they are between 3 and 5 years old.

These animals breed in summer and produce incredible numbers of eggs per female: frequently over one million eggs per breeding season. The eggs are fertilized in the water. Some species are hermaphroditic (the *O. edulis* of western Europe) and some, *C. virginia* (of the American east coast) are either male or female, with sex changes occurring several times in the lifetime of an individual.

All oysters form pearls, which are layers of calcium carbonate ($CaCO_3$) deposited around an irritant, such as a grain of sand. Only some pearls are lustrous; the best are from the species *Meleagrina vulgaris* of the Persian Gulf. Cultivated pearls result from an artificially introduced irritant. The oyster is then grown in a protected environment for several years before the pearl is taken out. This is a major industry in Japan.

Animals that are predators of oysters include the oyster drill (Murex), starfish, fish, birds, and sea mammals. See *mariculture, mollusk, murex, veliger.*

P

Pacific Bottom Water Cold water at the bottom of the Pacific Ocean. It originates in Antarctica and spreads north to become the major constituent of the Pacific bottom. It is detected as far north as Japan. Surface water sinks at the Antarctic and Tropical Convergences. At the latter, the equatorial countercurrent spreads water laterally, and as it sinks, this water moves poleward. Upwellings of cold water are well-known on both the North and South American coasts. See *Antarctic Convergence, Equatorial Currents.*

Pacific Ocean An ocean extending from the arctic in the north to the antarctic in the south, and from the Americas in the west to Asia and Australia in the east—it is the largest body of water on Earth. It is separated from the Atlantic Ocean by the Drake Passage, and from the Arctic Ocean by the Bering Strait. The boundary of the Pacific with the Indian Ocean runs across the Timor Sea to Cape Londonderry, Australia, and across the Bass Strait between Australia and Tasmania. The Pacific Basin may be considered as three distinct areas: east, west, and central. The eastern Pacific is the area stretching from the American coast to the Hawaiian Ridge; the central Pacific is a plain at average depths of about 5,000 m (16,500 feet). The western Pacific extends from the line of trenches (the Aleutian, Kurile, and Tonga) to New Zealand. Most of the islands in the Pacific are in the western Pacific.

The seas that are marginal to the Pacific Ocean are the Bering, Okhotsk, Yellow, and Timor seas and the Bay of California.

The water currents of the Pacific are wind-driven. In terms of the quantity of

water moved, the major current is the Antarctic Circumpolar Current. Others are the Kuroshio, North Equatorial, Equatorial Undercurrent, North Pacific, North Equatorial Countercurrent, California, and Peru currents.

The salinity of the Pacific varies with geography and temperature; it is highest in warm regions, where evaporation exceeds runoff and varies between 3.55 and 3.65%. It falls below 3.3% at the poles. The oxygen content is highest at the surface, drops with increasing depth, and increases again at depths greater than 4000 m (13,000 feet). The nutrient content, particularly of phosphates, is highest in the intermediate layer, where the oxygen concentration is lower. Upwellings along the American coasts bring nutrients up into the photic zone, and this results in the characteristic spectacular growth of plankton.

In addition to its great size, the basin of the Pacific is varied and geologically interesting. It has more oceanic bottom features (ridges, trenches, seamounts, mineral deposits, coral, hot spots, and islands) than do the Atlantic and Indian Ocean basins.

Voyages across large expanses of water in the Pacific began about 2,000 years ago with Polynesians in canoes using stick maps. A stick map is an accurate depiction of oceanic territory, with islands or reefs shown as focal points (i.e., conjunctions of several lines of rods). Information about winds, currents, and shoals was built into these maps. The adventurous, striking out over vast areas, were preceded by earlier coastal voyagers. Thus, stick maps came after a long period of sailing experience.

Others also sailed the Pacific. There were Chinese-Arab trade relations by the

1st century A.D. The lure of exotic treasures and fortunes to be made led many others on expeditions. While trade was the accepted reason for most Pacific travels, there were later voyages of colonization, missionary activity, and piracy. See *evolution of oceans, mid-ocean ridges, names of individual marginal seas and currents.*

Pacific Plate   The plate essentially coincident with the main or central Pacific Basin. This is a plain from which ridges and seamounts rise. It is cut by a series of fracture zones perpendicular to the ridge lines. The average depth of the plain is more than 5,000 m (17,000 feet).

The East Pacific Rise and its extensions form the eastern boundary of the Plate. The North American, Cocos, and Nazca plates are east of the Pacific Plate. The Plate edges are delineated by trenches and islands, frequently in double arcs. The inner one is seismically active, the outer one is an aseismic ridge line.

The southwestern edge of the Pacific Plate is the "andesite line," or the dogleg described by the New Hebrides and Tonga island arcs. There is some dispute over the composition of the crust under the marginal seas of the Plate, especially those of the southwestern Pacific. They have been called epicontinental by some; other researchers dispute this nomenclature. The Pacific is neither a "new" ocean nor is it the site of a "drowned" continent. See *andesite, crust, individual plates, Pacific Ocean.*

Pacific-type margin   The leading edge of a continent. At the convergence of two plates this leading edge is thrust upward. See *Atlantic-type margin, continental drift.*

pack ice   Any mass of floating pieces of ice, 3 m (10 feet) or less in diameter, driven together to form a solid layer. If there is a cover of pack ice in a waterway, it will stop all shipping except for those vessels that are icebreakers. See *Ice.*

Paleocene Epoch   The oldest epoch of the Tertiary Period of the Cenozoic Era. The predominant terrestrial rock of the Paleocene is sedimentary. The name for this epoch, signifying "ancient recent time," was suggested by the botanist W. P. Schimper in 1874.

The Paleocene epoch lasted about 10 million years, commencing about 70 million years ago. It saw two cycles of mountain uplift and the retreat of the sea. The Laramide orogeny—the upthrust of the Rocky Mountains—occurred in this period. The fauna of the Paleocene included some holdovers from the Cretaceous: cephalopods, pelecypods, gastropods, echinoids, and foraminiferans found favorable growing conditions. They repopulated many ecosystems after the very noticeable disappearance of the belemnites, ammonites, and fishlike dinosaurs. Reptiles moved onto land and became snakes, lizards, and crocodillians. The outstanding animal development during the Paleocene was terrestrial. See *Cenozoic Era, Tertiary Period.*

paleontology   The study of fossil plants and animals, and of the rock formations in which these fossils are found. The science was named by Ducrotoy de Blainville and Fischer von Waldheim in 1834. Before that, it was an area of natural science that had absorbed the interest of many professional scientists. Linnaeus, Buffon, Cuvier, Agassiz, Hutton, and others used paleontological specimens in their work before the official name was given to their area of study. See *fossil, individual scientists.*

Paleozoic Era   The period of earliest life on Earth. The name signifies "old life." Followed by the Mesozoic, the Paleozoic ended about 245 million years ago. It is divided into a Lower and an Upper segment. The Lower (or older) Paleozoic is further subdivided into the Cambrian, Ordovician, and Silurian periods; the Upper Paleozoic is subdivided into the Devonian, Carboniferous, and Permian periods.

The climate during this very long era seems to have remained fairly constant and warm. While warm seas overlaid many land areas, folding and downwarping raised mountainous regions above the water.

The flora of the period was dominated by algae in the water; on land, mosses and ferns had emerged. The fauna, however, was entirely aquatic. Trilobite fossils dominate the Cambrian remains. By the Upper Cambrian, cephalopods, brachiopods, and graptolites were present, and bryozoans, gastropods, and ostracods developed. By Silurian times there were fish.

The Upper Paleozoic saw several cycles of advance and retreat of the seas. The climate was much more variable, and deserts appeared in the Permian Period. The biota changed markedly as the climate changed. There were several ice ages in the late Paleozoic, and a major extinction at its end. See *Cambrian Period, Carboniferous Period, Devonian Period, Permian Period.*

palolo A polychaete worm of the genus *Eunice* that swarms in a pattern strictly geared to a lunar timetable. This worm is about 40 cm (16 inches) long and ordinarily a bottom dweller in warm tropical seas. It produces an enormous sac of eggs or sperm in its tail segment. This is synchronized with the third lunar quarter in June or July in the Atlantic Ocean near Florida, and in October or November near Samoa in the Pacific. The entire tail segment of the animal rises to the surface and ruptures, releasing the eggs or sperm. The tail portion resembles a worm, and even has eyes. It "lives" for a very short period, with the whole swarming period lasting only a day or two. The unruptured tail is considered a delicacy, and is caught and eaten on the spot. The female tails, which are blue, are considered the best. Male tails are greenish blue. See *Polychaeta.*

pancake ice New sea ice that is formed in autumn. The round pieces of ice are between 30 cm and 3 m (1 to 10 feet) in diameter. These are in turn coalesced into larger units. When they are pushed together, the relatively flat surfaces may become buckled, and the resulting floes are not as flat as the pancakes that formed them. See *floe, ice.*

Pangaea The supercontinent that comprised all of the continental landmass of the young Earth. See *continental drift; Gondwanaland; Panthalassa; plate tectonics; Wegener, Alfred Lothar.*

paper nautilus An octopus, *Argonauta*. The females manufacture a thin mucus envelope in which to brood their eggs, which hardens to an opaque, seamless cover, from which the name "paper nautilus" was derived. The males of the species are much smaller than the females; they were once thought to be a separate species. The sperm-transferring organ of the paper nautilus is a very specialized arm that in the 19th century was described as a species of worm. See *octopus.*

Panthalassa The primordial ocean that surrounded the supercontinent Pangaea before the latter's breakup in the Jurassic Period. See *continental drift, evolution of oceans, Pangaea.*

parapodia Projections on the segmented, marine annelid worms. Parapodia with cirri (curls) or setae (bristles) on their surface are characteristic of the polychaete worms. See *Annelida, Polychaeta.*

parasite An organism living in or on another organism (the host) and depending on the host for nourishment without giving anything in return. Many if not most marine animals may at some time in their lives harbor parasites. A great number of the known parasites are worms. Because parasites constitute an enormous group of often unrelated organisms, their mode of attack and their hosts are also unrelated.

Common Marine Parasites

Phylum	Parasite Class	Host
Platyhelminthes	Turbellaria	Internal parasites of crabs, mollusks, echinoderms
	Monogenera (fluke)	External on fish gills
	Cestoda (tapeworms)	Internal parasites of all crustaceans, and all vertebrates. Tapeworms are very specific about their vertebrate hosts.
Aschelminthes	Nematoda	All marine species
	Rotifera	External; crustacean gills
Acanthocephala		Internal; isopods, fish, birds, mammals
Annelida	Myzostomaria	External; crinoids.
	Hirudinea (leech)	Internal; fish
Arthropoda	Crustacea (Branchiura)	External; bony fish
Chordata	Cyclostomata (lamprey)	External; bony fish

parasitism   A relationship between two organisms in which one benefits at the expense and to the detriment of the other. See *commensal relationships, symbiosis.*

parrot fish   A group of tropical marine fish related to the wrasses. They are brilliantly colored and have oblong bodies with relatively small fins and tails. Their coloration and large beaky heads explain their name. Most parrot fish are herbivores, but some prey on coral fauna.

The largest parrot fish, *Pseudoscarus guacamia,* may grow to about 1 m in size and weigh about 14 kg (30 to 31 pounds). This species and several others are eaten, but on the whole, the parrot fish is not considered an object for commercial fishing.

pearl   A hard, smooth, lustrous, protective structure formed around an irritant by the mantle of a shelled mollusk whose shell interior is nacreous (mother-of-pearl). The coating applied to the irritant consists mainly of calcium carbonate and some proteins. A round or ovoid pearl forms in a part of the mollusk not subject to muscle pressure. If the pearl forms where it is affected by muscle action it is deformed (baroque) or flattened.

Gem-quality pearls are found in both marine and freshwater mollusks. They vary in color from white to cream, rose, lavender, green, gray, and black, depending on the locale and the producing organism. A cultured pearl is a pearl formed in a captive oyster by inserting an irritant into the oyster and then maintaining it in a protected environment for several years. This has become a major industry in Japan. The best natural pearls in terms of luster and color are supposedly those of the Persian Gulf. See *mantle, Mollusca, oyster.*

Peary, Robert Edwin (1856–1920)
An American naval officer and explorer who is credited with being the first man to deliberately arrive at the North Pole. In 1891 Peary organized his first attempt at attaining 90° North Latitude; it was unsuccessful. He explored the northern coast of Greenland in subsequent summers, and established the existence of water under the ice north of the island.

He designed an ice-withstanding ship, called the *Roosevelt,* in which in 1905 he sailed north to 87°6' before turning back. He tried again in 1908, and finally arrived at the Pole on April 6, 1909. This expedition established the Pole's existence on an ice platform: the North Pole is not on land, as the South Pole is.

Frederick Cook, a doctor on one of Peary's expeditions, claimed that he, not Peary, had reached the Pole first, almost a year earlier than Peary did, but eventually his claim was discredited. *See explorers and explorations.*

pelagic environment   The open ocean, excluding the ocean bottom and the shore. This is by far the largest area of "open space" on Earth. The organisms in this space are grouped into arbitrary categories by size and type. The categories frequently differ from one researcher to another. The main categories are plankton, both plant and animal; and nekton. Some biologists include fungi and non-photosynthesizing bacteria, along with the diatoms, dinoflagellates, green, brown, and red algae, coccoliths, and autotrophic bacteria as members of the phytoplankton. The familiar zooplankton include copepods, foraminiferans, tunicates, cnidarians, amphipods, pteropods, and chaetognaths. Other, less familiar groups of animals are also in this category of pelagic organisms.

The largest individuals in the open ocean are those capable of swimming against the current for some time. These constitute the nekton, which includes all open-ocean fish, sea mammals, large cephalopods, and some large crustaceans.

Pelagic populations in any one place depend on the availability of sunlight, current flow, nutrient concentration, salinity, and temperature. *See individual animals, nekton, plankton.*

pelagic sediment   An ocean-bottom deposit of material from seawater. The greatest contributions to this sediment are the remains of coccoliths and foraminiferans. Such deposits are calcareous, (calcium-containing). Diatomaceous and radiolarian remains also eventually drift to the bottom, producing siliceous, (silicon-containing) deposits. Phosphate deposits, which generally consist of apatite, originate from the skeletons of the bony fish.

Mineral deposits in the pelagic sediment are present as encrustations, usually consisting of oxides of iron and manganese. Minute grains of quartz and feldspar fall on the sea as dust, and eventually form part of the pelagic sediment. Some extraterrestrial iron and nickel, which come from meteorites, are also parts of this layer on the sea bottom. *See carbonate, coccolith, ocean floor, manganese, minerals, radiolarian.*

Pelecypoda   Bivalves; the class of Mollusca that includes clams, oysters, mussels, and scallops. They are found in

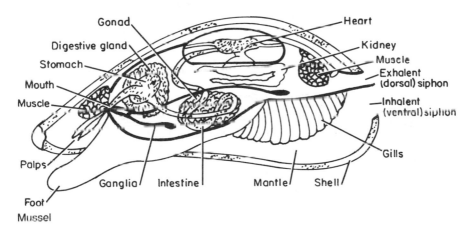

Mussel

all aquatic environments, fresh- and salt water, and at all depths. Their ancestors appeared in the Cambrian Period and continued to produce new forms on through to the early Ordovician.

The most obvious characteristic of bivalves is their shell, which is formed from secretions of the mantle. Some bivalves have nacreous (mother-of-pearl) shells which are of economic importance. Many bivalves are edible, and as such are sought after and economically important. Others are destructive: the shipworms, for example, are not worms at all but mollusks. See *clam, Mollusca, oyster, pearl, scallop, teredo.*

pelican   One of a small group of water birds of the family Pelicanidae. Pelicans live on rivers, lakes, and seacoasts in tropical to warm-temperate areas. They are related to the gannets and cormorants. Most pelicans nest in colonies. Their nests are most often in scrub trees that are found in wetlands. The young are fed on regurgitated food—mainly fish and mollusks—held in the parents' pouched throat. Ancient depictions of the pelican have it feeding its young on its own heart's blood—a fable based on inaccurate observation.

The American white pelican is found in the Everglades of Florida. It is the largest example of the group, and has an average wingspan in excess of 3 m (10 feet). The survival of the genus was in doubt in the early 1970s, but it has recovered from the effects of DDT poisoning, although its territory is still being encroached upon by human activities such as the channeling of the Everglades. This reduces the range of this bird and of other species. See *bird.*

penguin   A flightless, erect bird of the family *Spheniscidae,* order Sphenisciformes, that is adapted to swimming in cold seas and found mainly in the Southern Hemisphere. There are about fifteen species of penguin.

Penguins breed in colonies on cold, rocky coasts from California and the Galapagos to Africa, Australia, New Zealand, and South America. The emperor and

Adelie penguin are antarctic species, and the most "typical" penguins.

The group ranges in size from the small blue penguin, which is about 40 cm (16 inches) high, to the largest, the emperor penguin, which stands 120 cm (4 feet) high. All of the penguins are dark on their heads and backs. Their dark backs range from light brown to black. Penguins are extremely skillful swimmers and divers. Tagged emperor penguins have been found at depths exceeding 200 m (660 feet). They have white undersides, webbed yellow or orange feet, and flipper-like wings.

Both sexes of adult penguins look alike. In the breeding season, the parents take turns caring for the one or two chicks. Depending on the species, the young are fed regurgitated meals of fish, squid, or crustaceans. Thus, the diet of the emperor penguin, which is the deepest diver, consists almost exclusively of squid.

Penguin chicks are herded together in nurseries by the attending adults. This is part of the attempt to protect the young, which are preyed upon by skua (sea birds related to gulls) which steal the eggs. The predators that attack adult penguins are seals and killer whales. See *bird.*

peninsula   A landform that juts out into a body of water; an extension of a larger landmass. The Arabian, Iberian, and Florida peninsulas are examples. See *individual peninsulas.*

peptide formation   The linking together of amino acids to form polymers. The shorter polymers are known as peptides; the longer ones as proteins and polypeptides. The amino acids themselves are produced by the same cells that manufacture nucleic acids and carbohydrates.

The peptide linkage is represented chemically by the

$$H-N-\overset{\overset{\textstyle O}{\|}}{C}-$$

bond in the structure:

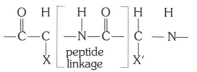

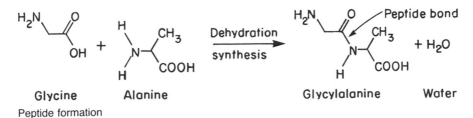

Peptide formation

This polymer chain has a repeating structure. The side chain, X or X^1, can be any atom or group. This means that considerable variation in the side chain and in the length or complexity of the main chain is possible, although only about 20 different side chains occur in nature. While the amino acids are common to all life, the particular peptide, polypeptide, and protein polymers they create are unique. Each such molecule is determined by the DNA and specific enzyme systems of a particular species of organism.

The enzyme systems that regulate protein formation are extremely conservative, and do not differ greatly from one species to another. The proteins in related species do have differences, but they are slight. The more closely related two species are, the more similar are their proteins. Evolutionary studies based on protein structure are now quite common. See *amino acid, compound, DNA, protein.*

perigee   That part of an orbit of a satellite (natural or artificial) at which the satellite is closest to the planet it is orbiting. See *apogee, lunar tide, perihelion.*

perihelion   That part of the orbit of a planet or comet at which the orbiting body is closest to its sun. See *aphelion, lunar tide, solstice.*

periwinkle   A small, widespread, marine snail of the genus *Littorina,* found on both sides of the Atlantic Ocean on intertidal rocks and flats, and feeding on algae. The periwinkle is usually gray and has a thick whorl. It is about 2 to 3 cm (0.8 to 1.2 inches) across, and known to anyone who has walked on a beach. Periwinkles are edible, and are used in Western European and Far Eastern cookery. See *Gastropoda, littoral, snail.*

Permian Period   The last period of the Paleozoic Era. This period is named for the Perm region of the western Ural Mountains in the Soviet Union. The best known North American formations that date from the Permian are in Texas. There were long periods of drought in the Permian Period that resulted in desert conditions on land. There were also several ice ages and glacial deposits, and scars of these exist on all of the Earth's southern continents. The presence of glacial scars, all from the same time period but on different landmasses, cannot be coincidental, and supports the continental drift theory. The continental landmasses must have been closer together than they are at present, and it is now postulated that the single supercontinent, Pangaea, was centered near the South Pole and broke up during the Permian.

The Permian Period lasted about 20 to 30 million years, and was known for mountain building: The Appalachians, Urals and highlands of central France were pushed up at this time, as were the northern Himalayas.

Marine life during the Permian was dominated by brachiopods, bryozoans, and goniatites, all of which were either extinct or greatly reduced in number by the end of the Permian, which was also a time of "great extinction." See *continental drift, extinction, Paleozoic Era.*

Peru-Chile Trench   Also known as Atacama Trench, this ocean deep is in the Eastern Pacific. It parallels the coast of South America for part of its length, then turns south. The deepest point is the Richards Deep at 7,635 m (about 25,000 feet). See *Pacific Ocean, trench.*

Peru Current   An eastern Pacific Ocean current also known as the Humboldt Current in honor of Alexander von Humboldt, who charted it and measured its flow in 1802. It is a cold, broad (900 km or 580 miles wide) surface current that flows north along the western South American coast, originating in the West Wind Drift in antarctic waters. The Peru Current separates from the major stream of the West Wind Drift as the latter heads toward the Drake Passage. The Peru branch moves north to about 4° South Latitude, where it turns westward as part of the South Equatorial Current. The cold Antarctic water is augmented by upwellings that bring cold water to the surface, along with nitrates and phosphates. The upwellings of nutrients provide the basis for an extremely rich and varied fauna.

The phenomenon known as El Niño is a disruption of the Peru Current. Its consequences are economically and meteorologically important even to regions far removed from the South American coast. See *Antarctic Water; currents; Drake Passage; El Niño; Humboldt, Friedrich Heinrich Alexander von.*

petrel   A widely distributed sea bird of the order Procellariiformes. The average petrel is a small black bird with a considerable wingspan. They are most often 25 to 90 cm (10 to 30 inches) long and weigh 100 to 400 g (0.2 to 0.9 pounds). Petrels have tubular nostrils and—uncharacteristically for birds—a good olfactory sense. Their beak is long and straight, with a hooked tip. They have webbed feet and swim well, and use the land only as a breeding ground. The average nest is a hole in a cliff, a spot on a cliff ledge, or a messy collection of seaweed, grass, and feathers on a rocky beach. Most petrel species produce only one or two eggs each year, and the young are slow-growing. The birds' diet is largely crustaceans, small fish, large plankton, and

squid. The bulk of the species spend some part of their lives in antarctic waters. While some petrels and fulmars are often associated with European, Mediterranean, or African shores, the most likely breeding ground for the average petrel is South America.

There are four families of petrels, comprising 92 species. They are the albatrosses, storm petrels *(Hydrobatidae),* diving petrels *(Pelecanoididae),* and petrels and shearwaters *(Procellariidae).* The storm petrel is also known as Mother Carey's chick or St. Peter's bird. It is about the size of a starling, and skims above the waves, taking whatever small creature or bit of debris it can find. The diving petrel is also a small member of this order. These birds use their wings as flippers, diving headlong into the waves like penguins. Their diet is more likely to include crustaceans and fish. See *albatross, bird, fulmar.*

petroleum   An oily, viscous flammable liquid, consisting mainly of hydrocarbons, which is refined to produce gasoline, naphtha, and other chemical products. It is found in dome formations on continental shelves and on continents. Offshore petroleum comes mainly from the Gulf of Mexico, the Persian Gulf, and the North Sea. Underwater domes are found by echo sounding and magnetometric techniques. One theory of petroleum formation holds that petroleum is formed in undersea locations by a compacting of organic materials that are brought together and nourished by upwellings.

There have been petroleum findings in depths of 3,000 to 4,000 m (10,000 to 30,000 feet). This would indicate a yet unexplained source of the organic material that produces this petroleum. Such deep deposits would be extremely difficult to tap commercially because the lack of a stable platform would render the recovery of such oil extremely expensive. The cost of petroleum crude oil would have to be quite high before such an operation would be a commercial success. See *minerals, oil.*

pH   A measure of the "hydrogen ion" content of water, and therefore of its acidity. Neutral water, which is neither acidic nor basic, has a pH of 7. A pH lower than 7 indicates acidity, and a pH higher than 7 indicates basicity. The pH of the primordial ocean was probably lower than that of today's seawater. This is theorized on the basis of a greater concentration of carbon dioxide in the atmosphere and therefore in the ocean.

The pH of the ocean is today about 8.2—slightly basic—and is remarkably constant despite the production of ammonia by animals, the dissolution of metallic oxides which would increase the water's basicity, and the production of acidic sulfur compounds by bacteria. This constancy is the result of the action of phosphate and carbonate ions, which buffer the ocean system, or "cushion" it against a sudden, large pH change by accepting large quantities of the hydrogen (H+) ion that causes acidity and giving up hydrogen ion in response to an increase in basicity, thus stabilizing the pH. See *ammonia, carbon dioxide, sulfur, water.*

Phaeophyta   Brown algae. These marine plants contain xanthophylls—pigments that give them their characteristic brown color. Examples of phaeophytes are the thallose seaweeds that inhabit the intertidal zones, the Fucales or wracks (e.g., *Fucus*), and the laminariales, or kelps (e.g., *Laminaria*). See *brown algae,* Fucus.

pheromone   A hormone that acts as chemical signal between animals of the same species. The scent trails that include the pheromones may be territorial markings, food locators, or sex attractants. Widely diverse organisms manufacture pheromones. In the marine biome they are produced by barnacles, by female crabs to attract males, and possibly by starfish to control egg deposition.

The typical pheromone is a fairly simple molecule that is produced in very small quantities. It is, however, picked up by the appropriate receptors on a given

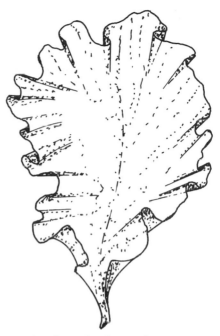

Ulva (sea lettuce), a brown algae

species of animal even when its concentration can be measured only in parts per trillion. Since artificial pheromones have been successfully used as pesticides by attracting insects into traps, there has been research in the use of pheromones to try to decipher the complex life cycle of the lobster, to act as shark repellent, and to direct the movement of schools of fish.

Philippine Sea   A marginal area of the Pacific Ocean south of Japan and surrounded by the islands of Taiwan, the Philippines, and the Ryukyus (Okinawa) on the west, and the Marianas on the east. The Kyushu-Palau ridge separates the Philippine Sea into two areas. Other delineating features of the Sea are the Kuroshio Current north of the Sea and the Equatorial Countercurrent, which forms its southern limit.

The Subtropical Convergence is at about 23° North Latitude. It runs through the Philippine Sea and alters the prevailing weather. The weather pattern for the area is Siberian in the northern portion, and the winter temperature near Japan is below freezing. Near Mindanao in the Philippines, the weather is tropical. Typhoons and other cyclonic storms occur, typically in September, and move into the Sea from the direction of the Caroline or the Marshall islands.

The Philippine islands are a mixture of volcanic and coral material. Many are too small to have formal names. The chain of islands starts south of Taiwan. Of the larger islands, Luzon is the northernmost. As the chain arcs to the south, it breaks into two. The islands of Mindoro and Palawan are in the western portion, while Leyte, Cebu, and Panay are in the east. Mindanao is the southernmost large island. The major fault line in the region runs between the two arcs and through the island of Mindanao.

The edge of the Philippine plate is the Philippine or Mindanao Trench, which lies east of the islands at this point. Depths greater than 10,400 m (34,500 feet) have been recorded in this trench.

When Magellan first sailed into the Philippine Sea in 1521, he found a population that consisted of people of Malay, Melanesian, Chinese, and Arab ancestry. The islands were named by Magellan for St. Lazarus. After they were formally occupied by Spain in 1564, the name of the islands was changed to the Philippines after King Philip II of Spain. The Spanish occupation continued until the war with the United States in 1898. During World War II the Philippines were occupied by Japan. The United States recovered the islands after several major battles in 1944. The Philippines became an independent nation in 1946. See *currents, island, Pacific Ocean, Pacific Plate.*

Phoenicians   The inhabitants of ancient Phoenicia (today's Syria and Lebanon); the biblical Canaanites. They first appeared in the 11th century B.C. along the length of the eastern Mediterranean coast. While not the first to sail into open water—Cycladic Greeks may have reached Spain by 2500 B.C. and Minoans had done extensive sailing and trading—

the Phoenicians took up where these other Bronze Age cultures faltered. They founded colonies along the North African coast, and were exploiting Spanish mines by 1000 B.C.

The Phoenicians' major scientific advance was celestial navigation. They were particularly likely to sight on the Pole Star, which the Greeks called the Phoenician Star for that reason. The ship these mariners used was the galley, usually a bireme (with two tiers of oars). Their longship, used as a warship, was long, slender and most often equipped with a ram and with sails. The traders' vessels were shorter, and wider, with rounded bows and sterns and no ram.

Long after their hegemony in the field had ended, Phoenicians were still being employed by others as experts in long-distance sailing and navigation.

Phoronida   A phylum of small worms, less than 1 cm (0.4 inches) long. They live in membranous tubes of their own making, on the bottom of shallow waters. Some species are borers. They can penetrate calcareous rock, coral, or bivalve shells by literally "dissolving" their way in, using an acid secretion that breaks down calcium carbonate.

Phoronids protrude a complex, retractible lophophore, their food gathering organ, which resembles that of the bryozoans and the entoprocts. The life cycle of the phoronids is complex. Their larvae are planktonic and metamorphose into the sessile stationary adults. See *Bryozoa, Entroprocta, lophophore, sessile.*

phosphorus   A nonmetallic chemical element that occurs widely in nature, especially as phosphates. It is present in seawater in the form of inorganic ions, as soluble deposits of complex salts, and as phosphorite on continental shelves, notably off the California coast.

Phosphorus is an essential element; phytoplankton remove it constantly from the water and use it to produce phosphoproteins, deoxyribonucleic acid (DNA), and phospholipids. The activity of the

PHOSPHORUS CYCLE

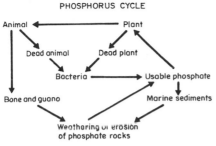

Phosphorus cycle

phytoplankton is, in fact, so great that it totally depletes the uppermost levels of seawater of all available phosphorus.

The concentration of phosphorus in the Atlantic Ocean is about 2 mg/m^3 at a depth of about 1000 m (3,300 feet). The concentration of phosphorus in the Pacific is about 2.2 mg/m^3 with that of the Indian Ocean falling between these two figures.

See *DNA, minerals, phosphorus cycle.*

phosphorus cycle The circulation of the element phosphorus in the biosphere. Through the erosion of phosphate rocks, or the use of phosphate fertilizer, phosphorus enters the soil and is taken up by plants, which are eaten by animals. The dead animals or plants are then decomposed by bacteria, and the released phosphorus, in the form of dissolved phosphate salts, re-enters the soil or is carried into the sea by runoff water. New phosphate rock is eventually formed from marine sediments.

See *phosphorus.*

photography, underwater The operation of cameras below the surface of the water. The use of cameras underwater requires modifications in equipment and lighting: the cameras used must be watertight or in housings that are, and the cameras or their containers must be designed to withstand water pressure. The lighting for underwater photography is supplied by tungsten or electronic flash equipment, which must also be protected.

The refractive index of water (the angle through which it bends incident light) is greater than that of air, and as a result the camera lens manipulates light differently when under water. The angle viewed is actually smaller than that in air, and distances appear shorter than they really are. Direct contact lenses can compensate for this difference, but they do not work as well under water as they would in air.

Water clarity is also a factor in gauging the availability of light for underwater photography. Since water absorbs the red, orange, and yellow end of the visible spectrum, a flash must be used that overcompensates for these elements of the spectrum; or the photographs that are taken will lose the reds and yellows and appear as washed-out greens and blues.

See *Marine archaeology.*

photosynthesis The process by which plants and some single-celled flagellated organisms convert inorganic carbon dioxide (CO_2), water (H_2O), nitrite ion (NO_3^-), and phosphate ion (PO_4^3-) into sugars and amino acids, using the energy in sunlight. This occurs wherever chlorophyll-bearing organisms can exist. Light is a key ingredient in the chemical reaction which can be represented by the following greatly simplified equation:

$$\times CO_2 + \times H_2O + Light\ (Energy) \rightarrow (C_nH_{2n}O_n)_x + \times O_2$$

where n is some small number and x is some other number, usually large. Photosynthesis can occur underwater to depths of about 100 m (330 feet), the photic zone, depending on the transparency of the water.

The extent of photosynthetic activity in any area determines the extent to which that particular area supports populations of organisms. The type of biome depends on the basic food supply manufactured by plants and plantlike protists. Thus, there can be a complex and well-populated food web in a seemingly inhospitable area. For example, phytoplankton blooms occur at the edge of ice shelves in both the North and South polar regions of the Earth. Upwellings of nutrients are an important factor in arctic water, where the surface water is nutrient-poor, but less so in the Antarctic, where it is not. Nevertheless, the phytoplankton bloom in antarctic waters extends 400 km (250 miles) into the Ross Sea. This bloom constitutes a major food source for krill, the tiny organisms which, in turn, are the food of baleen whales. Algae in melting ice may act as a starter population for this food cycle.

There have been many estimates and claims for the "fertility of the sea" and its capacity to be farmed as intensively as the land. Several factors argue against this theory. Phytoplankton do not normally feed terrestrial animals, including humans. They are eaten by zooplankton or die and fall through the sea to feed benthic organisms. The energy-storing compounds in marine organisms are oils rather than starches. This is exhibited throughout the marine food web. Only in shallow water and in warm weather is the photosynthetic capacity of the sea comparable to that on land on a square-meter to square-meter basis. The reason for this is that land plants are simply more efficient photosynthesizers. Ironically, the maximal photosynthetic activity in water occurs in those shallow, warm waters that are most likely to be contaminated by human economic activity.

See *algae, chlorophyll, food chains, mariculture.*

phycoerythrin   A distinctive red pigment of the Rhodophyta, or red algae.
See *algae.*

phytoplankton   See *plankton.*

Piccard   A family of Swiss scientists known for their underwater and stratospheric research and exploration. Auguste Piccard, (1884–1962), a physicist, designed a low-pressure balloon for stratospheric investigation. Jean Félix (1884–1963), his twin brother, a chemist who later became a naturalized American citizen, varied the construction of the balloon and both brothers made high altitude balloon ascents in 1931, 1932 and 1936 in order to study cosmic rays in the absence of atmospheric interference. Auguste also designed the bathyscaphe, a submersible vessel for deep-sea exploration. The vessel was built after World War II and made its first dive in 1958 off the coast of Senegal, West Africa. The support ship's captain was Jacques Cousteau. Other vessels based on the design of the bathyscaphe were built by Jacques Piccard, Auguste's son, who also designed a submarine, called the *Ben Franklin,* that utilized subsurface currents as its motive power.

See *bathyscaphe; Cousteau, Jacques-Yves; oceanic submersibles.*

pilot fish   *Naucrates ductor,* a pelagic, spiny-finned fish found in tropical to warm temperate waters. Pilot fish are best known for accompanying sharks, but will also follow other large fish or ships, to feed on the leftovers they discard. Pilot fish are distinctively banded, with vertical dark stripes on a blue-white body. The usual pilot fish is about 60 cm (2 feet) long and has a slender body.

See *commensal relationships, pelagic environment, shark.*

pilot whale   A small whale of the genus *Globicephala,* also known as the blackfish, that ranges in size from 4 to 6 m (13 to 20 feet). It lives in large groups in all but polar waters, and feeds mainly on squid. It is the most dolphin-like whale, and belongs to the family Delphinidae. These whales are trainable when captured, as are dolphins.

See *whale.*

pinniped   The name used to describe any of a group of fin-footed carnivorous aquatic mammals including the seals, sea lions, and walruses. The name means fin-footed. All pinnipeds are ungraceful on land and remarkably sleek and agile in the water. The differences between seals and sea lions are relatively minor ones. The former (Phocids) have no external ears and use only their forelimbs,

Pinnipeds

Common Name	Scientific Name	Habitat
	Sea Lions	
Stellar	*Eumetropias jubatus*	Alaska and the Pribilof Islands
California	*Zalophus californianus*	Western North America and the Galapagos Islands
South American	*Otaria flavescens*	Western South America
Australian	*Neophocus cinerea*	
Hooker's	*Phocartos hookeri*	New Zealand
	Fur Seals	
Alaska	*Callorhinus ursinus*	Northern Pacific
Juan Fernandez	*Arctocephalus philippi*	
Guadaloupe	*Arctocephalus townsendi*	
Galapagos	*Arctocephalus galapagoensis*	
Australian	*Arctocephalus pusillus doriferus*	
Antarctic	*Arctocephalus gazella*	
Subantarctic	*Arctocephalus tropicalis*	
	Phocids (True Seals)	
Baikal	*Phoca sibirica*	Lake Baikal
Ribbon	*Phoca fascinata* (white ribbon on black)	Siberian coast
Caspian	*Phoca caspica*	
Ringed	*Phoca hispida*	Arctic
Larga	*Phoca largha*	Northern Pacific
Hooded (also called bladder nose)	*Cystophora crustata*	
Bearded (large moustache)	*Erignathus barbatus*	
Harbor	*Phoca vitullina*	
Harp	*Phoca groenlandica*	
Gray or horsehead	*Nullchoerus grypus*	
Northern Elephant	*Mirounga angustirostris*	Western North America
Hawaiian Monk	*Monachus schauinslandi*	Protected
Mediterranean Monk	*Monachus monachus*	
Caribbean Monk	*Monachus tropicalis*	Probably extinct

Common Name	Pinnipeds Scientific Name	Habitat
Southern Elephant (largest phocid)	Mirounga leonina	South America
Weddell	Leptonychotes weddelli	Antarctic
Ross	Ommatophoca rossi	Antarctic
Leopard	Hydrunga leptonyx	Antarctic
Crabeater	Lobodon carcinophagus	Antarctic
	Walrus	
Walrus	Odobenus rosmarus	Arctic

See crabeater seal, harbor seal, harp seal, seal, walrus.

while sea lions (the otarids) hobble around on all four limbs. The fur seals are an intermediate group between the seals and sea lions: they are eared and move on four limbs, but have thick underfur. The walrus (an odobenid) has no external ear and moves ponderously on all fours.

plaice   Pleuronectes platessa, the European flounder. An important, commercially taken flatfish. It lives in North Atlantic waters, blending well with the bottom sands it lives on. The upper, exposed side of the plaice carries red, yellow, and orange speckles on a brown background. The plaice is a large fish, its maximum size being about 85 to 90 cm (about 3 feet).

The relatives of the European plaice are the North American right-eyed flatfish, the American plaice or sand dabs. They are also Pleuronectidae, but differ slightly in appearance from their European counterparts.

See fish, flatfish, flounder.

plankton   Tiny animals and plants that drift in the sea. The name comes from the same Greek root as "planet" and both terms denote wanderers. The term was first applied to organisms in the sea by Viktor Hensen, the director of the German expedition of 1889, known as the Plankton Expedition, whose specific charge was to find and systematize the organisms of the sea.

Plankton comprises a vast group of diverse aquatic organisms; some are freshwater, others marine. They are further separated into photosynthesizing organisms or phytoplankton, and non-photosynthesizing organisms, or zooplankton. This division is not made on a plant versus animal basis, since some chlorophyll-containing organisms are also flagellated and ingest other organisms, both of which are animal-like traits. The euglenoids are prime examples of protists that contain chlorophyll, and are basically phytoplankton. If, however, the light intensity in the water drops, photosynthesis

Name	Size in micrometers $(10^{-6}$ m = one millionth of a meter)
Ultraplankton	Less than 5
Nanoplankton	5–50
Microplankton	50–500
Mesoplankton	500
Macroplankton	5,000–50,000 (or 5 cm $\cong$ 2 inches)
Megaloplankton	larger than 50,000

declines, and the euglenoids must then hunt for their food rather than manufacturing it.

The plankton are also graded by size, ranging from the smallest, which are invisible without magnification, to those visible to the naked eye.

Some plankton are unicellular, others are colonial, and still others are multicellular organisms. Plankton are further categorized as those species that spend their entire lives floating with the motion of the water and those that are active rather than passive. The permanent floaters (i.e., those organisms that remain planktonic for their entire lives) are called holoplankton. Those that are planktonic only as larvae are called meroplankton. The meroplankton, or temporary plankton, may turn into adults that are benthic (fixed, or bottom-dwelling organisms) or nektonic (swimming organisms).

See *ecosystem, Euglena, food chains, nekton, Protista.*

plate tectonics   A theory of modern geology based on the movement of huge blocks, or plates, of the Earth's crust. Although the motion of continents was postulated many times, in this century, this concept was put forth most clearly by Alfred Wegener in 1915. He "fit to-

gether" the known landmasses using the best geographic information of the time, and called the primordial continent that had given rise to them Pangaea (all-Earth). However, Wegener could not explain the subsequent breakup of the continental mass of Pangaea, and simply concluded that the continents had drifted apart, without proposing a mechanism to explain their drift. For many years his theory remained an idea that awaited corroborating evidence. By the 1960s, there had been gathered a large body of supporting evidence for it. Fossils found in Antarctica proved to be related to Australian ones, and fossils found in both Antarctica and Australia were related to South American specimens. Magnetic anomalies in rocks of the ocean floor confirmed that there was motion of the Earth's crust, and indicated that Pangaea had indeed once existed as an equatorial landmass. The magnetic anomalies discovered by Vine and Matthews (of Cambridge University in England), and the nature of seafloor spreading explained by Robert Dietz, established Wegener's continental drift theory, now called plate tectonics.

The outer shell of the Earth is the crust and upper mantle. This is the sialic layer consisting of light rock rich in silica and

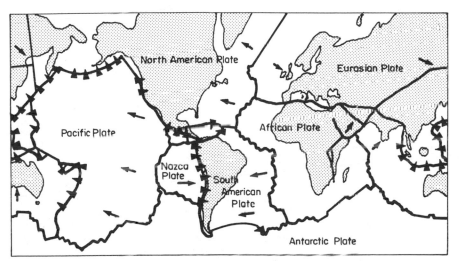

Major plates of the earth's crust and the directions in which they are moving

alumina, that floats on the denser sima, which consists of silicon and magnesium-containing rocks. The continents are sialic and therefore less dense than the lower mantle and asthenosphere (a semi-liquid layer). When the crustal plates move, they slide past each other (transform), move apart (diverge), or move toward each other (converge). When the plates converge and meet, one plate overrides the other, which then sinks into the asthenosphere. This action is called subduction.

The crustal movements have been slow but dramatic. Examples of visible evidence of these movements are oopholites and terranes. An oopholite is a piece of what was once ocean floor. Finding in the Himalayan Mountains a part of what was once an island chain in the Indian Ocean is, for example, evidence of incredible change in the shape of the Earth's continents and oceans. Terranes are "exotics," or pieces that do not belong to the continents they are surrounded by. A major piece of the Carolina Slate Belt, for instance, was accreted onto the North American continent after the Cambrian Period, when the primeval Atlantic Ocean opened and separated England from North America. Similar foreign inclusions on the west coast of North America are also known.

The single landmass of Pangaea, originally postulated by Wegener, existed during the Permian and into the Triassic periods from 280 million to 225 million years ago. By the end of the Triassic and in the late Jurassic, about 180 million years ago, it began to break up along what is now the Mid-Atlantic Ridge, a diverging zone where new material is introduced into the Earth's crust at the edges of the separating plates. According to one current hypothesis, this intruding material pushed the plates further apart. By the Cretaceous Period, 135 million years ago, the land masses were well separated. Since then the continents framing the Atlantic have continued to move away from each other fairly steadily.

The Australian, Indian, and Antarctic plates have not had unidirectional motion. The paths and movements of parts of the Indian Ocean, which includes Madagascar, and of the Caribbean and Mediterranean seas are not as well-defined as those of the Atlantic. See basalt; continental drift; crust; evolution of oceans; Ewing, William Maurice; Heezen, Bruce Charles; Hess, Harry H.; Wegener, Alfred.

plateau   A relatively level structure of the ocean floor, higher than the adjacent bottom, along a continent but beyond the continental shelf, or an extended flat portion of a mid-ocean rise. These marginal structures are extensions of the shelves they adjoin.

The Blake Plateau in the North Atlantic is probably a section of continent that has been subsiding since the Cretaceous Period. It is heavily overlain with limestone deposits and coral. The Bahama Banks are considered to be a marginal plateau, but exhibit the structure of a barrier reef and consist of coral. The underlying rock of Florida and that of the Bahamas islands is different, and not part of the plateau. The Campbell Plateau east of New Zealand is a large, shallow area with a continental bottom. It and New Zealand are both of continental origin. The Falkland Ridge, which geographically belongs to the same Devonian pattern found on the Falkland Islands, is a fairly sharp drop. The plateau to the east of Brazil may be a sinking portion of continental shelf, and therefore quite different from the structure of the Blake Plateau.

Some mid-ocean plateaus, such as the Coral Sea Plateau and the Bellona Plateau (about 1000 m long) are aseismic and coralline. The Melanesian Plateau has some volcanoes. The Flores Sea Plateau is coral in its base. The Mascarene Plateau in the Indian Ocean reaches from the Seychelles islands to Nazareth Bank and Mauritius. It is partly volcanic, and some of its oldest parts are Precambrian. It is postulated that this plateau once formed a part of the isthmus between

India, Madagascar, and Africa. See *Blake Plateau, coral, Cretaceous Period, Devonian Period, subsidence.*

Platyhelminthes   A phylum consisting mainly of parasitic flatworms. The freeliving members (Turbellaria) are of worldwide distribution and include fresh- and salt water types as well as terrestrial organisms. Those that are parasitic, flukes *(Trematoda)* and tapeworms *(cestoda),* may have marine hosts. The tapeworms in particular may have fish hosts, but are also known to infest humans. See *Turbellaria.*

Pleistocene Epoch   The first epoch of the Quaternary Period of the Cenozoic Era, usually referred to as the Age of Ice. It was when much of the globe was covered by ice sheets extending from the poles toward the equator. Modern land animals such as the horse, the elephant, and the cattle family made their appearance during the Pleistocene. The beginning of the Pleistocene is variously given as 1 to 4 million years ago, and it is agreed that it ended about ten thousand years ago. See *Ice Age, Quaternary Period.*

plesiosaurs   Extinct Mesozoic marine reptiles that lived about 190 to 65 million years ago. The plesiosaurs evolved into two groups: One had small heads and very long necks, the other was short-necked and had an enormous head. Both were worldwide in range. The former, the elasmosaurs, half of whose body length was neck and head, preyed on schooling fish. The latter, the pliosaurs, fed on large fish and cephalopods.

Many plesiosaurs—and particularly the elasmosaurs—were gigantic; some specimens are known to have been 13 to 15 m (43 to 50 feet) long. They were excellent swimmers whose limbs were paddle-shaped and allowed precise maneuverability. These animals literally "flew" through the water as penguins do now. See *reptile.*

Pliny the Elder (23–79 A.D.)   A Roman natural historian whose full name was Gaius Plinius Secundus. He came from an aristocratic Roman family, traveled extensively in the service of the Empire, and developed a flair for the collection of information. He specialized in art and natural history, and his two works, which are still entertaining reading, were a compendium of fine art and a book of natural history.

Unlike Aristotle, who flagged those entries in his compendium of natural history he considered suspect, Pliny recorded everything.

Pliny's interest in natural history was the cause of his death. While in the service of the Emperor Titus, defending the Bay of Naples from pirates, he went ashore to inspect an interesting cloud. It turned out to be the eruption of Vesuvius, and Pliny died from the noxious fumes produced by the eruption.

Pliocene Epoch   The last epoch of the Tertiary Period immediately preceding the Quaternary. The Pliocene lasted from about 10 million to about 4 million years ago. The British geologist Charles Lyell divided the epoch into the Older and Newer Pliocene. In the former, less than half of the remains of species from the former half resemble the species now on Earth, while more than half of the remains from the latter are like present forms of animal life. The newer Pliocene is now called the Pleistocene.

The major geologic change in the Pliocene was the retreat of the Tethys Sea, which occurred as the North Sea continued to creep north and the temperature dropped throughout the world.

The Pliocene fauna, particularly marine organisms, resemble the species of today. Echinoderms are common in fossil remains, as are other bivalves and gastropods. The colder temperature displaced corals and foraminiferans. On land, large mammals were dominant and the land bridges between America and Asia and between North and South America allowed the migration of these animals. See *Cenozoic Era, Pleistocene Epoch.*

Pogonophora A phylum of tube worms established in the 1950s. More than 100 species have now been identified, and there may be others yet unknown. These animals, which have no mouth, are sometimes called beardworms because of the featherlike tentacles with which they absorb food.

The chitinous tubes that the pogonophores inhabit are up to 0.5 m (20 inches) long, and contain the animal throughout its life. The sexes are separate, although the mode of reproduction is unknown, and the larval forms can swim. The blood of the tube worms contains hemoglobin. If tubeworms do have sense organs, they are well disguised.

Pogonophores live at great depths. They are usually encountered below the 1,000-m (3,300-foot) mark, and some are found below 7,000 m (23,100 feet). Fossil types have been identified, in both Cambrian and Ordovician deposits in Siberia and in other continental regions.

Some species of tube worms have been found in hydrothermal vent areas of the ocean floor, where hot sulfurous water bubbles out of a mid-ocean ridge. The worms that live near these hot vents concentrate sulfur for their bacterial symbionts, and are not poisoned by it. See *vent communities.*

poikilotherm An animal whose temperature varies with that of the environment. The blood of poikilotherms is often slightly above the ambient temperature, and these animals are therefore sometimes incorrectly called cold-blooded. Most animals other than birds and mammals are poikilothermic. See *homoiotherms, temperature of animals.*

polar biome The community of plants and animals found in the North and South polar regions of the Earth. The arctic supports more varieties of organisms, all ultimately dependent on a phytoplankton and zooplankton food-chain, than does the antarctic. Numerous predatory transients also use the territory for breeding colonies. See *bird, gull, penguin, pinniped, seal, tern, whale.*

polar front An atmospheric phenomenon that occurs at the latitudes near both poles, where polar cold water sinks below the surface water to form the intermediate layer of the world's oceans. Since there is a difference in the temperature of the air and the water in these areas, weather disturbances are frequently generated. The region, particularly the Arctic Convergence, is characterized by an almost constant cloudbank. See *Antarctic Convergence.*

Polaris The North Star, Pole Star or Phoenician Star (it has many names). It is part of the constellation known as the Little Dipper, or the Little Bear (these are only two of its names). It is at true north in the skies of the Northern Hemisphere. Because of its position and the fact that it is visible year-round, it has been used as a bearing in navigation. In the Southern Hemisphere the Southern Cross serves as the basis for celestial navigation. See *navigation.*

pollock A North Atlantic fish of the cod family *(Gadidae).* Pollock are carnivorous fish that are green or green-black on their upper surfaces and creamy white on the underside. The average adult individual is about 1 m (40 inches) long, and weighs up to 15 kg (33 pounds). Pollock are taken by both sport and commercial fishermen. Their meat is frequenty sold as cod. See *cod.*

pollution Contamination of the environment (air, water or earth) by waste material or any other product as a result of human activity. Water-polluting wastes can be industrial, resulting from the construction or destruction of buildings, equipment or materials, refinery or paper mill waste, or sewage; agricultural, from runoff containing fertilizers or pesticides; or urban, from sewage and litter. Airborne pollution is a problem in the sea as it is on land. The airborne pollutants are either small particles of metal, wood pulp, manufacturing waste, or any other material fine enough to be windblown, or are gaseous pollutants, such as sulfur and nitrogen oxides, that result from fossil fuel combustion. These affect seawater by altering its pH or by blocking out sunlight. By creating "acid rain," airborne pollutants also affect vegetation and human health.

Large cities are frequently major pollutors. They are almost always manufacturing centers as well as high-density population centers, and the local manufacturing processes use local waters to remove a variety of chemicals that are either end-products or byproducts of the manufacturing process. Cities also produce large quantities of garbage and sewage, both of which may enter the sea directly if the city is coastal, or indirectly if it is on a large river. Areas of the ocean that receive the effluent of cities are all too often "dead seas."

Besides rendering the shoreline aesthetically unappealing, floating garbage and chemical residues, including oil spills, prevent sunlight from reaching the phytoplankton in the water, and interfere with gas exchange between water and the atmosphere. Both factors decrease the available oxygen required by living organisms in the water. The chemical residues draw on the oxygen supply, since oxidation and eventual degradation of these chemical pollutants involves their reaction with oxygen. The decreased availability of atmospheric oxygen that is caused by floating pollutants means that natural cleansing of the water takes much longer.

Eutrophication is the result of an influx of phosphate or nitrate ions from either an agricultural area or an area of high population density. Phosphate and nitrate ions stimulate algal growth. For a while the algae increase the available oxygen supply, but as they grow they also cover the surface of the water, which means that other plants, down in the photic zone, begin to die. The death of some plants and the increase in the population of decomposing bacteria consumes more of the available oxygen, which means that there is less available for animals (e.g., fish), which then begin to die. Without intervention and aeration, the eutrophied area will become a foul-smelling and lifeless swamp.

Where coastal areas can be flushed by the action of waves, the effects of pollution are lessened but not removed. Better disposal methods and more effective wastewater treatment are essential, but require committment by governments and the investment of considerable funds. It is only recently that concerted efforts have been made to limit and clean up pollution.

In 1972 the first United Nations Conference on the Human Environment studied the effect of pollution, particularly industrial pollution. It was the Conference's conclusion that the effect of industrial wastes on water quality, and the resulting environmental impact, was most pronounced in enclosed marine areas, even very large enclosed areas such as the Mediterranean, Baltic, North Sea, and Caribbean. The mid-ocean effects of pollution were felt to be the least damaging.

In a 1982 follow-up report by the U.N. Conference on the Human Environment, it was stated that the open ocean was cleaner than it had been ten years earlier. The 1972 Conference in Stockholm had been very concerned with the concentration of chlorinated insecticides in all of the Earth's ocean waters. By 1982, as a result of the ban on the use of DDT and the decreased use of chlorinated insecticides in general, there were less of these compounds in the water. However, other pollutants remain a problem.

With increased awareness and investment in environmental clean-up, the water quality of the Northern Hemisphere is improving, even in those areas that are known problem spots, such as the Mediterranean. But as industrial development moves to the Southern Hemisphere, so will pollution. Heavy-metal dumping, acid rain, and nuclear wastes all pose problems of oceanic pollution. Since large concentrations of people live near the Earth's oceans, with an estimated 70%

of the world's population living less than 80 km (50 miles) from a seacoast, and humans also depend on food that is taken from the sea, international cooperation and study are required for continuing pollution control and environmental clean-up. See *estuary, Mediterranean.*

Polychaeta A class of the phylum Annelida, comprising the segmented worms. These animals range in size from less than 1 cm (0.3 inches) to longer than 3 m (10 feet). Among the polychaetes are animals with many varied life styles. There are free-swimming larvae that grow up to be free-swimming adults, other animals that can move but usually don't, and still others that don't move at all. The outstanding characteristic of the polychaetes is the presence of parapodia, or projections on the body, that bear setae (bristles). There is also distinct cephalization—the development of a head equipped with sensory organs.

Like other annelids, the polychaetes have pigmented blood. It is often red, although some have green blood. These animals have closed circulatory systems and live lives that may exceed ten years

in length. The different species are sometimes brilliantly colored. Although polychaetes are usually dioecious, with separate sexes, some species are hermaphrodites. See *Annelida, Nereidae, palolo.*

polymer A large molecule composed of repeating units of smaller molecules called monomers. Some polymers are built of repeating units of the same monomer, (e.g., repeating units of glucose form the polymers cellulose, amylose, and starch).

Proteins, polypeptides, and nucleic acids are also polymers, but are mixed polymers composed of more than one monomer. Amino acids are the constituent monomers of proteins. There are about twenty amino acids commonly found in naturally occurring proteins. The specific proteins are manufactured by cells according to the code contained in the DNA of their nuclei. See *DNA, RNA, sugar.*

Polynesia A collective name for the islands of the central and southern Pacific Ocean. The largest islands in this group

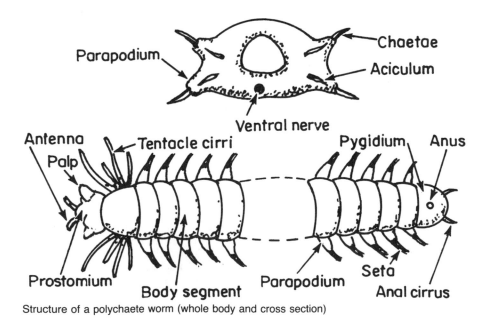

Structure of a polychaete worm (whole body and cross section)

are New Zealand and the Hawaiian Islands. Others are the Society Islands, (Tahiti), Fiji Islands, Samoa, Tonga, Easter Island (Rapa Nui), and the Tuamotus, Marquesas, and Line Islands (Christmas Island, Palmyra). See *atoll, island, Pacific Ocean.*

polyp A column-shaped cnidarian or part of a colony of cnidarians. The corals are typical examples. See *cnidarian.*

Ponce de Leon, Juan (1474–1521) A Spanish explorer who searched for the Fountain of Youth in the New World, de Leon was one of the gentlemen volunteers who accompanied Columbus on the discoverer's second voyage in 1493. He was attached to the military governor on Hispaniola and then sailed for Puerto Rico, giving it its name and conquering all of it by 1509. He was its governor until he was replaced by Diego Colon (the son of Columbus) in 1512. Ponce de Leon was then given an expedition to discover new lands, and sailed to the northwest in March 1513. Ponce de Leon was looking both for treasure and the "fountain of youth." This Eurasian myth had been transplanted to the New World; however, he seems to have been the only explorer to make serious attempts to find it.

The expedition moved north of Cape Canaveral before turning back south along the Florida coast and round it, finding the Florida Keys and naming them the Martyrs. Rounding the tip of the Keys he sailed north into the Gulf of Mexico and north on the Gulf Coast to roughly the latitude of Fort Myers, Florida. Possibly still looking for the "fountain of youth" he went south again and sighted the Yucatan Peninsula of Mexico. Ponce de Leon returned to Puerto Rico in October, and sailed for Spain in the next year (1514).

A second voyage to Florida in 1521 was to have planted a colony on Sanibel Island. Ponce de Leon was wounded in a skirmish with the local Indians, and died in Havana.

population A group of individual organisms belonging to a single species. On the whole, the population of a species tends to remain stable over long periods of time. The two most common reproductive strategies observed among stable populations are one resulting in small numbers of young that are protected and cared for by adults while they are juveniles, and another that produces large numbers of offspring that are left unprotected. The latter strategy is typified by the million or so eggs produced by the average female oyster. While the oyster's strategy does tend to be less stable, it does not lead to catastrophic declines in the number of individuals, which would point the species toward extinction.

Occasionally everything goes right for a particular population and one cohort, or age group, is very successful. The classic example of this phenomenon involved the 1904 herring spawn. The herring spawned in 1904 dominated the population of all herring catches for over fifteen years.

It is not only the availability of food or the lack of predation that will affect the make-up of a particular population. Temperature, weather conditions, storms, and currents will alter the biome and make one year better or worse than another. Some species respond in ways that are upon first appearance may seem rather peculiar. One of these responses is a lack of breeding behavior. In others, sex changes with variation in the environment are a possible and viable option. The driving evolutionary explanation for a sex change is that the two sexes have different metabolic demands. It is metabolically "cheaper" to maintain a population with a large percentage of male individuals.

In many animal populations, notably birds, seals, and other predators, not all adults breed. Furthermore, since reproduction is an energy-expensive function, adolescence may extend for years.

In fish populations, the high mortality among the young is due to both disease and predation. In any stable fish population, (i.e., one not perturbed by human fishing) adults tend to die as they age. The age of fish is determined by the annual rings on their scales. Some fish species can live to a great age, with some individuals having more than thirty rings not being rare.

In years of population "bloom," the rapidly increased population is one that has not had an accompanying increase in numbers of predators. While there is a large number of new individuals in the population, there is also greater competition for the same food supply and reproductive space. The initial effect is an increase in the die-off of young. This is followed by an increase in the predator population, and the end result is a much lower population in the next breeding cycle. Thus, a stable population is not one with no change, but one where there is both fluctuation and migration.

Populations vary in their behavior as well as in numbers. Some, such as those of whales and porpoises, are cooperative societies. In others there is active competition, as in the communities of elephant seals. Still others have parallel lives, such as the thousands of frigate birds that nest in the same place, all doing the same thing at the same time while seemingly oblivious of each other and without helping each other much. Penguins, on the other hand, share their babysitting chores with their neighbors. See *bird, fish, food chain, food web, migration, predator-prey relationship.*

porgy   A shallow-water food fish that is well distributed in warm and temperate waters in America and Europe. There are about 100 species of these relatives of the snappers. Their average size is about 30 cm (1 foot), but some tropical specimens may be more than twice that size. The European porgy is called a bream. See *fish.*

Porifera   A large phylum of multicellular, sessile (non-moving), primitive marine animals, commonly known as the sponges.

Sponges are found in all of the world's seas. There are about 5,000 species, with antarctic species particularly abundant. While most sponges are found in fairly shallow water, some are present at all depths, and a few individuals are hadal. A few species live in brackish or even fresh water.

The Porifera are distinguished by their very simple body plan, in which there is only one large opening, the osculum, and by the many small pores in their body wall. They feed by filtering water and its dissolved and suspended material through the small wall pores, then into and through the cells of their bodies, and expelling it through the osculum. The various genera of sponges are characterized by the type of spicule that is found in the body wall. These are the hard, support structures, and may consist of silicon compounds, calcium compounds, or a proteinaceous material, spongin.

The Porifera are usually tan or brown, although some living in shallow tropical waters are exotic greens, reds, or oranges as a result of their commensal algae.

Sponges range in size from 1 to 2 cm (0.2 to 0.8 inches) to giants more than 1 m (3.3 feet) in diameter. Their age range is considerable. They reproduce by budding, which is an asexual process, and by gamete production, which results in new, sexually generated individuals. The poriferans are quite undifferentiated; new individuals have grown almost from individual single cells produced by homogenizing an adult and straining the homogenate. An old method of reseeding sponge beds was to cut up sponges and strew the pieces around the ocean bottom.

The Porifera are a very ancient phylum that dates certainly from the early Cambrian Period. They do not fossilize well and have therefore left a very patchy record. See *chemical defenses, spicule, sponge.*

porpoise   A small, gregarious whale of the order Cetacea, family Phocaenidae. Dolphins are frequently called porpoises, but the true porpoise is smaller than a dolphin. It is generally about 2 m (7 feet) long, has a rounded nose rather than a pointed one, and a dorsal fin.

The common porpoises are the:

Harbor or common porpoise *(Phocaena phocaena)*, a cold water animal of the Northern Hemisphere.
Dall *(P. dalli).* the largest porpoise, which is found most frequently in the North Pacific.
Finless porpoise *(Neomeris phoca-enoidis)*, the individual found most often in South Asia and East Africa.
Spring finned porpoise *(P. spinipinnus)*, found near South America.
Spectacled porpoises *(P. dioptrica)*, which are antarctic.
California Gulf porpoise *(P. sinus)*, some-times placed in a separate category and sometimes lumped together with the spring fins.

The common porpoise plays before ships, as do dolphins. It has been known and trusted by man for centuries. The explosive sound porpoises make when emerging from the water is a well-known characteristic. Although they will eat fish and crustaceans, the standard meal of the porpoise is squid, which they prefer when they can find it. See *Cetacea, dolphin, whale.*

port (1) That side of a vessel that is to the left of an observer standing on the vessel and facing the bow or front end. (2) A harbor. (3) An opening in the side of a ship for loading or unloading cargo, for ventilation, or for the display and use of gunnery. See *starboard.*

Portuguese man of war A delicate, transparent, floating hydrozoan of the ge-nus *Physalia*, phylum *Cnidaria*, found in warm waters in and near the Caribbean Sea and in the Pacific and Indian Oceans. The Portuguese man of war is known for its gas-filled, floating body, which ranges from 10 to 30 cm (3 to 12 inches) long and stands about 15 cm (6 inches) above the water. The float acts as a sail, and is most often pale blue, but may also be pink or lavender.

The Portuguese man of war lives and travels in spectacular colonies, which trail a mass of polyps which can be several meters long, with some over 50 m (160 feet) long. The organism is differentiated into structures that capture prey, others that eat it, and still others that reproduce. The tentacles bear nematocysts that give a powerful, painful sting and can cause fatal shock in a human victim. The only known predator of the Portuguese man of war is the sea turtle. See *Cnidaria, toxin.*

posterior The rear or tail end of an animal. The word also denotes position. Thus, the tail of a fish is posterior to its pelvic fins. See *anterior, dorsal, ventral.*

potassium A chemical element of the alkali metal group occurring abundantly in nature. Potassium is an element nec-essary to animal and plant life. It is pres-ent in seawater as potassium ion, K^+, in roughly the same amounts as is cal-cium—about 0.04 grams per liter. The potassium argon dating of sediments is a fairly standard technique based on the degradation of radioactive potassium to inert argon in a process that takes thou-sands of years. See *dating, element, ra-dioactivity.*

prawn A large shrimp (genera *Pan-dalus* and *Peneus*), found in fresh, brack-ish, or salt water. Shrimp that are about 8 cm (3 inches) or larger are called prawns. The tropical freshwater species are often much larger, and individuals of 18 to 20 cm (7 to 8 inches) are not uncommon. See *Decapoda, shrimp.*

precession Rotation of the plane of orbit of the Earth (or the moon). This has an immediate effect on the Earth, causing variation in the tidal range. See *lunar tide, tidal range, tide.*

predator-prey relationship The re-lationship between a hunting animal and its animal food. The food web consists of a large number of plant organisms, which serve as food for a somewhat smaller number of small animals that in turn supply the food for a still smaller number of larger, carnivorous animals. There is frequently more than one predator for each small, herbivorous (plant-eating) animal. Furthermore, some very large animals are herbivores and some very small ones are carnivores. The very largest carnivores may even harbor plant populations on their bodies, such as the polar bear, which harbors algae in the hollow-stemmed hairs of its outer fur. Thus, the eater-to-eaten relationships are quite complex.

The relationship of a predator population to that of the prey shows that the predator population peaks after that of the prey. This in turn is echoed by the effect on the plant population fed on by the herbivores that form the bottom layer of the food web (or pyramid).

While there are organisms that do die of old age, those that are part of the regular food supply for others have nervous lives. Most prey organisms, however, do have some defenses, such as spines (with or without poisoned tips), a bad taste, mimicry or the ability to imitate other organisms so as to hide or appear potentially threatening, powerful partners (symbionts or commensals, such as in the case of the remora fish and shark), or chemical defenses. One example of the last type of protection is the toxic substance secreted by the Pacific sole, which repels other fish, notably sharks. See *carnivore, chemical defenses, commensal relationships, food chain, food web, herbivore, population, symbiosis.*

pressure   A force exerted on a surface. It is expressed in terms of force per unit area. The atmospheric pressure at sea level is that of a column of air exerting a force on a defined area. It is 101.325 kPa (kilopascals), where a pascal is defined as 1 newton per square meter, and a newton is defined as 1 kilogram-meter per second squared (kg.m/second^2). The

atmospheric pressure is sometimes given as 14.7 pounds/inch^2, or 1 torricelli, or 760 millimeters of mercury (mm Hg).

As an organism descends in water, it has to deal with both the air pressure on the water and the pressure exerted downward on itself by the column of water through which it has already descended. At a depth of 10 m (32.8 feet), the pressure is an additional atmosphere (101.325 kPa). At 20 m (65 feet) it is 2 atm more than at sea level. The pressure increases by 1 atm per 10 m.

Most surface- or near-surface-dwelling marine organisms deal with the pressure they encounter by using a swim bladder. Some hadal species also have swim bladders. Other organisms deal with the need to equilibrate their internal pressure and the external pressure on them by absorbing water into their tissues, increasing their internal pressure. This renders the organism almost transparent. Other mechanisms include the storage of low-salt fat, or oil, or air. See *decompression, diving, ocean-atmosphere relation.*

Priapulida   Roundworms, a class of the phylum Aschelminthes. Since the classification of this phylum is under revision, these worms may be reclassified as a phylum of their own. These slow-moving, sausage-shaped creatures are bottom dwellers. Their bulbous food-gathering structure, the proboscis, is covered with hooks and is used as a burrowing implement. The feathery gill structures form a "tail."

Priapulids range in size up to about 8 cm (3 inches) long. They are detritus feeders, but can occasionally capture a very slow-moving polychaete or another priapulid. See *Aschelminthes, detritus feeders, Polychaeta, proboscis.*

proboscis   The food-obtaining structure on the "head end" in invertebrates. It may be extendable, and is frequently equipped with suctorial or hooking structures.

The term "proboscis" also refers to

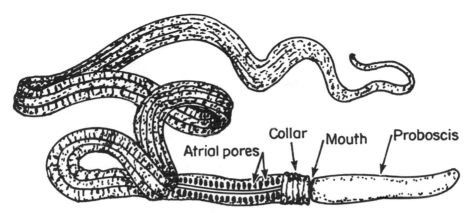

Collar **Mouth** **Proboscis**

Atrial pores

Acorn worm with elongated proboscis

the nose or nasal structures of an organism. Thus, the elephant seals' large nose sacs are called probosces. See *individual worm phyla, seal.*

protein A polymer of amino acids. See *amino acid, peptide formation, polymer.*

Protista A kingdom or group of simple organisms, including the bacteria, the blue-green algae and other algae, the fungi, and the protozoans. In some taxonomic classifications the Protista constitute a kingdom, but in others the prokaryotic bacteria and the blue-green algae are considered to be a separate kingdom, Monera. The protists consists of organisms whose nuclear material is surrounded by a nuclear membrane. This includes the ciliates; protozoa; dinoflagellates; and the red, brown, yellow, and green algae. See *Chrysophyta, Pyrrophyta, Sarcodina.*

protoplasm The substance of the living cell, enclosed by the cell membrane, (i.e., the nucleus and cytoplasm of the cell). Chemically, the protoplasm is an aqueous solution of nutrients, sugars, lipids, starches, oils and proteins. The pro-

toplasm is distinctive for each different kind of cell.

Protozoa A phylum of minute acellular or unicellular organisms living in marine, freshwater, and damp terrestrial environments. There are an estimated 30,000 living species of protozoans. They range from plantlike forms to animal types. The animal members of the Protozoa include ciliates, foraminiferans, radiolarians, amebae, some flagellates, and sporozoans. Some protozoans—the euglenoids—contain chlorophyll, are green, and conduct photosynthesis in sunlight. They become colorless and predatory in the dark. The Protozoa feed on detritus, bacteria, algae, and on each other. See *Protista.*

Pteropoda A group of small gastropod mollusks characterized by winglike organs with which they swim. The sea butterfly, actually a small snail, is a pteropod. These animals live on or near the surface. Their shell is a delicate, transparent one about 0.5 cm (0.02 inches) long. There are significant deposits of such shells in calcareous sediments. The shell, because it is so light and thin, is particularly susceptible to dissolution in cold, carbon dioxide-rich water and there are more pteropod remains in the waters of the North Atlantic and Mediterranean than in the Pacific. See *Gastropoda, sea butterfly.*

Ptolemy, Claudius (90?–168)

An Alexandrian Greek mathematician, astronomer, and cartographer. He also collected earlier geopgraphical works. Ptolemy's best known book was his *Geographical Treatise*—an eight-volume work, and an atlas of the then-known world.

A series of Ptolemy's maps showed curved meridians with only the central one, which he called the "prime meridian," depicted as a straight line. Coincidentally, this Ptolemaic line is reasonably close to the present line of 0° Longitude. Ptolemy drew his prime (or first) meridian as far west as he could, and placed it near the Canary Islands.

Ptolemaic maps were lost in the general decline of science and scientific interest in the West that followed the decline of the Roman Empire. The maps reappeared about 1400 A.D., when they were reintroduced by Arab scientists and Byzantine Greeks. They had become widely known by the time of the great voyages of the 15th and 16th centuries. The error in Ptolemy's maps resulted from his use of Poseidonius' (135–219, Syrian-Roman traveler, natural philosopher and geometer) figure for the circumference of the Earth, rather than Eratosthenes' figure, which was more accurate and larger. Like almost all other ancient cartographers, Ptolemy knew that the world was spherical.

Ptolemy was a much better geographer than an astronomer. His view of the Earth in relation to other extraterrestrial bodies was mathematically correct but inaccurate. He believed that the Earth, not the Sun, was the center of the universe. This geocentric depiction of the Sun and the known stars and planets was of extremely long duration. It became codified and was accepted by astronomers and geographers in many places, and it remained for Copernicus,

Brahe, Kepler, and Newton to disprove it.

The geocentric cosmos can very elegantly explain all observed astronomical phenomena. With it, stars can still be used to fix navigational points; however, the Spanish king Alfonso the Wise is supposed to have said, upon hearing the Ptolemaic exposition of the phenomenon of the eclipse: "If God had asked me how to design the world, I would have made it simpler." The story is probably apocryphal, but the statement points to the much later discovered reality of how the solar system works.

See *chart; Copernicus, Nicolaus; map; navigation.*

Puerto Rico Trench

The deepest point in the North Atlantic, at 8,385 m (31.330 feet), north of the island of Puerto Rico and almost on the 20° North Latitude line. The island is part of the Puerto Rico ridge, and the trench marks the edge of the Caribbean Plate.

The trench is a slip strike zone and shows continuations of the slip strikes seen on the island around it. It has been a feature on the surface of the Earth for more than 70 million years. It is partially filled with terrigenous sediment.

See *Caribbean Sea, slip strike, trench.*

puffer

Also known as the globefish, a spiny-skinned, mostly warm-water marine fish with the ability to puff itself up. Puffers are of the family *Tetradontidae,* a name indicative of their fused teeth. There are about 90 known species.

Puffers have characteristic tough, spiny skins. They are usually about 30 cm (12 inches) long, although some species are much larger. The most obvious trait of the puffer is the ability to inflate to soccer ball proportions.

The sharp-nosed puffer, *Canthigaster,* is smaller and more colorful than the usual gray-brown color of the puffers. Porcupine fish, another group of puffers, are thick-bodied animals that live in shal-

low water, Their spines become prominent when their body is inflated.

The spines of puffer fish are a defense
mechanism; any predator has to be will-
ing to put up with a mouthful of sharp
spines if it wishes to dine on these fish.
Some species, however, have poisonous
spines, and others such as the fugu have
poison-containing organs. When properly cleaned and prepared, however, the
fugu is a delicacy consumed by Japanese
gourmets. Because its organs are very
poisonous, however, any flesh that has
been touched by either the liver or the
reproductive organs is extremely toxic,
and several fatalities occur every year in
Japan from eating contaminated fugu
flesh. See fish, toxin.

puffin   A diving sea bird related to the
auk. It is somberly feathered in white and
black or gray, but dramatically colored
red, yellow, or blue on its legs and beak.
The birds are about 30 cm (1 foot) long,
and nest in large colonies on rocky cliffs
or islands. Both parents raise their single
chick in a burrow. The diet of puffins is
fish. Some representative species are:

purse seine   A type of net used for
trapping open-sea or pelagic fish. These
nets are usually set by two boats, and
can be drawn together like a purse around
a school of fish. The size of the mesh
determines the target population. Very-
fine-mesh purse seines are used in biological sampling. Purse seines for tuna
and other open ocean fish have recently
been redesigned so that dolphins acci-
dentally caught in these nets can escape,
thereby helping to preserve the dolphin
population. See fishing, nets.

pycnocline   A water layer that exhibits
a relatively rapid increase in density with
increasing depth. See Antarctic Bottom
Water.

pycnogonid   An arthropod that lives
on the ocean bottom, also called a sea
spider. The shallow water species are less
than 1 cm (0.4 inches) in diameter, while
hadal species may be larger than 1 m
(40 inches).

Pyrrophyta   A phylum of flagellated
algae. The dinoflagellates are in this group.
See dinoflagellates, flagella.

Common Name	Latin Name	Habitat
Common puffin	Fratercula arctica	North Atlantic
Horned puffin	F. corniculata	North Pacific
Tufted puffin	Lunda cirrhata	Temperate to warm Pacific area

See bird.

Q

Quaternary Period That part of the Cenozoic Era that began about 4 million years ago and extends to the present. It is in turn divided into the Pleistocene epoch and the most recent epoch, the Holocene. The major event of the Quaternary is the appearance and then disappearance of the vast sheet of ice that covered most of the Northern Hemisphere from the North Pole to about 40° North Latitude. The cooling trend that precipitated the glaciation started in the late Tertiary Period. The Quaternary is also marked by the appearance of humans. *See Cenozoic Era, Pleistocene Epoch, Tertiary Period.*

R

radar Radio Direction and Ranging: a device that uses ultra-high-frequency radio waves to measure the distance to an object by determining the time needed for the wave reflected by the object to return to some detection device. A directional antenna is used to determine the direction of the object, or the antenna can scan in continuous circles to locate any object in the path of its beam.

During World War II it was noted that rain and snow produced a particular signal on radar recorders. Using the Doppler effect, in which the frequency of a wave is shifted to a higher frequency if the source of the wave is moving toward the receiver, and to a lower frequency if the source is moving away from the receiver, it is possible to determine the amount of precipitation and air velocity of a storm. It is also possible to pinpoint the location of the precipitation and its intensity. This means that cyclonic storms can be mapped and tracked. Radar can also be used to track clear air turbulences (i.e., storms not associated with precipitation). A variation of radar, sonar, which employs sound waves, is used extensively in exploration, by measuring the time needed for the return of an echo. Given the medium, in which sound waves travel, the speed of sound is constant, and the time it takes to travel to and from some object is therefore easily translated into distance. See *navigation, wave, SONAR, sounding, storm, weather.*

radioactivity The spontaneous disintegration of an isotope of an element, accompanied by the release of alpha, beta, and gamma radiation. Radioactive isotopes occur both on land and in the world's oceans. Some radioactivity comes from the interplanetary bombardment of atoms, causing them to break up and release particles and energy, from the Earth itself, or from the human production of nuclear materials. The cosmic (extraterrestrial) contribution to the radioactivity of the oceans comes from the isotopes H^3, Be^7, Be^{10}, C^{14} and Si^{32}. These are "light atom" (meaning that they have smaller nuclei) radioisotopes. The principal radionuclides that date from the primordial earth are K^{40}, U^{235}, U^{238}, and Th^{232}.

In 1971, the Geochemical Ocean Sections Study began to inventory the contents of the oceans in order to establish a baseline, or "normal level" of radioactivity in seawater. The Study measured, among other items, the deepwater concentration of carbon 14 in different oceans. The rate of decay of carbon 14 is 1% per 83 years, and with this as a basic assumption, the turnover time in the various ocean deeps can be calculated. At depths greater than 1,500 m (4,800 feet), it is 510 years for Pacific waters, 250 years in the Indian Ocean, and 275 years in the deep Atlantic. Some deep water has stayed in the depths for more than 1,500 years, but the average turnover of all deep water is about 500 years. The outer limit of carbon 14 dating is about 40,000 years.

It is also possible, using ratios of isotopes (forms of a single chemical element that have different atomic mass) to determine the ocean of origin of certain bodies of water. The ratio of neodymium 143 to neodymium 144 is the means for determining the quantities of Atlantic versus Pacific water going through the Drake Passage.

Potassium 40 decays to argon 40 very slowly. It is assumed that all of the original argon that was present in the primitive earth is now in the atmosphere. Thus any argon found in rocks—either marine or terrestrial, is the product of the decay of K^{40}. Since some of this also escapes into the atmosphere, corrections have to be made for this. If it is not corrected for, rocks appear "younger" than they actually are. The potassium-argon system is accurate for rocks 400 million years old and, along with the rubidium-87 to strontium-87 (Rb^{87}-Sr^{87}) system, is applicable to the dating of most sediments and of continental rock as well.

Uranium decays to protactinium (U^{238} - Pa). This process is useful in dating sediments less than 60,000 years old. Another decay process that is used for assigning age to recent sediments is that of aluminum-26 to beryllium-10 (Al^{26}-Be^{10}). See *alpha ray, beta rays, element, isotope.*

radiation   The movement of energy away from a source. The energy from the Sun that actually arrives at the Earth's surface is less than half of the possible total. The rest of the energy is reflected by the atmosphere back into space. The sea acts as a great heat sink for the Sun's radiation, since the water in it is heated directly. This causes both convection currents, which bring warmer water to cold areas, and evaporation. The water that evaporates then enters into the hydrologic cycle, in which it goes from water vapor to cloud to precipitation (rain or snow) to ice melt to runoff to the sea.

Warm water also warms the air above it, and the air, like water, moves from regions of higher temperature to regions of lower temperature. Moving air is wind. Thus water air currents, and weather changes are dependent on the radiant energy received on Earth from the Sun. See *atmosphere, evaporation, meteorology, ocean, solar energy.*

Radiolaria   A group of symmetrically shaped marine protozoans with radiating thread-like pseudopodia. They date back to the Precambrian. Radiolarians are found in the photic sunlit zones of all oceans.

Their characteristic feature is a test, or external wall, of siliceous (silicon-containing) material. Numerous pseudopodia extend beyond the perforated test. Radiolarians reproduce largely asexually. There is an occasional production of gametes, but the usual reproductive mode involves the test splitting in half, with one segment becoming part of each daughter cell. Upon the death of the animal, the tiny test (from 50 to 500 micrometers to 10^{-6}m) settles to the ocean bottom. Extensive deposits of tests are a sign of large populations of radiolarians, and are found in regions of upwellings. The rising of nutrient material from the depths of the ocean supports the large populations of radiolarians that mark the Equatorial Divergence, the Antarctic Convergence, and the coast of South America. See *Protista, Protozoa.*

rat-tail fish   See *Macruridae.*

ray   A cartilagenous, winglike fish of the order Batoide (or Rajiformes). They live in almost every ocean, and some species even move up rivers. Most are sluggish bottom dwellers living on mollusks. The manta ray, however, will occasionally break the surface. It is also one of the largest of the sea's creatures, with a 6m (20 foot) "wingspan."

Other members of this group are the stingrays, electric rays, and guitar fish. All are characterized by a long, spiny, and sometimes venomous tail, and highly modified pectoral fins that resemble wings. The gill slits on the underside are part of the very "streamlined" bodies of the rays. Water is brought into the body by a series of spiracles on the upper surface. Most rays bear live young. See *Chondrichthyes, ovoviviparity.*

red clay   An oceanic sediment found at great depths. It consists of windblown particles of continental origin. Red clays dominate the sediments of the southwest Pacific Ocean. See *ocean floor, sediments.*

Red Sea   A narrow, marginal sea of the Indian Ocean separating Africa from the Arabian peninsula. It is about 300 km (190 miles) wide at it widest point, and 1900 km (1200 miles) long. The Sinai peninsula juts into the northern end of the sea, dividing it into the Gulf of Suez on the west and the Gulf of Aqaba on the east. The eastern portion is very deep (over 1800 m or about 6,000 feet), while the Gulf of Suez is shallow. The entire region is geologically interesting, since it is part of a rift valley system extending from the Dead Sea through East Africa into Kenya. The northern end of the Red Sea is a downfault. The central and southern portions show complex rift activity, with new magma emerging and rising to spread the African and Arabian plates even further apart.

Water enters the Red Sea from the Gulf of Aden over the fairly shallow sill (125m; 412 feet) at Bab el Mandeb. Practically no water enters from the surrounding lands except during the spring torrents, which bring sediment. The high rate of evaporation due to the hot, dry climate makes the Red Sea one of the saltiest bodies of ocean water on Earth, with a salinity of over 4.0% in summer. It is also the warmest. Some anomalous water, even hotter and saltier than the very warm, salty water to be expected in the Red Sea, has been retrieved from the bottom. Samples of water taken in the central channel from depths of 2,200 m (7,260 feet) and below have shown temperatures above 40°C, and have more than a 20% salt content. The brine and the sediment associated with it are very rich in minerals, and the deposition of iron, manganese, zinc, lead, copper, silver, and gold is thought parallel the processes of a billion years ago. The anomalous, warm deep water is thought by some to be connate or fossil water that contains dissolved minerals that originated elsewhere and are now appearing in the Red Sea.

The winds associated with the Red Sea are northerlies in the northwestern portion of the waterway and southerlies in the southeastern region. This makes sailing on the sea difficult. The technological answer was the dhow—a small, lateen-rigged ship. The triangular sails of this vessel are hung on yards that are suspended from the mast, making it possible to beat almost directly into the wind. See *dhow, hot springs.*

red tide   Seawater covered or discolored by a large population of dinoflagellates; often fatal to many forms of marine life. Red tides are produced by the sudden growth, or bloom, of a dinoflagellate or other flagellate species. The Gulf of Mexico has red tides that are caused by *Gymnodinium brevis.* The result is often a gigantic fish-kill. Walvis Bay on the Atlantic side of South Africa is similarly affected. The red tides caused by the genus *Gonyaulax* result in human deaths because these dinoflagellates are ingested by clams and mussels, which then concentrate the paralyzing neurotoxin produced by these dinoflagellates. See *dinoflagellate, toxin.*

reef   A mass or ridge of rock in either a freshwater or marine waterway. A reef is unlike a bar, which consists of unconsolidated sand and is thus subject to removal by wind, current, or both. A reef is either submerged coast, and therefore structurally like the rock of the nearby landmass, or a biogenic creation. A biogenic reef is constructed by corals and coralline algae. See *barrier reef, coral.*

reflection   The return of a ray or wave of energy toward its source. When water waves (or light, radiation, or sound waves) strike a surface, the surface acts as a barrier and redirects the waves in the direction of their origin. See *diffusion, refraction.*

refraction   The apparent bending of an energy wave on its passage from one medium (e.g., air) into another (e.g., water). Refraction is caused by the differing densities of the two media through which the wave passes, and the resulting change in velocity as the wave moves from one medium to the other.

relative humidity The ratio of the quantity of water vapor actually present in a sample of air to the quantity that the atmosphere can hold at that temperature and pressure.

The relative humidity is measured using two thermometers, one wet, the other dry. The temperature shown by the wet bulb will be some number of degrees lower than that of the dry one because the evaporation of water from the wet bulb requires energy (heat) and therefore the bulb will be cooled and show a temperature lower than that of the dry bulb. If the atmosphere is saturated with water vapor, no more water can evaporate and the two thermometers will have the same temperature reading. The relationship between the wet-bulb and dry-bulb readings is not a simple one and complex charts are used to determine the relative humidity on the basis of the two temperature readings and the barometric pressure. Electric hygrometers measure the water content of the air directly by measuring the electrical conductivity of the air. See *air, fog.*

relict beach An archaic beachfront that is no longer at the seashore. A rising sea level has submerged some beaches, which are discernible on continental shelves. See *beach, blue holes, continental shelf, subsidence.*

remora Any member of a small group of specialized fish that can attach themselves to large fish or ships by means of a suction disk on the top of the head. They are anywhere from 30 to 90 cm (1 to 3 feet) long and dark gray or green in color. The remoras are most often associated as commensals with sharks. Like sharks, they live in warm waters worldwide. Their usual diet is detritus or anything that falls in their direction.

Remoras seem to feed on the stray bits that the shark neglects to snap up, although some seem to clean the larger animals, thus "paying" for their free ride.

Remoras do not depend on larger fish for movement; they can and do swim well on their own. See *commensal relationship, fish, shark, symbiosis.*

reptile Any member of the class reptilia, comprising air-breathing vertebrates characterized by scaly skin, internal fertilization followed by the production of shelled but not always calcareous eggs, and in most species a three-chambered heart. The reptiles include the snakes, lizards, turtles, alligators, crocodiles, and their near relatives. Sea snakes and turtles are the most prominent marine reptiles, although some crocodilians live in brackish waters. The fossil record shows many extinct aquatic reptilian forms, such as the plesiosaurs and ichthyosaurs. See *ichthyosaur, plesiosaur, snakes, sea turtle, vertebrate.*

***Research,* H.M.S.** A ship used in 1827 by the English sea captain Peter Dillon, to search for survivors of La Pérouse's expedition to Australia in 1787–1788 on the *Boussole.*

The name *Research* was also given to a British surveying ship commissioned in 1889. Another vessel named *Research* was a nonmagnetic ship built to avoid mines, and launched during World War II, which was scrapped after just six years at sea. See *La Pérouse, Jean-François Galoup comte de*

respiration The process by which an organism uses organic material (food) as an energy source. See *metabolism, oxygen, photosynthesis.*

Rhine River This western European river rises in central Switzerland and flows roughly northwest along a 1,300 -km (820-mile) course to the North Sea. More than half of this course is navigable, which means that the Rhine is a very important river commercially.

The Rhine is one of Europe's significant historic, legendary, geographic, and political features. Its many falls and attractive mountainous surroundings have made it a center for tourism and vine-

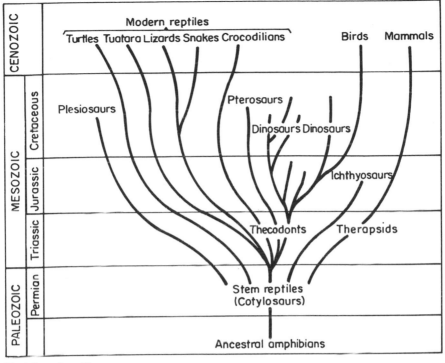

Evolution of reptiles

yards. In other parts of its course, such as near Bonn, in the Federal Republic of Germany, it is an industrial and therefore often polluted waterway. The city of Cologne, Germany, is one location that has been known for its "smelly river" since Coleridge's visit in the last years of the 18th century.

The delta region of the Rhine is one of fragmented streams. The Neder Rijn, or Lower Rhine, and the Waal are smaller rivers that rise from the breakup of the Rhine proper. These in turn divide again, forming the network of small streams that thread through the delta, which constitutes most of the land area of The Netherlands.

Current engineering works are changing the course and bottom topography of the Rhine throughout most of its length. The objectives are to control flow, improve navigation, and control pollution. The major mouths of the Rhine will all

eventually be protected with a series of "superdikes" designed to control storm surges. Many of these have already been completed in the region around Eindhoven, The Netherlands. See *North Sea, storm surge.*

Rhodophyta   The red algae, a division of the kingdom Plantae in some taxonomies. Their chlorophyll is hidden by the pigments phycoerythrin and phycocyanins. Rhodophyta are usually found at depths greater than those at which green or brown algae flourish. It is believed that their red pigment enhances the feeble light that reaches these plants, and enables them to photosynthesize despite the low intensity light at these greater depths. See *algae.*

Rhône River   A great river 800 km (500 miles) long in southeastern France, that rises in Switzerland, flows through

Lake Geneva, and then flows south through France into the Mediterranean Sea. The delta of the Rhône extends over 40 km (25 miles) from Arles to the Mediterranean, and is the major undersea feature of the western portion of this sea. The river is divided in its lower reaches into the Grand and Petit Rhône, with the marshy Camargue between them. See *delta, Mediterranean.*

ribbon worm   Any of a group of worms in the phylum Nemertea; they are marine forms of the free-living flatworms. These animals may be as long as 30 m(100 feet). The phylum contains both benthic (bottom dwelling) and pelagic (open-sea) forms. They are not to be confused with ribbon fish or Venus's girdle. See *benthic ecosystem, Ctenophora, Lampridiformes, Nemertea, pelagic environment, Venus's girdle.*

rift valley   A depression in the Earth resulting from the collapse of a large section of the crust. Rift valleys are often part of the mid-ocean ridge system. They occur where crustal segments split and move away from the upwelling magma.

In some regions, a rift valley is the obvious separation of two plates. The African Rift Valley is an extension on land of the division of the African from the Arabian plates. It runs down the Red Sea and continues on what is the eastern portion of Africa, through Ethiopia and into Tanzania and Kenya. See *fracture zone, mid-ocean ridges.*

right whale   A large baleen whale with an extraordinarily large head. Adult right whales are 15 to 18 m (50 to 60 feet) long, about 30% of which represents the head of the animal. The name was originally applied to the Greenland or bowhead whale *(Balaena mysticetus)* because it was "right" for whalers. The resulting overfishing almost exterminated these animals, and they are still an endangered species.

The right whale is an attractive species, with blue-black superior surfaces and light undersides. It feeds on tiny crustaceans. Related species are the black right whale of the North Atlantic *(B. glaciales)*, the pygmy whale *(Caperea marginate)* of South America, and *B. sieboldi* of the North Pacific. See baleen, Cetacea, whale.

Rio Grande Rise   An aseismic ridge in the South Atlantic off the southeastern coast of South America. It continues the line of the Rio de la Plata and divides the Argentine and Brazil basins. This ridge is analogous to the Walvis Ridge, also in the South Atlantic and off the coast of Africa. See *aseismic ridge.*

rip tide or rip current   A rapid, narrow, surface current moving from shore out to sea at an abnormally rapid speed. These are not tides at all. While their rate is unusual, their direction away from the beach is normal. They occur on long, unbroken coasts, where waves approach parallel to the shore. Because each wave pushes water shoreward, periods of high wave activity lead to the buildup of a mass of water near the shore, raising the level of the sea. This condition continues until the weight of the water breaks through the nearshore water at a low point and surges seaward for some period. Once the pressure is released, the wave activity and seaward flow of water returns to normal. See *currents.*

ripple marks   Characteristic wave patterns on the ocean bottom. They are the result of wave action. They separate the bottom sediment into crest-and-trough patterns that mirror the crest-and-trough parts of waves. These bottom features are found in areas that have a bottom of sand or other finely divided sediment. Ripple marks may fossilize, to later reveal a sea floor that bears the marks of ancient wave patterns. See *turbidity currents.*

river bore   A river tidal wave with a high, well-defined crest. It occurs at spring high tides (high tides at a new or full moon) if the river channel leading to the sea is constricted. The constriction concentrates the body of water rushing in with the force of the tide, and as this water sweeps upstream it has a well defined, abrupt crest. The height of the crest varies; it is greater at the banks than at midstream. It also varies from one river to another and from one spot in a given river to some other location. See *Bay of Fundy, tide.*

RNA Ribose nucleic acid, the substance that transcribes the information encoded by DNA into proteins in the living cell. The sugar unit in RNA is ribose, and the bases are adenine, guanine, cytosine, and uracil; the uracil replaces the thymine that would be the fourth base in DNA. RNA is a single-stranded chain. It is formed as a complement to DNA, which acts as a template for it. Where there is a guanine (G) on the DNA chain, the RNA base that pairs with it is a cytosine (C). The complement of adenine (A) is uracil (U). While DNA is a very large, double-stranded, much more complex molecule than RNA, there are more RNA molecules in cells.

RNA is present in several forms. Messenger, or m-RNA, carries the encoded message of the DNA from the nucleus of a cell to the ribosomes—the organelles that manufacture the species-specific protein encoded by the DNA and whose amino-acid sequence is established by the order of the base arrangement in the DNA.

Messenger RNA is constructed as a complement to only a part of one strand of a particular DNA, (i.e. the segment of the DNA that is the instruction, or specific sequence of bases, for a particular series of amino acids). The m-RNA then moves out of the nucleus to form the ribosome, where its own sequence of bases, complementary to those on the DNA, is translated into ribosomal RNA (r-RNA). This last complement of the DNA's complement is therefore identical to the structure encoded by the original DNA section being modelled. The ribosomal RNA is acted upon by transfer, or t-RNA, which brings the appropriate amino acids for the par-

ticular protein sequence being assembled, and lines them up on the ribosomal RNA to form the protein chain. See *DNA, protein.*

roaring forties An old sailing term referring to the prevailing westerly winds— winds moving from west to east in the 40 to 50° latitude range. Since there is less landmass in the Southern Hemisphere, the effect of these strong winds, which range up to gale force (thus giving them their name), is normally encountered there. Sailing ships such as clippers moved very quickly before such winds. See *clipper, westerlies, wind.*

rock louse An isopod member of the class Crustacea that lives on the shore above the splash line (highest point of the high tide), and is not a louse at all. See *isopod, littoral.*

Romanche Deep See *fracture zone, Pacific Ocean.*

Rondelet, Guillaume (1507–1566) A French physician, traveler, professor of anatomy, and chancellor of the University of Montpellier (France). Rondelet was one of the "Encyclopedic Naturalists," and is a significant figure in the history of marine science because of his book on marine fish published in 1554. It was translated from the Latin four years later with the title, *Complete History of Fish.* In it, Rondelet discusses freshwater as well as marine life; thus he includes everything in aquatic environments from beavers to fish to dolphins. Almost all of his 300-odd entries are illustrated.

Rondelet analyzed the various systems of aquatic creatures in much the same manner that Aristotle did: with primary consideration of the why and way in which a structure would work. He noted the swim bladder in freshwater fish and then looked for it in marine specimens. He dissected dolphins, noting their relationship to both pigs and humans. His description of the sea urchin is excellent and a worthy complement to that of Aristotle, and Rondelet's drawing of this creature is the earliest extant depiction of an invertebrate. See *anatomy; Aristotle; Belon, Pierre; Gesner, Konrad.*

Ross, John (1777–1856) A Scots arctic explorer. Ross sought the Northwest Passage in the arctic islands of the Canadian Northwest Territory. During his second voyage his nephew, James Clark Ross, located the North Magnetic Pole. A lasting body of work produced by Ross was his extensive charting of the islands of the arctic. See *Magnetic Pole, Northwest Passage, Ross, James Clark.*

Ross, James Clark (1800–1862) A Scots explorer who discovered the Ross Sea in the antarctic in 1841. He is also noted for magnetic survey studies of the polar regions. He was present on an earlier expedition led by his uncle, Sir John Ross, during which he located the North Magnetic Pole in 1831. Subsequently, in the 1839–1843 expedition, he sailed with the ships *Erebus* and *Terror* to find the South Magnetic Pole. In the antarctic, he charted Graham Land, the Weddell Ice Sheet, and Victoria Land. This expedition established him as a notable scientist-explorer. See *Antarctica, explorers & exploration, magnetic pole, Northwest Passage.*

Ross Sea A sea arm of the South Pacific below the Antarctic Circle, approximately due south of New Zealand. It is bordered by Victoria Land on the east, and it merges into the Ross Ice Shelf south of 77° South Latitude. The closest landmass (other than Antarctica) to the Ross Sea is New Zealand. The water of the Ross Sea is fairly constant in temperature ($-2°C$ to $0°$) and salinity (3.35 to 3.47%). The pattern of circulation is part of the East Wind Drift.

The sea covers a continental shelf that is about 300 to 450 m (1,200 to 1,800 feet) deep. Its depth is attributed to pressure from the weight of the ice shelf to the south. Ridges parallel to Victoria

Land's coast show signs of Tertiary volcanic activity. The Ross Sea is relatively free of ice when compared to other antarctic waters. Admiral Byrd's *Little America* base is on its western coast. See *Antarctica, ice shelf.*

Rotifera Microscopic worms of the phylum Aschelminthes. Most are freshwater animals, but there are about 15 species that are marine. Rotifers have a "head" with a ciliated "mouth," and many are parasitic. *Seison* is a parasitic species that lives on the gills of crustaceans. See *cilia, parasite.*

roundworm See *Aschelminthes, Annelida.*

Royal Society The Royal Society of London for the Promotion of Natural Knowledge. Founded in 1660, it is the oldest scientific society in Great Britain and one of the oldest worldwide. Current membership in the society numbers about 1,000 and includes some members who are not English. The Society started with an informal group of people interested in scientific subjects. Some of the founding members included John Dryden, Robert Hooke, Christopher Wren, and John Evelyn. The group was given a charter by Charles II in 1662, but no financial support.

Beginning in 1665, the Society published its *Philosophical Transactions,* a periodical that survives today. Notably, the Society actively encouraged interest in a variety of fields and recognized achievement by both Englishmen and foreigners. Members of this period included Joseph Banks, Isaac Newton and Edmund Halley. Benjamin Franklin was a colonial member. The enterprises the Society sponsored ranged from voyages of exploration, the first of which was the 1768 voyage of Captain James Cook, to the 1919 expedition to photograph the solar eclipse, a voyage that was intended to and did prove Einstein's theory of relativity.

S

Sagitta The arrow worms, members of the phylum Chaetognatha. They are about 0.5 cm (0.2 inches) long and form an important part of the zooplankton. They are carnivores, feeding on smaller organisms. The arrow worms are exclusively marine animals and have been used as indicator organisms. The species of Sagitta in a particular area are specific to that region; thus, seawater entering the Atlantic from the Mediterranean will bring with it the arrow worms of the Mediterranean. The movement of the water can be traced by the species of sagitta in it. See *carnivore, Chaetognatha.*

Sahul Shelf A continental shelf off the Australian shore, under the shallow Timor Sea. It lies along the northwest coast of Australia, running from the arbitrary divide with the Arafura Shelf to the southwest. The Sahul Shelf extends to the Rowley Shelf and then almost to the Sunda Shelf, whose edge is marked by the Timor Trough.

The Sahul Shelf exhibits evidence of ancient erosion and geologically recent subsidence. The coral islands on the shelf are not oceanic, volcanic atolls, but were once part of the Australian continental mass, and the land between them and Australia was once above sea level. The canyons in the Sahul Shelf are analogous to the ["Sunda River"] system. See *island, subsidence.*

sail A cloth set on masts or vertical supports, and yards or horizontal supports, that crosses these structures and is intended to catch the wind and propel a ship. In general, square sails are set on yards, triangular ones directly on the masts themselves. Ships carrying only triangular sails are also referred to as fore-and-aft rigged ships. However, some large square-rigged ships also carry triangular sails. Such sails on the bowsprit or leading sail support are jibs. If the triangular sail is hung on the mizzen (rearmost mast) it is a stay or spanker. See *individual names of sailing ships.*

sailfish A tropical fish *(Istiophorus)* that is a relative of the swordfish and the marlin; it is sought by commercial and sport fishermen. The latter prefer it because it is both a fighting fish and a beautiful one. The sailfish is dark blue above, with a silvery ventral surface. It has a long, pointed snout and a distinctive "sail," its large dorsal fin. The average size of the sailfish is 3.4 to 3.5 m (11 to 12) feet), and it weighs about 90 kg (200 pounds). The diet of the sailfish consists largely of smaller, schooling fish.

Saint Lawrence, Gulf of A large, roughly triangular area, about 150,000 km² (90,000 square miles) of the North Atlantic Ocean along eastern Canada between Quebec, Newfoundland, New Brunswick, and Nova Scotia. The Gulf is connected to the North Atlantic by the Straits of Belle Isle between Newfoundland and Labrador and the Cabot Strait between Newfoundland and Nova Scotia. The southerly portion is fairly shallow, with depths of about 50 to 80 m (180 to 250 feet). The dividing line is the Laurentian Channel, which runs northwest to southeast with an average depth of 300 to 450 m (1,200 to 1,500 feet).

Cold water of the North Atlantic enters the area via the Strait of Belle Isle and then moves west, where it encounters St. Lawrence river water, which flows southeast between the Gaspé Peninsula and

Anticosti Island. The salinity in the Gulf is low, particularly in summer, when it is about 2.6%.

The Appalachian Ridge line bends at the Gaspé and then wanders through the Gulf. The current shape of the Gulf and of the St. Lawrence River channel was created during the last Ice Age. The sediments in it are due to current and recent geological glacial activity. See *Atlantic Ocean, Labrador Current.*

St. Peter and St. Paul   A cluster of rocks protruding above the surface of the South Atlantic Ocean between Brazil and West Africa, about 1,000 km (600 miles) northeast of Natal, Brazil. These rocks are seamounts east of the Ceara Plain and the Mid-Atlantic Ridge. They were landmarks in the day of sailing ships. See *Atlantic Ocean, seamounts.*

salinity   The amount of dissolved inorganic minerals (salts) in seawater. The standard measure of salinity is in grams per kilogram of water. The saltiness of the oceans has been exploited for millennia. Neolithic man harvested salt in salt pans and used the product as a trade good. Explanations for the salt in the ocean's waters have been put forth for almost as long as salt has been used.

By the 17th century, careful experiments to determine the physical and chemical composition of seawater had begun. Robert Boyle, an English scientist well known for his work on air pressure, compared the specific gravities of seawater samples taken at different depths. He concluded that the salt content at all depths was essentially the same. Excep-

tions were the water of bays and of areas near undersea springs. Upon Boyle's suggestion for further research, others measured the specific gravity of seawater at different latitudes, and established that there were variations in its saltiness.

It remained for the chemists Antoine Lavoisier of France and Tobern Bergman of Sweden to separately do successful analyses of seawater. Lavoisier's samples came from Dieppe, France, accounting for their less-than-average concentration of salt. He identified as constituents of seawater, sodium, calcium, and magnesium chlorides; sodium, magnesium, and calcium sulfates; and calcium carbonate. Edmund Halley (of comet fame) realized that Lavoisier was right in assuming the terrestrial origin of the salt in the sea. He then theorized that the dissolution and transportation of these minerals could be used as a measure of the age of the Earth. Because the ocean was in a steady state of salinity, said Halley, it had to have achieved its present state of saltiness long ago in very early geologic times.

The work of William Dittmar on the H.M.S. *Challenger* voyage established the composition of the sea. (See table below.)

The total weight represented by the ions above is 99.95% of that of all ions present in seawater. See Challenger, *H.M.S., individual ions and salts.*

salmon   Any of several large, anadromous food and sport fish of the order Salmoniformes. The name salmon is used for the Atlantic Ocean species *(Salmo salar)* and also for the red, pink, coho, king, chum, and Pacific salmon of the genus *Oncorhynchus.*

Most Common Ions Present in Seawater

Ion	% of total salt (by weight)	Ion	% of total salt
Cl^-	55.04	Na^+	30.61
SO_4^{2-}	7.68	Mg^{2+}	3.69
HCO_3^-	0.41	Ca^{2+}	1.16
Br^-	0.19	K^+	1.10
H_3BO_3	0.07		

The Atlantic species is present in both European and American waters. It is carnivorous, feeding on crustaceans. The average adult fish weighs between 5 and 9 kg (11 to 30 pounds), although larger ones have been recorded. The color of the fish varies with its age and sex, the season, locale, and availability of food. Most adults, however, are silvery below and speckled brown on their upper surface. They may have red splashes of color on the brown surface. The adults, having matured in the ocean, ascend their native rivers to mate. The timing of the "salmon run" or movement of these fish upstream in any particular river, is a function of the temperature. Thus, Atlantic salmon are found in Irish rivers at an earlier date than in Norwegian rivers because the latter are colder. The female burrows a shallow depression in the river bottom, and the eggs are fertilized in the rudimentary nest. When they hatch, early in the spring, the young salmon feed on insects. They look like trout at this stage and are called parr.

The adolescent salmon is called a smolt. At this stage it is about two years old and ready to go to sea. There it feeds on small crustaceans and grows rapidly. The average adult is about five years old when it ascends its native river to mate. It is an exhausting trip during which the fish do not feed. Many of the salmon die after mating, others, in their weakened and exhausted state, are caught by predators; some return to the ocean and repeat the trip about two years later.

The largest species of salmon are king salmon of the Pacific Ocean. Specimens of more than 50 kg (110 pounds) are known, but not common. The pink salmon are the smallest salmon species, at about 2.5 kg (6 pounds). These species differ in their appearance and the distance they travel to spawn, but are in general more alike in pattern and more colorful than is the Atlantic species.

When they are ascending their native river, the physiologic changes in Pacific salmon, and particularly the king salmon, make it impossible for the adults to feed as they move upriver. If a given fish does not turn back almost immediately, it is committed to the upriver direction.

The steelhead trout (Salmo gairdneri) is close to the salmon in color and behavior and is related to the salmon. There are also other species that are migratory and called salmon, such as the wall-eyed pike, squawfish, and yellowtail, but these are unrelated to the true salmon.

Commercially, salmon are caught in a variety of nets or weirs. Because of serious overfishing, the range of many varieties, particularly in the Atlantic, has been severely curtailed. Extensive restocking has been going on and some species are becoming more plentiful.

salps   Free-swimming tunicates of the class Thaliacea, family Salpidae, and ge-

Varieties of Salmon

Generic name	Common name	Habitat
Salmo salar	Atlantic (parr and grilse are juveniles)	Scandinavia, North America, British Isles, Iceland, etc.
Oncorhynchus tshavytsha	Chinook, king, Colorado River, Sacramento	Central California to the Yukon
O. garhusha	Pink or humpback	California to Alaska, Japan
O. nerka	Red, sockeye, blueback, Fraser River	Oregon—North Alaska
O. keta	Dog, chum	Alaska, Siberia,

nus *Salpa,* most often found in warm tropical seas. These organisms are solitary or elongated colonial animals that float on the surface of the sea. Their most distinguishing characteristic is their primitive reproductive system: salps reproduce most often by budding (i.e., as an outgrowth of the parent). See *Tunicata.*

salt   The common substance called salt is sodium chloride (NaCl). It is found in seawater and also as the mineral rock salt. Pure salt consists almost entirely of chloride, colorless, cube-shaped crystals. Dissolved salt is the most prevalent substance in seawater. Salt is obtained commercially by a series of processes from underground mines, by evaporation from salt pans, and others. Salt is used in the preservation and preparation of food, in commercial refrigeration, and in manufacturing other chemicals.

There are, however, other kinds of salts than NaCl. Chemically, a salt is defined as a compound which on dissolving produces negative and positive ions. The solubility of salts varies enormously. See *mineral, salinity, solubility, water.*

salt domes   Geologic formations consisting of evaporites, largely or entirely composed of sodium chloride, salt domes

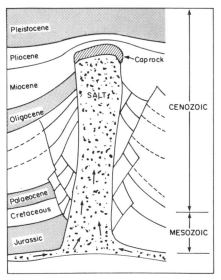

Salt domes

are most often vertical and cylindrical columns at least 1 km in diameter, which are thought to have developed in sedimentary basins and later been overlain by sediment. The entirety of the the the formation may then have been folded by upthrusting of the Earth's crust.

Salt domes are economically significant, since oil and gas deposits are often associated with them. The presence of a series of salt domes is often an indication of an oil or gas field in the same vicinity. When hollowed out, salt domes are also considered very safe storage units for "hot" radioactive waste material. See *oil.*

salt flats   Formed by the evaporation of a body of water. They usually occur in arid areas where an inflow of salt-containing water is dammed naturally and evaporates. This may result where isolated bays are separated from the sea by coral growth or the creation of barrier islands, or where a sedimentary basin or shallow sea subsides. A shallow shelf may also become a site for the deposition of more insoluble salts, sulfates being the first salts to precipitate. Non-marine salt flats may result where a river deposits its load of dissolved material in a dry region. The Dead Sea is an example of a current, building salt flat. Some other such inland areas, such as the Bonneville Flats in the United States, were once marine. See *coast, estuary.*

salt marsh   An area of brackish water usually built on a continental shelf. Only recently has the importance of the salt marsh ecosystem been recognized as vital to both esthetic and commercial interests. In earlier times a marsh was "waste land" and considered to be an area fit only for outcasts. Civilization meant drainage and conversion of the area to farmland, such as was done centuries ago with the fen country of the east coast of England.

The typical salt marsh is the result of a silt-carrying river. The outflow to the sea is impeded either by a barrier island or by the slow subsidence of the region. The area of the marsh is a series of hillocks separated by water channels. It is not a permanent area, since the continued accretion of silt builds up the land and effectively dries it. This progression from marsh to dry land can be seen in the change in vegetation with time. The dominant plants are at first sea grasses, then salt-tolerating plants, then terrestrial ones. This in turn creates a similar change in the local fauna.

In general, the enclosed nature of a salt marsh means that the water temperature is somewhat higher than that of the nearby ocean. A salt marsh is also a region of shallow water. Both of these factors mean that there is increased photosynthesis, and the greater, resulting plant production in turn supports larger animal populations than those that would be found in a similar volume of ocean water. New soil brought by the river system increases the fertility of the region and so do the droppings of the animals that feed on the plants. The resulting quantities of shrimp and birds—to name just two groups of animals—are very large.

Many animals of the continental shelf region, especially some fish and crustaceans, spend some part of their lifespan in the marsh, as larvae, juveniles, feeding adults. Many marsh animals can spend some part of their lives out of the water or in water of low salt content. The clams of the littoral can exist in sand for hours awaiting the high tide that inundates them. Mussels and limpets cling to occasionally exposed rocks. Bird communities, both passarine and indigenous, use the grasses and local shellfish of the marsh as vital ingredients in their diets. Since a salt marsh, unlike a coral atoll, is not an isolated system, it is a more difficult ecosystem to study, and the interrelationship of many species in salt marshes is still unknown.

Marshes are ecologically fragile areas. They are sensitive to pollution, which may be chemical from industry or agriculture, and thermal pollution. The latter is frequently the result of river water being used as the coolant in industrial plants.

The tropical and semi-tropical version of a salt marsh is a mangrove swamp. Like any other salt marsh, it is vital to the ecosystem of the region.

See *delta, estuary, individual animals, pollution.*

salvage   The saving, in navigable water, of a ship or its cargo. Salvage may involve towing a vessel that has mechanical problems, moving a stranded vessel to open water, or raising a sunken vessel. Salvage vessels are usually very large, powerful tugs equipped with various heavy-duty pumps, winches, and cranes. Salvage techniques may involve one or several methods of towing, lifting, and pumping out a damaged ship. In maritime law, the vessel to be salvaged, and its goods, belong to the owner unless abandoned. The owner of the vessel to be salvaged, who may pay by prearranged fee or a share of the material retrieved, contracts for the rescuer's service in salvage. The fee depends on the success of the salvage mission. See *marine archeology.*

San Andreas Fault   A slip strike series of faults along the coast of California where the American and the Pacific plates move past each other. It is analogous to other faults in the Pacific basin. The San Andreas is a line leading from the Gulf of California, where the East Pacific Rise dips under North America, northwest past San Francisco.

It has been theorized that all of the Pacific faults are rotating the land on their seaward sides in a counterclockwise direction. This would mean that those parts of California to the west of the fault are gradually sliding toward Alaska. See *California, Gulf of, crust, Pacific Plate, slip strike.*

sand   Granular mineral particles formed by the erosive activity of ocean or river currents, waves, wind, or glacial movement. Most sand is terrigenous and brought to the ocean by rivers. The mineral most commonly found in sand is quartz, although it may contain inclusions of volcanic rock, coral, or gypsum. These inclusions account for the variety of colors of sand, and for its grittiness. Coral sand is powdery.

Sand is used commercially as an aggregate in construction. Concrete is sand in a lime matrix. Sand is also mined for use as an abrasive. See *mud, sediment, silt.*

sand crab   A crab of the genus *Ocypoda,* that lives on beaches in temperate to tropical habitats. There are about 200 species of these gray- or buff-colored animals which are sometimes called ghost crabs. They live in burrows and are seen running on beaches above the high-tide line.

The American version of the sand crab is found on the Atlantic coast from about 40° North Latitude to about 40° South Latitude. This crab, *O. quadratus,* is about 4 to 5 cm (1.5 to 2 inches) in diameter. *O. ceratophthalmus* is found in the Indian Ocean and the eastern Pacific; *O. saraton* lives on the coast of East Africa and along the Arabian peninsula. The principal diet of the sand crab is insects. *O. quadratus* lives almost exclusively on sand fleas. See *backshore, crab.*

sand dollar   A thin, disk-shaped echinoderm belonging to the class *Asteroidea.* The sand dollars are most often about 7 to 8 cm (3 inches) in diameter. The upper surface is marked with a characteristic five-petalled shape which is a modification of the water vascular system common to echinoderms. In some species, the petaloids are cut-outs, leaving the disk perforated.

The most often encountered species of sand dollar is *Echinoarachnius parma,* which is found in North American and Far Eastern waters. See *Echinodermata.*

Sarcodina   A class of unicellular organisms in the kingdom Protista. There are over 11,000 species in the Sarcodina, including the Foraminifera and Radi-

olaria. See *Foraminifera, Protista, Radiolaria.*

sardine   A food fish of the herring family (Culpeidae) found in dense schools in almost all warm ocean waters. The adults are 25 to 30 cm (10 to 12 inches) long, dark green or blue on their upper surfaces and silvery-white on the underside. Their backs are frequently marked with black. Sardines, like herring, are primitive bony fish with unscaled heads and no lateral line.

Sardines spawn in the spring, when the plankton population that they feed on is at a maximum. They rise and fall from more shallow to deeper waters on a daily basis following their prey, the zooplankton. Commercial fishermen take advantage of this behavior and net sardines at night, when both they and the plankton rise to the surface to feed.

The method used to preserve sardines as food depends on the locale. Some of the techniques in use have been used for a very long time. Sunken ships that were part of the commercial fleet of Rome carried pickled sardines in amphorae as cargo. Sardines preserved in salt can be found in the markets along the Mediterranean coast, as were those fish in the amphorae of long ago. Canning is today's most common process of preservation. The species of sardine caught depends on availability. Besides being used as a food item, the sardine is also a source of oil. The oil is used in the manufacture of paints and varnishes, other chemical raw materials, and as an edible fat. Sardine meal, or what is left after the fat is extracted, is incorporated into animal feed.

Some common sardine species and their locales are:

Pilchard *(Sardinia pilchardus)*	Southern and southwestern Europe and North Africa
Pacific *(Sardinops caerulea)*	Western North America
South American *(S. sagax)*	Western South America
Japanese *(S. melanosticta)*	Japan
S. neopilchardus	Australia
S. ocellata	South Africa

The so-called Maine sardine is not a sardine, but rather a small or juvenile herring. See *fish, herring, marine oil.*

Sargasso Sea   A relatively still area in the North Atlantic, between 25° and 35° North Latitude and 40° to 70° West Longitude. It is the "eye" in the anticyclonic series of currents of which the Gulf Stream is one. Although eddy currents wander into it, the Sargasso is relatively warm and still. Its name comes from the large quantity of floating algae, or *Sargassum,* found in it. These branched algae support an ecosystem of small animals that use the "weed" as both a food supply and structural support.

The Sargasso Sea has been called an "aquatic desert," implying that little or nothing lives in it. It is a highly saline body of warm water floating on colder water. This thermal gradient keeps nutrient-rich water from welling up to the surface. Thus, the plankton production in the sea is less than it might be otherwise, but it is not altogether lacking.

The Sargasso is the spawning ground for most of the European and American species of eels, and for fabulous legends. The eels are real, the legends entertaining. See *eel, Fucus.*

Sargassum   See Fucus.

Sars, George (1837–1927)   A Norwegian marine-life scientist known for his work on the Crustacea; the son of Michael Sars. Like his father, he was born in Bergen and later moved to Christiania (Oslo), where he spent the rest of his life. He was originally a doctor, but eventually specialized in the Crustacea and wrote a nine-volume work on the crustaceans of Norway, which occupied him for almost forty years (1890–1927). He was the outstanding expert of his time on the Crustacea, and as such was the examiner of the *Challenger* Expedition's crustacean samples.

In 1864 Sars was appointed by the government (at that time the government of Norway was Danish) to investigate the cod. He discovered that cod eggs float, rather than sinking, as had been theorized, and proved that this is true of almost all food-fish eggs. This fact is very important in the study of the life history and migration patterns of the oceanic fish that are to Denmark and Norway vital elements in the industry of the region. See Challenger, *H.M.S.; Sars, Michael.*

Sars, Michael (1805–1869)   A Norwegian naturalist and marine zoologist. He was born in Bergen and educated as a pastor; natural history was his hobby. As an amateur naturalist, he published a work on natural history in 1829. Another work, on the littoral fauna of Norway, appeared in 1846. By then Sars was no longer an amateur, and accepted an appointment as professor of zoology at the university in Christiania (now Oslo), Norway.

Sars made a number of important discoveries in marine zoology. The discovery of the sessile (stationary) stage in medusal development of the *Scyphozoa,* the development of mollusks from free-swimming ciliate larvae, and the first record of a living crinoid are among his contributions. In conjunction with his son George, he worked assiduously to disprove the prevailing theory of an abiotic zone in the ocean. See *Crinoidea,* Forbes, Edward; *Medusa.*

satellite   In astronomy, a celestial object that revolves with or around a larger celestial object. Artificial satellites are manmade objects placed in orbit around the Earth for scientific, technological, and military uses such as reconnaissance, surveying, meteorological observation, navigational assistance, and communications, for relaying radio and television signals.

The Space Age began in 1957 with the Soviet Union's *Sputnik,* which made possible the first real look at Earth. It became quickly apparent that satellite sightings from several places on Earth at exactly the same time would make triangulation, and therefore exact mapping, possible.

The American *Landsat* spacecraft created a photographic record of Earth. This was the first proof of Columbus' accurate guess. He postulated that the Earth was egg (or pear) shaped. The Northern end is more pointed; the Southern is rounder. Used with computer enhancement, different terrestrial structures, soil conditions, land usage, and subsurface conditions became visible. In water, plankton concentrations and their concomitant fish predators produce a distinctive color when viewed by satellite. Although the individuals cannot be seen, the entire population can be mapped and migrations of fish can be tracked.

Satellite cartographers map volcanoes, currents, geothermal hot spots, and other energy resources. They also map other planets, using the data produced by the *Voyager* series of satellite photographs. See *maps, Moon.*

scallop   A bivalve mollusk of the family Pectinidae; a common genus is *Pectin.* There are over 400 species of scallop, and their range is worldwide, extending from the littoral regions of the coast to ocean depths of over 4,000 m (13,000 feet). Scallops are an important food and are taken commercially off the New England and Canadian coasts, usually from depths of about 40 to 120 m (130 to 390 feet).

The scallop has an almost hemispherical, ribbed shell with winglike projections at the hinges. The hinge is controlled by the animal's large adductor muscle. Scallops range in size from 2.5 to over 15 cm (1 to 6 inches); the smaller ones are found in bays or estuaries, and their color varies from white to yellow, red, or purple. The outer edge of the mantle has numerous blue eyes. These are visible just inside the partly open shell of the undisturbed scallop as it rests on the ocean bottom.

Scallop eggs are fertilized in the water. They hatch quickly to release free-swimming larvae, which may develop a byssal extrusion (a thread by which they attach themselves to a surface). Scallops are filter feeders and live in sand or on gravel in clear water.

While some scallops are sessile (stationary), the average individual can and does move by rapidly opening and closing its shell. This jet propulsion mechanism moves the scallop along.

The normal predator of the scallop is the starfish. Man has known the scallop as food for thousands of years. The pretty shells have found their way into decorative motifs since the very beginnings of human history near the sea. During the Middle Ages, the scallop shell was adopted as a religious pilgrim's sign. See *byssal threads, filter feeders, mantle, mollusk, shell.*

Scaphopoda   A class of primitive marine mollusks. They are commonly called tooth or tusk shells, or elephant tusks. There are about 200 known species. Many species of these bottom-dwellers are found in oceanic deeps, some at or below the 4000-m (13,000-foot) level.

The scaphopod's headlike proboscis is surrounded by tentacles that probe the surrounding mud, searching for food. Both water and nutrients are moved through the mantle cavity by means of muscle action, the usual means of circulating material in other mollusks also. See *Mollusca, tusk shells.*

scarp   A sharply falling undersea slope. It separates plateaus or regions of gentle slope. See *canyon, continental slope, ocean floor, plateau.*

scattering layers   See *deep scattering layers.*

scavenger   An animal that feeds on dead organisms. The decomposing bacteria are the ultimate and smallest scavengers. See *bacteria, decomposers, detritus.*

schooling   A behavior of fish related to the herd behavior of many terrestrial organisms. A possible rationale for schooling is that it reduces predation. Since a predator can eat just so much, it cannot devour an entire school of fish. Some individuals are bound to survive.

This ensures a breeding population in the school. Socially, this instinctive response to predation is involuntary, and thought by some biologists to be individual. Thus, as each individual fish moves to escape predators and maximize its feeding, all others do the same, resulting in a large number of individuals in one area at one time doing the same thing. See *fish, pelagic environment*.

schooner   A sailing vessel with two or more masts carrying fore-and-aft sails. A number of vessels of this type were used in voyages of exploration. The *Fram*, a vessel famous in arctic exploration and now in a museum in Oslo, is a schooner. See *explorers and explorations, Fram, sail*.

Sclerospongiae The coralline sponges, a small class (there are only seven species) of the phylum Porifera. These animals have relatively large and complex skeletons composed of calcium carbonate or of spongin, (a proteinaceous material), or they are siliceous (glassy). See *Porifera, sponge, spongin*.

Scoresby, William (1789–1857) An English scientist and explorer of the arctic. His father was a whaler, and Scoresby went to Greenland and its surrounding waters on numerous voyages with his father. The northernmost point he reached was 81°30', in 1806.

Scoresby retired from the sea in 1822 and became a clergyman. Like many other country vicars of the time, he was a natural scientist and author who wrote of his arctic experiences and on magnetism.

Scotia Ridge   A part of the mid-ocean ridge that connects the southern end of South America to Antarctica by way of a long loop of raised seafloor that runs through the South Sandwich Islands. The ridge ends abruptly in the Scotia Trench east of the South Sandwich Islands. Its deepest point is the Meteor Deep, 8,260m (more than 25,000 feet) below sea level. See *mid-ocean ridges*.

Scotia Sea   The area of the southern Atlantic Ocean bounded by the Falkland Islands and South Georgia, South Sandwich, South Orkney, and South Shetland islands. It was first explored by Sir Francis Drake in 1578.

The conformation of the sea bottom, particularly at the western end of the Scotia Sea, channels the water pouring through the Drake Passage north and east across the Atlantic. There is a bend in the Antarctic Convergence that reflects this.

About 30% of the seafloor consists of foraminiferan calcareous remains, and about 30% consists of the siliceous remains of diatoms. The rest is sand and silted sediment, with a considerable occurrence of manganese nodules.

The warm, deep water of the Scotia Sea moves south, bringing nitrates and phosphates toward the Antarctica. This accounts in part for the incredible plankton populations south of the Antarctica Convergence. See *Antarctic Convergence, Drake Passage*.

Scott, Robert Falcon (1868–1912)   An explorer and officer in the British navy. Scott was given command of the British National Antarctic Expedition, which was charged with the scientific exploration of Victoria Land (Antarctica). He sailed in the H.M.S. *Discovery* in 1901 and spent two seasons (two southern summers) in Antarctica. During the first summer season (1902–1903), Scott, Wilson, and Shackleton reached 82°16'33" South Latitude, which was then the record for southernmost travel on Earth. He sailed for the antarctic again in 1910 with the intention of reaching the South Pole before another explorer, the Norwegian Roald Amundsen.

Scott and his party set off in Antarctica in November 1911, and he and four others reached the South Pole on January 12, 1912, only to find Amundsen's markers already there. The Scott party's trek back to their supply depot was disastrous: one man died of injuries, another committed suicide, not wishing to be a burden, and Scott and the two remaining men were trapped by a blizzard and died. The tragic end of the Scott attempt to reach the South Pole caught the public eye, especially in England, where it was viewed as a national loss. See *Amundsen, Roald; Antarctica; explorers and explorations.*

Scripps Institution of Oceanography   A research facility founded in 1903 by Edward Scripps and his half-sister as the Marine Biological Station of San Diego, California. It is now part of the University of California at San Diego, and operates as a center for research in marine sciences. Its major subdivisions are marine geology, sedimentation, air-sea interactions, ecology, and paleontology. The Institution maintains research vessels as well as satellite receiving equipment. Scripps Canyon, a submarine canyon off San Diego, was explored by the Institution and its research ships.

SCUBA   An apparatus used for breathing while swimming underwater. Scuba diving is probably the best-known diving technique, and is the one that made undersea exploration accessible to the nonprofessional diver. It has made it possible for man to enter into the marine environment. With SCUBA, both the biologist and the archeologist could for the first time observe an undisturbed site. While nets, seines, and hooks have their uses in studies of undersea phenomena, accurate undersea observation and photography became available as adjuncts of the SCUBA diving method.

The Self Contained Underwater Breathing Apparatus (SCUBA) was developed in the 1940s and rapidly became widely exploited. Almost any healthy person can use this apparatus, given reasonable precautions. The depth limit of SCUBA equipment is set, rather arbitrarily, at 50 m (160 feet), and for a short period. The length of stay at this or at any depth depends on the individual diver's physiology.

The SCUBA equipment most often used by amateur divers (all non-helmeted divers) is of the "open-circuit"

type. This consists of a compressed air tank that is carried by the diver. Its mouthpiece is a regulator supplying air at normal atmospheric pressure into a face mask. The diver exhales into the mask and the characteristic stream of bubbles is produced. Closed circuit SCUBA apparatus involves either pure oxygen or an oxygen-helium mixture. Neither technique is safe without extensive training and preparation. Semi-closed SCUBA circuits recycle the exhaled oxygen. See *Cousteau, Jacques-Yves; diving; marine archeology.*

Scyphozoa   A class of coelenterates comprising the jellyfish or true medusae. Their outstanding characteristic is four-part radial symmetry. Both the hydroid (polyp) and medusal generations of the Scyphozoa can be quartered into four identical units.

All Scyphozoa have a bell- or umbrella-shaped body with tentacles in the subumbrellar area. Nematocysts or sting-

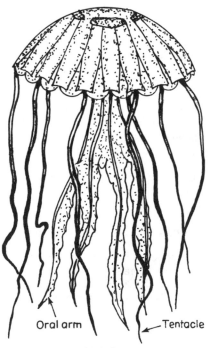

Pelagia, a type of jellyfish

ing cells cover the tentacles. The mesogleal jelly that forms the interstitial material of the jellyfish is fibrous. These fibers, which stiffen the umbrella, are unique to the class Scyphozoa. The interior of the jellyfish has a central, four-pouched stomach and primitive digestive system. Jellyfish are dioecious, with a distinct separation of the sexes.

Jellyfish may vary considerably in size. Some—the Stauromedusae—are nonswimmers and sessile. Some scyphozoans may live several years. They can be any color, and range from colorless to brown, green, pink, or violet. Several species are luminescent. The range of the class is from the ocean surface to almost any depth except the hadal zone, and from tropical waters to the subpolar regions.

The oldest fossil that bears evidence of being a scyphozoan is a Jurassic medusa, but some researchers believe that Cambrian rocks have traces of what may be medusa arms. See Cnidaria; medusa; nematocyst; Sars, Michael.

sea A term used to mean the ocean in general, or a large marginal part of a particular ocean, or a large landlocked body of water that may have an oceanic outlet. The Caribbean is a marginal sea, as is the Sea of Japan. The Aral and Caspian seas are really lakes. See individual names of seas.

sea anemone A sea polyp that resembles a flower, of the phylum Cnidaria, class Anthozoa, and order Actiniaria. Sea anemones are common littoral animals whose general body plan is a ring of tentacles surrounding an oral disk. These animals' stinging cells or nematocysts, which are arranged in multiples of six, are on the tentacles and rarely on the columnar body. There is usually a pedal disk by which the animal attaches itself to a solid surface. Some species are floaters. The anemone has strong longitudinal muscles that can move it along the bottom. Even the most seemingly sessile (nonmoving) individuals can move, albeit slowly.

Most anemones live in rock pools near the surface, although some species are found at great depths. They are most abundant in tropical waters, where they are also more likely to be very large. The giants of the Great Barrier Reef can be up to 1 m (40 inches) in diameter. These animals are attractively colored—specimens have been found with colors ranging from delicate ivories to yellows, pinks, reds, and browns. Some species are hermaphroditic, with a single organism exhibiting both male and female sex characteristics, while others are dioecious. Like many other sessile animals, the anemones can be quite long-lived. Aquarium specimens have been known to live 50 years and longer.

Anemones will eat anything that they can trap with their tentacles. They spit out the non-food components of a meal. Their central cavity is almost filled with a giant digestive gland.

Symbiotic and commensal relationships with other animals are well-documented phenomena in the anemone's life. Wrasses and other fish live in the tangle of the anemone's tentacles, luring prey and catching crumbs from the host's meal. Hermit crabs "adopt" a particular anemone and attach it to their shell. When the crab outgrows the shell and moves into another one, it takes "its" anemone with it to the new shell. See chemical defenses, Cnidaria, commensal relationships, littoral, sessile.

sea bass The common name of the perchlike, mostly marine, fish of the family Serranidae. This family numbers over 400 species, which are found in tropical and semitropical waters. They are the groupers, coneys, grasbys, jewfish, and sea bass. The individuals range in size from 3 to 5 cm (1 to 2 inches) to very sizable fish over 2 m (7 feet) long.

Members of the subfamily Moronidae live in colder waters, and include the familiar temperate-water fish, the striped bass (Morone saxatilis) and the white perch.

The entire group of sea bass and related fish are fished by both sport and commercial fishermen from tropical to temperate regions of the Earth. The group is largely carnivorous, feeding on crustaceans and mollusks. Their breeding patterns vary considerably. Some species are hermaphrodites, others are dioecious, and some have rather exotic behavior: the groupers, for example, change sex in maturity. See *fish, grouper, hermaphrodism.*

sea butterfly   A pteropod marine snail. See *Pteropoda.*

sea cave   A wave-carved cave in a cliff facing the sea.

sea cucumber   A warty, stationary, marine animal of the class Holothuroidea (phylum Echinodermata), which resembles a squash or cucumber. Aristotle referred to an animal that he called a holothurian, but it is not known if he meant the sea cucumber. This common name comes from Pliny's *cucumis marinus.*

Holothurians are found in almost every marine environment. They have elongated bodies, most often 18 to 30 cm (7 to 12 inches) long. There is a tentacle-surrounded mouth at one end and an anus at the other. While the holothurians do have the five-part body that is typical of the echinoderms, the sea cucumber is not quite symmetric. It lies preferentially on one side, and that side is usually lighter in color than the other. It can be white, pink, salmon, or yellow, while the upper surface is frequently much more drab in shades of gray, olive, or brown. The exterior of the animal is a warty and knobby surface that produces a sticky mucus. This sticky material entraps the prey, which is then conveyed to the mouth by the tentacles. The spiny body wall of the sea cucumber contains calcareous ossicles (calcium-containing bone-like structures) that are unique to a species of these creatures and are used as a means of classification. Both the tentacles and the tiny podia operate by way of the water vascular system, which is a common feature of all the echinoderms. When the animal moves, it does so slowly, and looks like a giant, sluggish caterpillar.

The reproductive pattern of the sea cucumber is also fairly typical of the rest of the echinoderms. Some are hermaphroditic (with both male and female characteristics on a single organism) while others are dioecious. Some species, particularly arctic ones, brood their eggs.

Sea cucumbers manufacture a toxic substance with which they appear to kill small animals and stun larger ones such as fish. The toxin doesn't seem to interfere with the use of these animals as food for humans. Dried sea cucumbers are

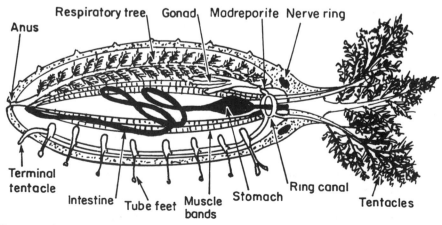

Sea cucumber

often used as basic ingredients in fish sauces and stews in the Far East.

As is the case with many long-lived and fairly sessile organisms, sea cucumbers have commensal organisms living on or near them. The sea cucumbers are also parasitized, particularly by marine snails. See *chemical defenses, commensal relationships, echinoderm.*

sea fan   See *gorgonians.*

sea floor   See *continental drift, ocean basins, ocean floor.*

sea gooseberry   See *comb jellies, Ctenophora.*

sea hare   A shell less gastropod of the genus *Aplysia.* Some species have a rudimentary shell hidden by the mantle. The common name of these animals is based on the appearance of the anterior tentacles, which resemble the ears of a hare, and the arched back.

These hermaphrodites live in warm to tropical seas and produce strings of eggs that when dry look like seed cases. The diet of the sea hare is largely seaweed, but these animals occasionally capture small crustaceans, annelids, and other mollusks.

Aplysia produce a purple liquid as camouflage. This liquid is a bile protein and serves as a chemical defense for this rather slow-moving creature. See *chemical defense, Gastropoda, hermaphrodite, mantle.*

sea horse   One of a group of small marine fish of the family Syngnathidae, order Gasterosteiformes, that have a distinctive horselike head and neck and a prehensile tail. They are poor swimmers, and therefore hold onto seaweed. The sea horse moves vertically using its swim bladder as a float. It feeds on tiny crustaceans and other zooplankton.

The singular structure of the sea horse is its brood pouch, which is beneath the male's tail. The female departs after depositing the fertilized eggs in the brood pouch, and the male tends the eggs.

When they hatch, the young emerge from the male.

sea ice   Any ice, floating or fast, that is formed from freezing ocean water. See *fast ice, ice, pack ice, pancake ice.*

sea lettuce   Several species (especially *Ulva*) of green algae found in temperate to cold ocean water. *Ulva* is found in small sheets that are only two cells thick, which float in the open ocean. The tropical counterpart of *Ulva* is the spongeweed. See *algae, chlorophyll, green algae, spongeweed.*

sea level   The difference in the height of the sea from some arbitrary point on land; in reporting sea level, the value used is an average between the water level at high and low tide.

Changes in sea level depend on a number of factors, such as atmospheric pressure, salinity, and most noticeably, temperature. The volume of cold or high-salinity water is less than that of an equal weight of warmer or fresher water. A high atmospheric pressure tends to flatten the water surface; atmospheric lows have the opposite effect.

The effect of the Earth's rotation is evidenced in the Northern Hemisphere by an accumulation of water on the land to the right of a stream, whereas the accumulation is to the left in the Southern Hemisphere. Thus, oceanic gyres will create differing mean sea levels, a phenomenon that was observed by early explorers of the Caribbean. The mean sea level is highest for the North American continent at Nova Scotia.

Long-term effects of changes in sea level are being carefully studied. Previous advances and retreats of the continental ice sheets have altered the level of the sea to a mark about 100 m (330 feet) below the present mean level. The mean sea level has moved up about 12 cm (4.5 inches) since 1900. The change has been directly correlated with a global warming trend. This in turn has been ascribed to the combination of the general warming that has been going on since the last glacial period, and the greatly increased use of fossil fuels in the last 100 to 150 years. See *greenhouse effect, subsidence.*

sea lily   See *Crinoidea.*

sea monsters   Very large sea creatures that may have been the nucleus for some legends about sea monsters. There are giant pelagic (open sea) and benthic (bottom-dwelling) animals. A few that may have been the foundation for the monster theory could be whales, giant worms, ribbon worms, ribbon fish (Lampridiformes), giant clams, squid, crabs, and even the walrus. See *giant clam, Lampridiformes, ribbon worm, squid, whale.*

sea pansy   A soft coral cnidarian belonging to the class Anthozoa; it consists of a flat plate with polyps on one side. The genus *Renilla* is common to both coasts of North America. See *Anthozoa, Cnidaria, coral.*

sea pen   A cnidarian of the genus *Pennatula*, belonging to the class Anthozoa, and which looks like a feather. There are about 300 species of this colonial animal. They live in shallow to deep water, and their range is almost worldwide. The central stalk or peduncle of the sea pen bears the "branches." Many species are luminescent. See *Anthozoa, bioluminescence, Cnidaria.*

sea robin   A marine fish of the family Triglidae, with an armor plated head. Some species have bony plates over their entire bodies. Sea robins have long, trailing feelers that explore the bottom, searching for mollusks and crustaceans.

The sea robins are often bright green or blue. They can, and frequently do, make audible sounds. The largest species is about 70 cm (28 inches) long.

These fish are occasionally used by humans as food. In the earlier years of the 20th century, sea robins were considered "trash fish" and were dumped overboard if they became entangled in lines set for other fish. See *fish.*

sea snake   A venomous, tropical marine serpent of the family Hydrophiidae. There are about fifty species. Their distinguishing feature is a flattened body; they swim using their tail as an oar. The usual individual is 1 to 1.3 m (3 to 4 feet) long, but the Japanese varieties may be twice that length.

Sea snakes may be found in tropical ocean water anywhere, but the largest number of species are found in Asiatic or Australian waters. Some species lay eggs on the beach, but most bear live young in the water. The sea snakes are fish eaters. They are most often solitary animals, although large groups of them are occasionally sighted. As a whole, sea snakes have especially lethal venom. The venoms are neurotoxins, which attack nerve centers that control breathing and heartbeat. See *reptile, toxin.*

sea spiders   Small, long-legged arthropods that live on ocean bottoms. Also known as whip scorpions, they belong to the class Pycnogonida, and represent about 600 species. The individuals range in size from 2 to 3 mm (0.01 inches) to 50 cm (20 inches) or larger. The small ones are found in shallow water, the larger specimens are deep-sea dwellers.

Sea spiders have four to six pairs of walking legs. They are carnivorous animals, some feeding on other invertebrates by sucking them dry, while others tear food apart with powerful appendages and pass it onto a proboscis—a food gathering mouth structure. While their reproductive mode is internal fertilization, the males incubate the eggs. See *proboscis, pycnogonid.*

sea squirt   A sessile marine animal that looks like a lumpy potato; a member of the class Ascidacea, subphylum Tunicata. Sea squirts live in all seas, with a range from the intertidal flats to oceanic deeps. Their bodies are covered by a thick skin or tunic, which has two large openings, one of which sucks water in while the other expels it. The sea squirts feed by filtering debris, small animals, and plankton out of the water.

The sea squirts are on the whole small; individuals range in size from less than 1 cm to about 6 cm (0.4 to 3 inches) in diameter. Some species are colonial, others solitary. The animals are hermaphroditic, each individual having male and female gonads. When the eggs are fertilized (externally), they hatch into free-swimming larvae that look like tadpoles. As the larvae mature, the notochord does not develop into a spinal column. Instead, the larval organism attaches itself to some likely spot and loses its vertebrate-like structures. In addition to their relatively common form of sexual reproduction, squirts also reproduce by budding.

In addition to their curious, almost vertebrate status, ascidians are peculiar because their blood and tissues contain vanadium oxides, compounds which no other animals concentrate in their systems. See *notochord, Tunicata*.

sea turtle   Any of several quadruped marine reptiles characterized by a typical shell made up of a series of plates, the shape and size of which are species specific. The shell covers both the dorsal and ventral surface of the animal. The head, limbs, and tail may be retracted completely into the shell. The sea turtles range in size and feeding habits but all are strong swimmers that rarely come to shore. Females do this only to lay eggs on their native beaches, Thus, a male turtle may never have been on a beach after hatching. The sea turtles are classified as the leatherback (Dermochelyidae), green, loggerhead, hawksbill, and redleg (Cheloniidae).

The leatherback is the largest sea turtle. It is an omnivore whose shell is often 2 m (6.5 feet) long and it weighs about 500 kg (1,100 pounds). The name refers to the a tough covering of floppy skin, with the characteristic shell beneath it.

The green turtle is also omnivorous. It is an endangered species, and several attempts are underway to preserve it in the wild, restock it on native beaches, and keep it alive in zoos. The green turtle is about half the size of the leatherback.

It is brownish green, but the name is probably due to its green fat. This species is most often hunted for its meat. It lives in tropical waters worldwide.

The loggerhead is about the same size as the green turtle. Its shell is red-brown and its feeding pattern is also a carnivorous one.

The hawksbill is much smaller than the other sea turtles. The average individual weighs less than 50 kg (110 pounds). This species is also taken by hunters; its shell is used to make ornaments. It is a worldwide creature and an omnivore.

The redlegs—the Atlantic and somewhat larger Pacific species—are primarily tropical animals. They are 60 to 70 cm (25 to 30 inches) long, and are omnivores.

The exploitation of the sea turtles for meat and jewelry has greatly decreased their number. Since reproduction is a very energy-intensive activity, these animals do not breed in lean years. The removal of eggs from beaches by humans (the eggs have long been considered as aphrodisiacs in the Caribbean), and the destruction of the beaches used for egg-laying, has further decreased the likelihood of the turtles' survival without the active intervention of conservationists. One project for this purpose involves taking eggs from nests scooped out of the sand by female turtles and moving some of the clutch, which may contain as many as 100 eggs, to another beach where that species of turtle had appeared in the past. It is hoped that the hatchlings who make it back to the sea—despite the attempts the shore birds to pick them off—will at maturity return to the new "native" beach.

While the sea turtles are primarily animals of tropical waters, some are blown off course and end up in cold or temperate-cold water. This puts them into thermal shock, since their reptilian circulatory system is not equipped to deal with temperature drops (which is why reptiles in the temperate zone hibernate). Since sea turtles are an endangered group, many locales post a "turtle watch." In New York waters, for example, sea turtles

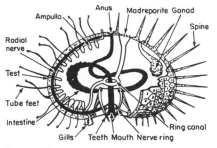

Radial nerve • Ampulla • Anus • Madreporite • Gonad • Spine • Test • Tube feet • Intestine • Gills • Teeth • Mouth • Nerve ring • Ring canal

Sea urchin

that come ashore are routinely brought to the N.Y. Zoological Society's aquarium, which arranges transport for them back to the waters around Florida. See *extinction, poikilotherm, reptile.*

sea urchin   An echinoderm (class Echinoidea), enclosed in a thin brittle shell covered with movable spines; it looks like a dark red, blue, or purple pincushion. The name is derived from the Greek for hedgehog. The sea urchin has been known for millennia, and has left a fossil record since the Ordovician Period, 500 million years ago.

The urchin is a regular echinoid (symmetric) with a roughly globular aboral (top) and a flattened oral (bottom) surface. Its mouth is on the oral surface and is equipped with hard teeth. Almost all of the urchin is covered with a calcareous, spiny test or shell. In addition to the spines, five sets of podia (feet) protrude from the test. The urchin moves using both its feet and its spines.

Urchins are ubiquitous in tropical and warm temperate waters. They are used as a food by many people, and have a place in the lexicon of both the French and Chinese cuisines. The sea urchin's eggs, which develop rapidly after external fertilization, are primary tools for embryologists; the development of the eggs is a "textbook" illustration of the staging of cell division. See *Aristotle, Echinodermata.*

seal   A flippered marine mammal of the families Phocidae and Otariidae, of worldwide range but prevalent in cool to cold coastal waters. Seals vary in size from 1 to 8 m (3 to 25 feet); they have sleek, streamlined bodies and paddle with their flippers, which are modified forelimbs. Their bodies are well insulated with fat, which may serve as a pressure regulator as well as a thermal insulator. Both functions are necessary, since seals dive rapidly to depths of over 250 m (775 feet) and stay submerged for 20 to 30 minutes at a time.

Seals are divided into earless and eared groups. The ear, where it exists, is very small. The earless seals, or Phocidae, number 13 genera and 18 species. These are generally small, and their major means of propulsion is their tail. The eared seals constitute the family Otariidae. They consist of seven genera, which comprise 12 species. This group is the better known of the two, and includes the sea lions and fur seals. They have longer flippers and propel themselves forward with these forelimbs. The eared seals also have much more mobility on land. While their motion on the shore is a comical waddle, it is much more directed than what the earless seal can manage. It is the eared seal that is the circus and aquarium clown.

The average seal has a dense, lustrous, brown pelt, the shade of which varies considerably, from almost black in some species, to a sandy buff.

The bull male of the sea lion group grows much larger than the female, and these enormous animals collect large breeding harems if they can. The young males, or bachelors, live in groups on parts of the beach out of the territory of any bull, and constantly try to effect "harem takeovers."

The young of seals are born after a gestation of about a year. They feed at first on a milk that is extremely rich in both fats and proteins, then progress to a regurgitated diet provided by the mother, and, in a few weeks, are swimming and catching their own fish and crustaceans. Infants and their mothers have strong scent markings—a factor that is vital for the survival of the pup, since it permits the pup and its mother to seek and find each other in the crowded rookeries.

The dynamics of a seal colony have been extensively studied using elephant seals as subjects. These animals are somewhat larger than the average seal, and also have huge noses, from which they derive their name. They were hunted for fur to near extinction in the 19th century. Effective protection by the United States and Mexico has re-established a large population of elephant seals on the islands off California. A measure of the size of the population was attempted by banding the pups. The protection was effective, since there were six pups born in 1911 and 25,000 in 1982. The extent to which the area can support the population growth is still not known. See *mammal, pinniped.*

Sealab A U.S. Navy project designed to determine the ability of humans to live at relatively great depths (180 m or 600 feet) for long periods. The Sealab habitat was occupied for over a month at one point in its use. The breathing mixture pumped into it was one of oxygen and helium. See *diving.*

seamount An undersea, relatively isolated mountain rising from the seafloor to a height of about 1,000 m (3,300 feet). If it has a pointed summit it is called a sea peak; if it is flat-topped, it is called a guyot.

The shape of most seamounts is elliptical. Only the very small ones have steep sides; large ones have gently sloping sides. These features are true for the seamounts present in every ocean. Hundreds of these mountains have been mapped, particularly in the Pacific, and it has been estimated by oceanographers that this number is only a small portion of the total. Although they are separate entities, seamounts occur in groups or chains, as do island arcs.

Seamounts seem to be of volcanic origin, and may result from eruptions along crustal fissures. In most cases the volcanic cone shows signs of weathering (erosion). There may be sedimentary accretion on a seamount, or a coral reef that has used the dormant or dead volcano as a base for growth.

Some seamounts, including a significant number with coral or limestone tops, are below the level of coral growth. This suggests subsidence by an unknown mechanism. See *continental drift, crust, guyot, hot spot, island.*

seawater Water in or from the sea. Seawater comprises about 97% of all the water on Earth. It is of remarkable worldwide consistency in terms of content. This was confirmed by the careful analyses done by William Dittmar, the chemist on

Some Elements in Seawater

Strontium (Sr^{2+}) ion	8 ppm (parts per million)
Oxygen (as O_2)	4.6–7.5 ppm (this varies with depth; the greatest concentration is at the surface)
Silicon (Si^{4+}) ion	3 ppm
Fluorine (F^-) ion	1.3 ppm
Nitrogen (N_2, NO_2, NO_3^-, NH_4^+)	0.5 ppm
Argon (A)	0.5 ppm
Lithium (Li^+) ion	0.17 ppm
Phosphorus (HPO_4^{2-}, $H_2PO_4^-$) ions	0.07 ppm
Iodine (I^-) ion	0.06 ppm
Carbon (in CO_2)	Traces

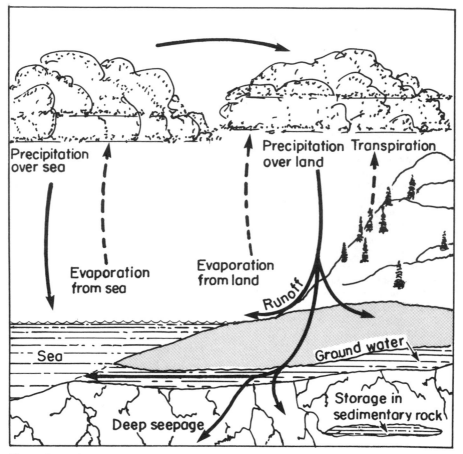

The water cycle

the *Challenger* Expedition of 1872–1876. Dissolved materials, other than sodium and potassium chlorides, comprise only about 0.05% of all of the dissolved material in seawater. This small percentage includes traces of practically every element. But while they are present, most elements are present in amounts too small to be of commercial interest.

The carbonate and halite (chloride, bromide, iodide) deposits found on land are of marine origin, and date from all geologic periods, from the Precambrian to the present. Since they are fairly consistent, it is assumed that the composition of seawater has been essentially the same through time. The composition of Pre-

cambrian water is not known, but is assumed to be similar to the seawater of today. While the composition of seawater is relatively uniform, this does not mean there is no difference.

The patterns of water flow in various oceans have been established by core sampling. The study of foraminiferan remains shows that not only is there local variation in the temperature and salinity of seawater, because land formations hinder its circulation, but that these formations maintain the individual character of whole oceans. Thus, deep water flows from the Atlantic to the Pacific, but the water in the two bodies have slight but significant differences:

Atlantic Deeps	Substance	Pacific Deeps
Relatively rich	$CaCo_3$	Relatively poor
Relatively poor	SiO_2	Relatively rich
Relatively rich	O_2	Relatively poor

The surface of the Atlantic is about 2° warmer and 1 ppm more saline than that of the Pacific. The difference in the waters dates roughly from the point about 12 million years ago when the Panamanian isthmus closed. Similar events isolated the Mediterranean. A comparison of the levels of the isotopes C^{13} and O^{18} was used to establish this point. The assumptions made were that there was unimpeded circulation prior to the change in landmass, and that the rate of decay of the radioisotopes is known. See *antarctic currents, carbon dioxide*, Challenger, *H.M.S., circulation, meteorology, minerals, radioactivity, salinity.*

Sedgwick, Adam (1785–1873) An English geologist who studied and named some of the earliest segments of the Earth's history. Although he trained in mathematics, he was appointed professor of geology in Trinity College, Cambridge, in 1818, and proceeded to become a geologist. In 1827 he and Roderick Murchison investigated rock formations in western Scotland and then in southern England. They gave the Cambrian Period its name, using an old name for Wales. This team also examined Devonshire and published their work in 1839. A profound professional disagreement led to both the breakup of the team and the incorrect naming of some Paleozoic strata. This was not resolved until after the deaths of both Sedgwick and Murchison. See *Cambrian Period, Devonian Period, Ordovician Period.*

sediment   Matter that settles to the bottom of the sea. Sediments were first seriously sampled by Matthew Maury, and systematically studied by the *Challenger* expedition (1872–1876). They are divided into two large groups, the terrigenous and the pelagic. The former is largely of terrestrial origin and consists of rock, gravel, sand, and mud—with volcanic intrusion, and coral debris. The pelagic sediments are those that "rain down" from all levels of the water and finally land on the bottom. These are the shells or skeletal remains of plankton.

The *Challenger* sampling showed the divisions of the types of sediment, with calcareous sediment disappearing at about 4,000 to 5,000 m (13,000 to 16,000 feet). The red clay that Murray described as soft, plastic, and greasy is of volcanic origin and terrigenous. At still greater depths radiolarians were found. Prior to the *Challenger* sampling, these were unknown. The sediment at these greater depths is a pale gray to yellow, claylike material.

	Ocean Area Covered, Millions of km²(mi²)			
Sediment	Atlantic	Pacific	Indian	Percent
Calcareous				
Globigerina	40.1(15.5)	51.9(20)	34.4(13.3)	35
Pteropod	1.5(0.58)			
Siliceous				
Diatomaceous	4.1(1.58)	14.4(5.56)	12.6(4.86)	10
Radiolarian	uncertain	6.6(1.55)	0.3(0.116)	
Terrigenous				
Red Clay	15.9(6.14)	70.3(27.13)	16.0(6.17)	5

Also found on the sea bottom are minerals precipitated from the water, and material of extraterrestrial origin. Although they constitute only a small part of the total sediment, cosmic dust, meteor fragments, and tektites are found in all old sediments and present-day bottom samples.

The location of sediments is a function of their deposition rates or their movement to the ocean by rivers, glaciers, and wind, and of the rate of generation of various materials by volcanic or biogenic activity. The sediments are then carried along and finally dropped in the ocean by the various currents that move solids horizontally and vertically. See Challenger, H.M.S., Foraminifera, ocean currents, ocean floor, Maury, Matthew F., Radiolaria.

seiche   The occasional, rhythmic rise and fall of water in enclosed lagoons or bays. It is not tidal. The phenomenon was carefully observed on Lake Constance (Switzerland) in the 16th century. The oscillation of the water is the result of a disturbance that sets up "waves." The time period of these "waves" is described by Merian's formula,

$$T = \frac{2L}{\sqrt{gh}}$$

L = length of the body of water
h = depth $\quad$ " $\quad$ " $\quad$ " $\quad$ " $\quad$ "
g = gravity

The so-called wave will bounce back and forth until its energy is dissipated through friction. The seiche is dependent on the shape of the basin in which it occurs.

Coastal seiches are derived from air-water interactions or long-period waves coming in from the open ocean. These are caused by strong winds or large atmospheric disturbances, or by earthquakes either on the coast or undersea. The "surfbeat"—a phenomenon of an embayed area—has also been named as the origin of seiches.

The effect of seiches can be good. They are the means of mixing water in a constricted area, and of distributing nutrients, dissolved gases, and heat. On the other hand, moored ships have been rammed into docks, or even broken up,

when hit by a node of a strong seiche. See estuary, tide, wave.

seine   A net that hangs in the water with one edge held up by floats and the opposite edge held flush against the bottom by weights. When the sides of the net are brought together, they enclose the catch. The net and its contents are then drawn ashore or onto a fishing boat. A purse seine is one that is drawn shut by means of an interwoven rope. See fishing industry, net, purse seine.

seismology   The study of earthquakes and of artificially produced vibrations of the Earth, such as those caused by nuclear explosions. When earthquakes or landslides occur under the sea, they set long waves in motion. These waves are incorrectly called tidal waves; they are properly called tsunamis or seismic surges.

Whether naturally occurring or set off by explosives or by high-energy electrical discharges, these long waves have a measurable velocity. When this data is combined with gravity and magnetic measurements, and with much higher frequency echo soundings, the result is a profile of the bottom. This profile is quite detailed since different rocks and sediments "reflect" signals differently, and this permits the composition of the bottom to be determined. See earthquake, oceanography, surveys, tsunami.

semidiurnal tide   Tides that occur in a cycle of two high tides and two low tides each tidal (lunar) day. See lunar day.

Sepia   A genus of cuttlefish. It also defines the black fluid or ink produced by some cephalopods as a protective device. See cuttlefish, octopus, squid.

sessile   A term used to describe an organism that is attached to another structure by its base (i.e., unable to move freely); a description of animals that do not move, or whose movement is slow enough that they need other means of avoiding predation. Many different species are sedentary, and these creatures range through almost every phylum. Some are motile for a part of their lives and sessile during the rest. The oyster is an example of a sessile adult that comes from a motile juvenile.

setae   The small bristles on either arthropod (crab, lobster, etc.) exoskeletons, or the paired projections on the segments of annelid worms. On polychaete worms, the setae are borne by the parapodia, the "false feet" that characterize this group. See *Annelida, Polychaeta.*

sferics   Also known as atmospherics; the electromagnetic radiations that are end-products of atmospheric electrical discharges. They are the principal causes of radio static.

Since sferics are the result of discharges created by storms, they are used to track the paths of storms or to predict their activity. Different storms produce different discharge patterns. See *electromagnetic spectrum.*

shad   See *herring.*

shark   Any of a widely diverse group of carnivorous, cartilaginous fish most prevalent in warm seas, where they are found at varying depths, with some found at great depths. The sharks belong to the families Carchariidae, or sand sharks; lamnidae or mackerel sharks; Carcharhinidae, or requiem sharks; Squalidae, or dogfishes; and many others. Most sharks kill their prey by biting chunks out of it. Some species have poisoned spines anterior to the dorsal spine, whose function may be connected with feeding and also may be a defense mechanism. The outstanding characteristics of the sharks are the five sets of exposed gill slits, an asymmetric tail (the top part is larger), numerous deciduous teeth (they fall out and are replaced), and placoid scales. These scales are not overlapping; each is small, uniform in size, and bears a tiny barb in it that points toward the tail.

Sharks are varied in size, ranging from the small dogfish of about 1 m (3.3 feet) to the basking sharks and the great white sharks, which are among the largest fish and largest of all sea creatures excepting the whales. The greatest number of shark species are relatively small, at less than 3 m (10 feet). The whale shark, the largest species, has been given an estimated maximum measurement of about 10 m (35 feet). It eats plankton and small fish. There is much conjecture (and hysteria) about the size and ferocity of sharks. There are, however, shark repellants. Some fish that sharks prey on manufacture toxic substances as protection.

Sharks have been intensively studied, putting some favorite myths to rest although perhaps only temporarily. Like other predators, they kill for food and do not eat for days or weeks afterwards. They rest and sleep without moving.

Many sharks rise through the ocean in the daylight and descend at night in a diurnal pattern that mirrors the movement of their prey. Since they do not have a swim bladder, it is thought by some that the shark's enormous liver, whose volume consists largely of oil, may be an equilibrating device. Sharks also store urea (nitrogenous waste), which may also be a depth-equilibrating substance.

Sharks comprise a large number of ovoviviparous species with eggs that hatch within the female's body, as well as viviparous ones. The period of gestation is quite a long one: dogfish pups may spend two years in the oviducts of the mother.

Sharks are caught commercially for food and for their livers. The liver is used as an oil source. Large quantities of squalene can be produced from shark liver oil; this compound, when purified, is used extensively in the preparation of pharmaceuticals and cosmetics.

Because there are so many gaps in what is known about sharks, the possibility of overfishing these animals may be a very real one. See *Chondrichthyes, fish, individual sharks, marine oil.*

shearwater   A member of a group of oceanic, web-footed birds of the family Procellariidae, that glide along wave troughs. The puffins are typical members of this group. These birds are almost always flying or in the water. They come to land only to make nests and raise their chicks. They do so in dense colonies on offshore islands. Shearwater colonies are found from cold temperate to almost polar regions on the eastern side of the Americas. The colonies also are found off the coasts of Australia and New Zealand.

The single egg hatches in about seven weeks, and is fledged in about 10 to 12 weeks. The typical shearwater is a slender bird about 30 to 60 cm (12 to 25 inches) long. Its feathers are gray-brown to black on the upper surface; the underside is white. The body is long and the wingspan distinctive, being at least 1.5 times as long as the body. Its hooked bill makes the shearwater an effective predator.

Widespread and typical species are the sooty shearwater (*Puffinus griseus),* the common shearwater, also known as the Manx *(P. puffinus),* and Newall's shearwater, which is found in Hawaii (*P. newelli).* The short-tailed shearwater (*P. tenuirostris)* is found in Australia or Alaska, and is hunted for its meat. See *bird.*

ship   An oceangoing vessel, usually of some size. Using this somewhat arbitrary definition, the ocean voyages of the Polynesians and some of the Norse voyages were made in boats, not ships. The first significant ships were Phoenician galleys, which eventually became the ships used throughout the entire Mediterranean Sea.

Sails as a reliable means of propulsion appeared about 1000 A.D., and the original sailing ships had one mast, and later two and three. The rudder was developed about 1200 A.D. and allowed for greatly increased maneuverability.

Merchant vessels were usually not different from warships. There were a few exceptions: war galleys had rams with metal sheathings on them and defensive or offensive fortifications were constructed on these vessels to give the fighting crew a platform from which to oper-

ate. Gradually the difference between trading and fighting ships became more pronounced, and trading vessels became broader than warships and carried less sail. The breadth of the beam gave trading ships increased cargo space, and the decrease in sail meant that a smaller crew could work the ship; however, these factors meant that merchantmen were increasingly slower than warships. The size of the ships by the 15th and 16th centuries is significant: they were not large by our standards, the caravel averaging only about 250 tons and the largest carracks 1,500 to 1,600 tons.

The peak of success of the sailing merchant ship was the Age of the Clipper. These beautiful vessels were economically significant because they could handle small, expensive cargoes at great speed. However, they became uneconomical once steamships became reliable.

The first steamship, a merchant ship, crossed the Atlantic in 1838. The paddlewheeler came and disappeared within a decade, so rapid was the technical change that brought about a revolution in ocean travel. The largest ship of the time, the *Great Eastern* (19,000 tons) was a commercial failure, but the beginning of a total change in transportation: before this point ocean travel had not been commonplace, whereas by the end of the 19th century it was. See *individual ships by type.*

shipworm   An elongated, boring marine clam that resembles a worm; also known as pileworm. There are about 14 genera of this family, the Teredinidae. They are of economic importance since they cause damage to wooden piers, pilings, and ships. See *borers, clam, Teredinidae.*

shore   The seacoast, from the lowest low-water line to the highest high-water line. See *beach, coast.*

shrimp   A group of edible, commercially important decapod crustaceans belonging to the order Natantia, with a semi-transparent body and a flexible abdomen that ends in a fantail. There are roughly 2,000 species of shrimp, ranging from freshwater to deep-ocean dwellers, and of sizes from less than 7 cm to more than 20 cm (3 to 8 inches).

These delicate animals move rapidly in a backwards direction by flexing their abdomen and tail. Most shrimp are omnivores, some are scavengers. They are rather typical of the decapods, living in almost every range of marine environment. They produce thousands of eggs upon a single mating, and these are carried on the swimmerets of the female as they develop. In some species the juveniles live as a separate, segregated group.

Shrimp are frequently hatched in the shallow waters of estuaries, and the continuing study of wetlands and marshes produces fresh information about the life cycle of this group of animals. This is of major interest not only to ecologists, but to commercial groups, since shrimp are an important part of the overall fish and shellfish catch. Many, many tons of these crustaceans are taken annually in the Gulf of Mexico, the Mediterranean, the Baltic, the Sea of Japan, and other areas of the world's oceans. See *Arthropoda, Decapoda, mantis shrimp, marsh.*

Sicily The largest Mediterranean island. It is situated off the coast of southern Italy, to which it was geographically connected in the Pleistocene epoch, and possibly earlier. It has one of the world's most active volcanoes, Mount Etna. The whole island and the surrounding sea are subject to frequent seismic activity. See *island, Mediterranean.*

Sicily, Straits of A passage about 500 m (16,000 feet) deep between the island of Sicily and the Italian mainland. It separates the Mediterranean into two large areas. The cold Atlantic water that enters the Mediterranean at Gibraltar flows along the north shore of Africa and enters the eastern Mediterranean by passing through the Straits of Sicily. Geologically,

this passage is old. It has certainly existed since the mid-Pleistocene, and was probably a water channel before that time. See *Mediterranean, strait.*

silicoflagellates Small phytoplankton that engage in photosynthesis and, like diatoms, have a glassy frustule or shell. They range in size from about 10 to 150 micrometers. (1 micrometer = 1 millionth of a meter). More members of this group existed in earlier geologic times than do now. Their fossil remains are found in sedimentary rocks dating from the Cretaceous Period to the present. See *diatom, phytoplankton, plankton.*

silicon A reactive, nonmetallic element, which in combined form is, after oxygen, the most abundant element in the Earth's crust. Its compounds are found in the sea, and are both biogenic and nonbiogenic in origin. Silicates predominate in the nonbiogenic sediments of the sea; quartz, feldspar, and other clay minerals are either carried into oceans by rivers, or are the result of undersea volcanic activity. Sand is silicon dioxide (SiO_2).

Biogenic silicates come from the tests (shells) of diatoms, silico flagellates, and radiolarians. Differing depths have characteristic concentrations of silicates, as do different oceans. Deeps in the Indian Ocean and Antarctic deep water are particularly rich in silicates. There is obviously a rapid turnover in the use of silicon, but the exact mechanism by which this occurs is unclear. See *diatom, element, minerals, Radiolaria, sediment, silicoflagellates.*

sill An underwater ridge that separates ocean areas. It may be at the mouth of a fiord or between two oceanic basins. The Caribbean, for example, is separated into several basins by sills, and a sill separates the Mediterranean from the Atlantic Ocean at Gibraltar.

silt Finely divided particles that are the intermediate form of particulate matter between sand and clay. Particles of silt range from 0.01 to 0.05 mm in diameter. Mud is within this range.

Silurian Period　The period following the Ordovician Period of the Paleozoic Era and preceding the Devonian. This time is usually given as 430 million to about 395 million years ago. The name Silurian was devised by Murchison, who chose it for the Silures, a tribe living in pre-Roman Wales.

The Silurian was an age of mountain building: The events of this time raised the Scottish Highlands in Britain and the Taconic Mountains in eastern North America, and closed the proto-Atlantic Ocean. Canada, Scandinavia, and South Africa were probably equatorial landmasses, Japan and the Philippines were near the North Pole, and South America and Australia were near the South Pole.

Generally speaking, the climate throughout this time was mild to tropical and the seas were shallow, a combination that encouraged the growth of calcareous invertebrates, which in turn produced vast limestone deposits. As the tiny animals died, their shells piled up in layers many meters thick. Niagara Falls is a prime example of a massive limestone formation of Silurian origin.

The first fish that had true jaws appeared in the Silurian, as did land plants. Fossil trilobites, graptolites, crinoids, brachiopods, and mollusks are also frequently found in the deposits of Silurian rock systems. These remains are of worldwide distribution and are used to date these rock formations. See *fish; geologic time; index fossil; Murchison, Roderick Impey; Ordovician Period; Paleozoic Era.*

Siphonophora　An order of the Cnidaria that forms free-swimming or floating pelagic colonies of both polyp and medusal forms. *Physalia* is an example. See *Cnidaria, pelagic environment, Portuguese man of war.*

Siphunculida　A phylum of sausage-shaped worms. These animals are small,

ranging from 1 to 2 cm (0.5 to 1 inches long). Their prominent physical characteristic is a long proboscis equipped with a circle of small tentacles. The proboscis collects plankton and debris. Siphunculids are usually burrowers in the ocean bottom; they live in all habitats, from intertidal zones to hadal depths down to 4,000 m (13,000 feet). Those that inhabit the depths are usually larger, some growing to lengths of 50 to 60 cm (18 to 25 inches).

Siphunculids live in commensal relationships with other animals. Their association with entoprocts and coral has been investigated. See *benthos, commensal relationships, intertidal, plankton, proboscis.*

Sirenia　The sea cows; a group of large, herbivorous, aquatic mammals that includes the manatee and the dugong. Most of them are extinct or on the endangered species list. The largest sirenian, Stellar's sea cow, was about 8 m (25 feet) long and weighed several hundred kilograms. It was hunted to extinction about 100 years after its discovery in 1741.

Dugongs live in warm, coastal Indo-Pacific waters; manatees live in the coastal waters of the Atlantic. They will enter estuaries, where they are frequently injured or killed by the propellers of motor boats.

The manatee is a fairly representative sirenian: it is 3 to 5 m (10 to 16 feet) long, rotund, slow-moving, and cumbersome. Manatees have strong flippers and are skillful swimmers. They live in herds, tend their single calves communally, and eat enormous quantities of *Zostera* and water hyacinths.

The sirenians are related to elephants. There are many more representatives of this group in the fossil record than exist now: 20 genera have been found from the Eocene to the present. See *extinction, mammal.*

sirocco　A hot, dry, sand-laden wind blowing north out of Africa. It is a regular feature in spring in the Mediterranean region. Since moisture is picked up by the wind as it passes the sea, the sirocco weather in Europe is humid. Sand is frequently precipitated with the rain brought by the wind.

In Egypt, the same winds, blowing more easterly than northerly, are called the khamsin. See *wind*.

Skaggerak   A body of water in the northeastern part of the North Sea, between Denmark and Norway. It is the connecting waterway between the North Sea and the Kattegat, which leads to the Baltic. The two passages are ice-free year-round, and are vital links in the shipping of the Scandinavian countries, the German northern ports, and all of the Baltic. Because of high volumes of river runoff, the salt level of both the Skaggerak and the Kattegat is low, and the mean sea level of both is high. See *Baltic Sea, North Sea*.

skate   A flat, cartilaginous fish with distinctive large pectoral fins which extend from nose to tail, giving the fish a diamond shape. The upper surface resembles sharkskin. Some skates have an electric organ in their tail and fend off predators with electric charges. There are 99 genera in this suborder, Rajoidea, of the order Batoidei. The skate has a worldwide range, and a very wide depth range as well. Skates have been found in depths beyond 2,700 m (9,000 feet).

Skates are bottom dwellers. They range in size from the Atlantic little skate (50 cm or 19 inches) to the big (2.5 to 8 m or 8 to 26 feet) Pacific skate. They are taken by both commercial and sport fishermen. See *Elasmobranchi, ray*.

skua   A powerfully built brown seabird of the arctic and antarctic. The skua nests in either of the two polar regions. The great skua *(Catharacta skua)* is about 60 cm (24 inches) long. The three jaeger species, of the family Stercorariidae, are sometimes called the lesser Skuas. All skuas prey on other birds, taking eggs and young gulls, penguins, terns, and petrels. The skua, like gulls, will eat garbage, and in dire need will eat their own young. See *bird, carnivore, jaeger*.

sloop   A sailing vessel with one mast, rigged fore-and-aft. The name is also used today for smaller naval escort vessels.

snorkel   A breathing tube used by an underwater swimmer in conjunction with a mask in breath-hold or free diving. See *diving, SCUBA*.

SOFAR   Sound Fixing And Ranging, a triangulation technique based on the mode of travel of sound waves underwater. Sounds created in deep ocean waters (e.g., by an explosion) move great distances with little attenuation. Land-based listening stations in several different locations that receive such sound signals can then pinpoint their origin. See *navigation, SONAR, sounding*.

soft coral (Alcyonacea)   Corals that, unlike the hard or true corals, do not surround themselves with a distinctively shaped calcareous (hard, calcified) skeleton. Instead, they are related to the gorgonians, and with them constitute the subclass Octocorallia of the class Anthozoa, phylum Cnidaria, whereas the true corals constitute the subclass Hexacorallia. The names indicate the number of arms about the mouth of the animal; and the number of symmetric segments of the body: six or multiples of six in the Hexacorallia and eight or multiples of eight in the soft corals.

Soft corals are the dominant corals in the Indian and Indo-Pacific oceans; the gorgonians are predominant in the Atlantic and eastern Pacific.

Alcyonaceans are usually found in symbiotic relationships with Zooxanthellae—the dinoflagellates. The dinoflagellates are photosynthesizing organisms, and the result of their pairing with soft corals is that the two animals can live in water that has no or little plankton). As long as there is light, the coral does not starve. See *chemical defenses, coral, dinoflagellate, gorgonian, symbiosis*.

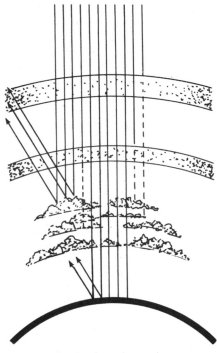

Sunlight reflection from the earth

solar energy   Energy given off by the Sun in the form of electromagnetic radiation. It is the ultimate source of most of the energy on Earth. Solar energy is, overall, fairly constant, and the energy directed at the Earth, known as the solar constant, is 2 g cal/cm^2 minute or about 1.4 kW/m^2. About 35% of this energy is radiated immediately back into space by the top layer of the atmosphere. Only about 50 to 60% of the Sun's energy (i.e., only about 0.8 kW/m^2) reaches Earth either directly or indirectly via reflection from clouds.

The topmost layer of the ocean (top 10 m or 33 feet) absorbs heat and light. Some energy is radiated back into the atmosphere and some is directed downward to a layer 10 to 20 times deeper (100 to 200 m or 330 to 650 feet). Some solar energy is also used to evaporate surface water. This energy—the latent heat of vaporization—is the driving force for oceanic winds, which in turn generate ocean currents and waves. The ocean currents are, in turn, the means of equilibrating water temperature. Through this mechanism, energy is transported from low latitudes to high ones at the rate of about 10^{15} (a quadrillion) calories per second.

Since there is considerable vertical temperature variation in the Earth's oceans, particularly in areas of high insolation (sunshine) and low latitude, large-scale experiments to produce energy by using thermal gradients in ocean waters (from 10° North Latitude to about 10° South Latitude) are being carried out. See *atmosphere, electromagnetic spectrum, insolation, reflection, tide, wind.*

solar tide   A partial tide caused by the gravitational pull of the Sun on the Earth. The Earth is closer to the Sun at perihelion because the Earth's orbit is an ellipse. Perihelion occurs in winter in the Northern Hemisphere, while aphelion (the farthest distance of the Earth from the Sun) occurs in summer. (The moon also moves closer to and farther from Earth respectively, during these two seasons.) Because of both of these movements, the tidal ranges are greater in the Northern Hemisphere during the winter than they are in the summer.

The Chinese, by the 2nd century B.C., recognized the correlation of both the Sun and the Moon in determining the tide. See *aphelion, lunar tide, perihelion, tide.*

sole   An edible flatfish of the family Soleidae. There are about 100 species that live in temperate to tropical waters. The sole lies on the ocean bottom, on its white or blind side. Sole are reasonably large fish, averaging about 50 to 60 cm (1.5 to 2.0 feet) in length. The young sole swim upon hatching, and have an eye on each side of the head. By the time the hatchling is a few days old, it has drifted to the bottom, and its left eye (usually) migrates to the right side of its head.

Sole are fished commercially, particularly in the Atlantic and Mediterranean.

The Dover sole *(Solea solea)* is one of the most highly prized of these fish. See *flatfish.*

Solomon Basin   A part of the Coral Sea in the southwestern Pacific Ocean, east of Papua New Guinea and west of the Solomon Islands. It lies at about 150 to 160° East Latitude and 5 to 12° South Latitude. The region was first approached by Europeans in 1567, when Alvaro de Mendoza entered the area as he traveled west from Peru. However, the Polynesians, Arabs, and Chinese had preceded him. Several other European voyagers also claimed to be the first on the scene. The *Challenger* Expedition charted the area.

The northern reaches of the area, south of New Britain, constitute a geologically young geosyncline (depression) that is variously called the New Britain, Bougainville, or Solomons Trench. The deepest point of this depression is the Planet Deep, 9,140 m (29,988 feet) below the surface—a greater distance below sea level than Mount Everest is above it.

Seismic activity is frequent in the Solomon Basin, as are undersea landslides. This, and the high rate of island weathering, produce a good deal of terrigenous sediment inshore. The sediment is largely from *Globigerina* in the southeast part of the Basin. This part of the Pacific seems to have subsided in the late Tertiary, drowning terraces on some of the larger islands.

The climate in the area of the Solomon Basin is hot and humid most of the year. There is a considerable cloud cover that keeps evaporation and therefore salinity down. The temperature of the water is about 27°C (80°F) year-round.

During World War II there were major battles in the Solomons, particularly on Guadalcanal. See *explorers and explorations, Globigerina, Pacific Ocean, subsidence, trench.*

solstice   The two days each year when the polar axis is inclined 23°30' toward the sun. The summer solstice in the Northern Hemisphere is on June 21 (except in leap years)—the day on which the noon Sun is directly overhead at the Tropic of Cancer, 23°30' North Latitude. In the Southern Hemisphere, this day is the winter solstice.

On December 21 (except in leap years), the noon sun is directly overhead at 23°30' South Latitude—the Tropic of Capricorn. This is the summer solstice in the Southern Hemisphere and the winter solstice in the northern.

On the day of the winter solstice (in either hemisphere) the sun never rises above the horizon for an observer at the 66°30' parallel of latitude (the Polar Circle) or any point closer to the pole than the Polar Circle. See *equinox.*

solubility   The extent to which a substance will dissolve in a solvent. In the context of marine science, the solvent is water.

The extent to which a substance dissolves is a function of the properties of both the solvent and solute (the substance being dissolved). Solubility is determined by the difference in the intermolecular forces between the molecules of the pure solvent, the molecules of the pure solute (the initial state), and the molecules of the solvent and the solute in the solution (the final state).

Almost every substance has a measurable solubility in water, including those that are often thought of as insoluble. The solubility may be expressed either as milligrams per liter (mg/l) or millimoles per liter (mmoles/l). The solubility of any substance is variable, changing with changes in temperature and pressure. Calcium carbonate is a significant example of this: it is sparingly soluble in ocean water and with increasing depth precipitates out of the water as solid particles of salt; however, at still greater depth and pressure it redissolves. See *minerals.*

solution   A homogeneous molecular mixture of two or more substances, of which one is usually a solid (the solute) and the other a liquid (the solvent). In the context of the marine environment, water is the solvent—the ingredient present in the greater quantity—and the other ingredient(s) are the solute(s) [the substance(s) dissolved]. Solutes may also be gases such as carbon dioxide, oxygen, or nitrogen; or they may be solids such as sodium chloride and various other salts, or metallic atoms. See *minerals, solubility, water.*

Somali Current   A strong, north-flowing current along the East African coast. It is part of the Indian Ocean gyre. The velocity of the Somali Current averages about 4 km/hour, and has been clocked as high as 7 km/hour. The Current flows northeast for about 1,500 km (1,000 miles) and then turns east at the Horn of Africa and continues as the Monsoon Current. The stream exists only in summer. See *current, gyre, Indian Ocean.*

SONAR   The acronym for Sound Navigation And Ranging, a technique used at sea for detecting and determining the position of underwater objects (e.g., sunken ships, schools of fish), and for determining the depth of the water under a ship. See *echo sounding, LORAN, SO-FAR, sounding.*

sounding   An old nautical term for the measurement of the depth of a body of water. Many mechanical devices were proposed for this purpose, but none could replace the seaman's weighted line. Bonnycastle first attempted echo sounding in the 1830s, but the best readings were those of Sir James Clark Ross, who used a leaded line when he charted the Arctic in 1839–1843.

Matthew Maury corrected depths erroneously reported by earlier surveyors. These were incorrect because it was not realized that the soft sediment through which the sounding weights fell constituted the sea bottom.

The *Challenger* Expedition's work was meticulously careful. This group used weighted, thin piano wire for both sounding and sampling. See Challenger *Expedition; echo sounding; Maury, Matthew Fontaine; Ross, James Clark; seismic sounding, SONAR.*

South American Plate   One of the larger elements of the Earth's crust. It is approximately continuous with the western and Caribbean coastlines of the South American continent. On the Pacific side it abuts the overriding Nazca Plate. On the Atlantic side, the South American Plate spreads westward from the Mid-Atlantic Ridge. See: *Atlantic-type margin, continent, crust, Mid-Atlantic Ridge, mid-ocean ridges, Nazca Plate, plate tectonics.*

South China Sea   Part of the warm, complex ocean area at the joining of the Indian and Pacific oceans, bordered by the mainland of Asia, the Indonesian islands, Japan, and the Philippines. The area is pinched into a large gulf by Thailand. The South China Sea is a deep basin (over 5,000 m, or 15,000 feet), with the circulation of its bottom water impeded by the Bashi Sill between Taiwan and Luzon and the Karimata Sill between the China and Java seas and the Malacca Strait, the connection to the Indian Ocean. The sea is rich in sediment brought down the Red and Mekong Rivers. The shoreline and nearshore undersea landforms of both of these exhibit the typical structures of drowned river valleys. The major climatic feature of the South China Sea is the monsoon. See *Mekong River, monsoon.*

Southeast Indian Rise   A part of the mid-ocean ridge system which divides the Australian from the Antarctic Plate. See *mid-ocean ridges, plate tectonics.*

South Equatorial Current and Countercurrent   Two fairly well defined streams in the Atlantic; the current flows east-to-west and the countercurrent flows west-to-east. In the Pacific however, island groups change the direction of the current, with the result that the continuous gyre is not as apparent. The South Equatorial Current enters the western Pacific and is turned southward through Melanesia into the Coral Sea and the Fiji Basin. See *currents, Equatorial Currents, gyre.*

Southeastern Pacific Plateau The portion of the Pacific Ocean floor closest to South America. Except for some notable deeps, it is continuous with the Pacific Antarctic Ridge, and rises to an average 2,000 to 3,000 m (6650 to 9900 feet) above the mean level of the ocean bottom. See *mid-ocean ridges, plateau.*

spanker A fore-and-aft sail on the mizzenmast (the mast behind the mainmast) on a sailing ship. See *mizzen, sail.*

spar A sail-supporting timber. It can be a mast, a yard, or a boom. See *boom, mast, sail, yard.*

specific gravity The ratio of the density of a substance to the density of water, at 4°C and 760 torr, the normal atmospheric pressure at sea level.

sperm whale A large blunt-nosed whale *(Physter catodon)* noted for its enormous squared off head containing a large closed cavity filled with a fluid mixture of spermaceti wax and sperm oil. The sperm whale, or cachalot, has small flippers, a lumpy back, and teeth. It is dark gray to brown. Males have an average length of about 20 m (65 feet); females are smaller. The sperm whale can dive rapidly to depths of 350 to 400 m (1200 to 1300 feet), and has been found at even lower depths. The whale lives in small herds in temperate to tropical waters. Its principal food is squid, including giant squid.

Spermaceti is a white, waxy substance obtained from the oil in the head of the sperm whale, while ambergris is a material obtained from the whale's digestive tract. Both substances are used in the manufacture of cosmetics. Because of these valuable materials and its meat and blubber, the sperm whale was and is hunted, and is found today in ever-decreasing numbers.

Ambergris is a fatty substance that is supposed to protect the sperm whale's interior from the hard beaks of the squid it eats. It is, however, not found regularly in sperm whales, and is therefore thought by some to be the result of an intestinal blockage. Its exact origin and function is quite uncertain.

Sperm whales have been found with scars of giant squid suckers on their hides. It was not until the 1970s that enough of a giant squid's body was found in a sperm whale to give credence to the existence of these squid. Up to that time many respectable zoologists believed that the huge squid scars were the result of battles between normal-sized squid and baby whales, and that the scars grew in size as the whales grew.

The sperm whale has a pygmy relative, the kogia. It is a dolphin-like creature, 4 m (13 feet), long and is not hunted. See *ambergris, extinction, squid, whales.*

spicule A small, calcareous, siliceous or protein support structure found in various invertebrates, such as sponges, echinoderms, chitons, and radiolarians. See *sponge.*

spider crab A thick-bodied scavenger crab with extremely long legs, of the family Majidae. The spider crab decorates its shell with bits of debris, which it uses as camouflage. The giant crab found near Japan, probably the world's largest anthropod, is a spider crab. Its maximum length, from claw to claw, frequently exceeds 4 m (13 feet). See *crab, sea monster.*

Spirogyra A green alga that lives in fresh water.

spit A narrow bank or small point of land usually composed of sand. Spits are the creations of longshore currents, and are most often perpendicular to the mainland. See *beach, coast.*

sponge A marine animal (phylum Porifera) characterized by a porous structure and a skeleton of interlocking, thornlike fibers that may be calcareous (Calcispongiae), siliceous (Hyalospongiae), or proteinaceous (Demospongiae). The last is the largest group of sponges. There are thousands of known species of sponge of which 20 or so live in fresh water. Marine sponges range from intertidal ones

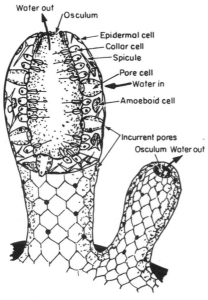

Colonial sponges

to hadal ones living at depths 8,500 m below sea level (26,000 feet). Their color can vary from pale yellow to orange, red, brown, green, and black. The glass sponges are found in deep water. These very pretty specimens, such as the Venus' flower baskets, are taken as collectors' items.

The sponge body is a series of many pores (ostia) and a large, single exit, the osculum. The exterior is usually a layer of flat cells called pinacocytes. Water enters the central cavity of the spongocoel because of the current created by the cells lining it, the choanocytes. The middle of the body wall is the mesoglea, a gelatinous material in which live the amebocytes—cells that move about and engulf and digest bacteria and debris brought in by the water current.

The amebocytes also produce motile egg cells which, when fertilized (most sponges are hermaphroditic), float around until they find a suitable spot for anchorage and a sessile adult life. Sponges also reproduce by budding, and can regenerate completely: Sponge tissue can be ground up and passed through a sieve,

and the tiny fragments will, if protected, grow into sponges.

As an industry, sponge fishing has all but disappeared.

While most sponges are relative harmless, long-lived, sessile animals, they do have some chemical defenses. They apparently manufacture some toxic substances that kill bacteria. They can also produce offensive tastes and smells that discourage predators. Some sponges, such as the burrowing sponge, actually attack other organisms. These organisms dig into live coral or mollusk shells. Most burrowing sponges *(Cliona)* attack empty shells or dead coral. *Siphonodictyon* species attack live organisms and use toxic materials to stop coral growth. By this means, they prevent the coral from overgrowing the sponge's osculum. The result is a killed coral patch or a mollusk dead in its shell. See *chemical defenses, hermaphroditism, Porifera, spicule, Venus flower basket.*

spongin   A hardened protein that forms the spicules, or support structures, of some sponges. See *Porifera, spicule, sponge.*

sponge weed   A green alga that inhabits warm water. It grows in a branched colony, which may reach 2 m (6.2 feet) long. See *algae, Chlorophyta, sea lettuce.*

spring tide   A maximal tide—either the highest high tide or the lowest low tide. Spring tides correlate with the new and full moon of the lunar cycle. See *lunar tide, solar tide, tide.*

squall   A wind of rapid onset. The winds in a squall may quickly reach velocities exceeding 70 km/hour (58 mph), and do not last long. They occur most often in connection with violent thunderstorms or cyclones.

In areas where there is cold mountain air that can quickly descend into a warmer valley, squally weather occurs frequently. The fiords of Norway and New Zealand are typical of such areas. Mountain passes that channel the wind are another. The *foehn* in the Alps and the *chinook* in Alaska are examples of warm-air squalls.

A squall line is an approach line of a large system of unstable air. See *meteorology, storm, wind.*

square rigger   A vessel bearing square sails that are set into the wind, as opposed to fore-and-aft sails. The latter are most often triangular. See *clipper, lateen, sail.*

squid   A member of the large order, Teuthoidea, of marine mollusks of the class Cephalopoda. Squid have a long body, a distinct head, and ten arms, two of which are much longer than the others. These are used as tentacles. All the arms have undersides covered with suction disks. These in turn are controlled by giant neurons, and the whole is coordinated by a sizeable brain. The squid has exceptionally good eye-to-tentacle control. Like the octopus, it is teachable in an aquarium setting.

Squid range in size from 1 cm (0.2 inches) to the giants, 20 m (66 feet) long. The mantle totally covers the "pen," which is all that remains of the molluscan shell in these animals. The animal moves by drawing water into the mantle cavity and then forcibly ejecting it. The movement is a backward jet propulsion. Squid move quickly, and capture fish and other prey easily by grasping them with their tentacles and tearing the prey apart with their beaky jaws. Camouflage is produced by ejecting either ink (a blue-black fluid) or a phospholuminescent material.

Squid are either male or female. In mating, the male inserts a sperm packet into the mantle cavity of the female. When the eggs are developed and fertilized, they are released into the water, where they develop into miniature squid. There is no separate juvenile form of this creature.

Squid have long been taken for human food in the Orient and the Mediterranean, and are now a commercial catch in the U.S. They are also prey for whales, birds, and bony fish.

A live but dying giant squid caught near Bergen, Norway, in 1982 weighed more than 200 kg (480 pounds) and had tentacles that measured 8 m (25 feet). The blood of such an animal cannot carry enough oxygen in warm or shallow water to keep it alive. Such giant squid are rather sluggish animals and poor swimmers (for squid). See *mantle, mollusk, Sepia, sperm whale.*

starboard   The side of a vessel to the right of an observer standing on it and facing the bow (front end). See *port.*

starch   A high-molecular-weight polymer composed of glucose monomer units. Starch is a carbohydrate storage compound of many plants, particularly terrestrial plants. See *sugar.*

starfish   Creatures of the class Asteroidea of the phylum Echinodermata, that are common inhabitants of all seas and are not fish. They are also known as sea stars. The starfish is most commonly encountered in a littoral rock pool. It is 15 to 30 cm (6 to 12 inches) in diameter, although some that are 65 cm (26 inches) across are known.

The body plan of the starfish is a five-part arrangement, with five arms disposed in radial symmetry. The upper surface consists of horny plates. The undersides of the arms have tube feet running along their lengths. The tube feet are a series of small, extremely flexible suction cups. The light sensitivity of starfish is not good. Whatever light receptors these animals have are at the ends of their arms. A water vascular system controls their arms, their movement, and their response to hard surfaces.

They take in water through the madreporite, a porous structure on their aboral (upper) surface, and this intake affects the degree of suction that the starfish can apply to whatever its tube feet are gripping.

Primitive species of starfish feed by engulfing their food, others by everting their stomach onto a meal and digesting it on the spot. The resulting digested mass is then "vacuumed" back into the

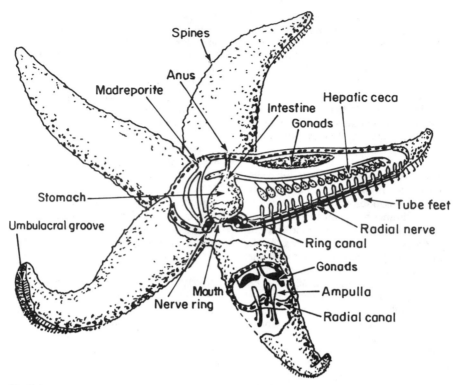

Starfish

body of the starfish along with the stomach. Starfish feed on mollusks, coral, and other echinoderms. See *Asteroidea, Echinodermata.*

stay   A rope that supports a mast or spar. Staysails, which are most often triangular, are hung on stays. Frequently the sails themselves are referred to as stays. See *sail.*

Stomatopoda   The mantis shrimp, an order of the class Decapoda, that resemble the praying mantis, an insect. There are about 200 species of these arthropods living in tropical to warm seas. They are found from shallow water to depths of about 500 m (1,650 feet); however, most are creatures of the littoral. The mantis shrimp are about 4 to 35 cm (2 to 13 inches) long, but most species are in the 15 to 20-cm range. They are frequently brightly colored, with green and

brown predominating, but red and blue are also seen.

These slender carnivores wait for prey (small worms, other arthropods) in rock crevices or burrows. They then pounce, in a move very similar to that of the praying mantis. The carapace (relatively hard shell) of this shrimp is most distinct, and serves to intensify its insect-like appearance. See *Decapoda, shrimp.*

stone crab   A large, edible, Caribbean crab *(Menippe mercenaria).* It is taken commercially by crabbers, who commonly harvest its large chela, or pinching claw, and leave the crab in the water to regenerate a new one. See *crab.*

stony coral   A small marine invertebrates of the class Anthozoa, phylum Cnidaria, order Madreporaria or Scleractinia. This is the most familiar and widespread class of coral. There are over

1,000 species, only a few of which are solitary creatures. The range of the stony coral is from the intertidal regions to depths of 6000 m (19,000 feet), and their color varies from the palest shades of yellow and pink to black. The color depends on the algae that live on the calcareous skeleton of the coral, which is white.

Brain, mushroom, star, and staghorn are among the names of the different formations of specific colonial varieties of stony coral. The living coral grows about 0.5 m (20 inches) annually if it has optimum conditions of food, salinity, temperature, oxygen, and exposure to light. See *Anthozoa, coral, Cnidaria, polyp.*

storm A meteorological disturbance involving a change in atmospheric pressure, high winds, high tides, and precipitation. If this condition moves over a large area and travels in a circular path it is called a cyclonic storm.

Extratropical cyclones start at the Polar Convergence, where there is cold air to the north and warmer air to the south (in the Northern Hemisphere, with the reverse holding true in the Southern Hemisphere). As warm air is lifted by an inrush of cold air, it is given a spin by the Earth's rotation. This is anticlockwise in the Northern Hemisphere and clockwise in the southern. As the air pressure decreases, the wind speed increases. This frequently leads to a warm front developing. If a cold front catches up with this warm front, a semi-stationary belt of bad weather occurs. This is very typical at the Polar Convergence.

Tropical cyclones develop in an analogous manner, usually between the 10 to 15° parallels of latitude. They are most likely in August and September in the Northern Hemisphere, and in January and February in the Southern. The most commonly observed path of these storms is west. This is a result of the trade winds, which peak and move from east to west. Tropical cyclones are hurricanes in the Atlantic and typhoons in the Pacific or Indian Ocean. See *Beaufort Scale, meteorology, sirocco, squall, wind.*

storm surge A rise in the normal water level in a particular area of the ocean. Storm surges are caused by wind and by low pressure and are augmented by high tides. See *storm.*

strait A narrow, natural waterway between two large bodies of water. See *individual straits.*

sturgeon A group of large fish (genus *Acipenser*) widely distributed in northern Europe and North America. They live in the sea and spawn in rivers, to which they return more than once in a lifetime.

Name	Range	Length	Weight
Common (*Acipenser sturio*)	Western Europe	3 m	225 kg
Atlantic (*A. oxyrhynchus*)	Eastern North America	3–4 m	250++ kg
Russian (*A. guldenstatii*)	Black and Caspian Sea	3–4 m	250+ kg
Sterlet (*A. ruthenus*)	Black and Caspian Sea	1 m	—
White, Oregon or Sacramento (*A. transmontanus*)	Western North America	3 m	130 kg
Chinese (*A. sinensis*)	China Sea, Yangtse	2 m	150 kg
Japanese (*A. kikuchi*)	Japan	2 m	130 kg

Some species of sturgeon have abandoned the trip back to the sea and spend their entire lives in rivers. These fish are related to the paddlefish.

The sturgeon have several features of primitive fish, including an unequally lobed tail, laterally located bony plates (scutes), and no teeth in the adult. They are long-lived, individuals of more than 200 years having been tagged. Since they continue to grow, some individuals are very large.

Sturgeon are prized for both their flesh and for their eggs, which are considered the best caviar. They are therefore chronically in danger of being overfished.

Several anadromous species of sturgeon are listed on the previous page. The wholly fresh water animals, such as the shovelnose sturgeon, are omitted.

subduction   A process that occurs when one edge of a crustal plate dives beneath the edge of a converging plate. Usually the plate subsiding is an oceanic one, and it dives beneath a continent-bearing plate. The reason given for the sinking of oceanic plates is that they are denser than the granitic continental plates. See *crust, plate tectonics, trench.*

submarine canyon   A valley or system of valleys cut into the ocean floor. These may be of several types. They may be narrow, v-shaped clefts that are the visible edges of faults, or broad valleys with flat floors that look like river valleys on land; or they may be cuts in front of delta cones.

There is considerable discussion concerning the origin of the v-shaped canyons. They occur near the mouths, or at the sites of former mouths, of some large river systems. It is therefore very tempting to conclude that they are the drowned seaward ends of those river systems. Sediment is carried along the canyon's length, which can extend along the entire width of a continental shelf, then down the continental slope, and finally into fan-shaped patterns of valleys on the ocean floor.

Other gashes and seams in the ocean floor may be the result of glacial action or volcanic activity. Land and mudslides

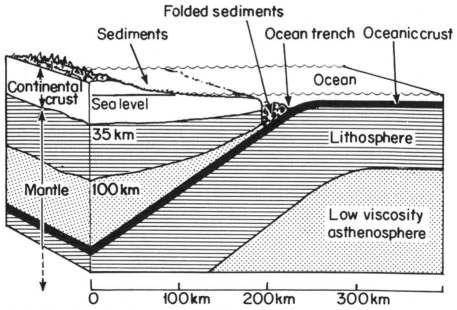

Section through crust and upper mantle showing a subduction zone where a continental plate is advancing over an oceanic plate

occur frequently. These change the seascape considerably, particularly in areas of high sediment build-up such as river deltas. See *continental slope, delta, ocean floor, sediment.*

submersibles Small vessels that can dive and remain submerged for long periods. They are designed to be easily moved, and are battery-powered and capable of withstanding the great pressure of ocean depths. They are intended for the detailed study of small areas or the movement of equipment on the sea bottom. The submersible is usually moved from one location to another by tenders or mother ships.

The earliest submersible—the bathysphere—was used by Barton and Beebe. They descended to a depth of 1,000 m near Bermuda in 1934. The next real work on the use of the bathysphere at great depths had to await the 1950s.

Submersibles have been used in the inspection and repair of pipelines, cables, and drilling platforms. They are also used extensively in research. The *Alvin* and its tender *Lulu* have been involved in several very deep sea dives. One explored the East Pacific Rise and its vent communities. Another, in 1974, the Mid-Atlantic Ridge. The latter was a joint Franco-American project that included the French submersibles *Archimede* and *Cyana*. Very recently, the wreck site of the R.M.S.

Titanic which sank in 1912 in the North Atlantic, was explored by a submersible.

See Alvin; Archimede; *Beebe, Charles William; East Pacific Rise; Mid-Atlantic Ridge.*

subsidence The dropping of dry land into the sea. This can have several causes. One might be a rise in the mean sea level, another the collapse of an isolated guyot, or coral island. The latter may be the result of volcanic activity. If a volcanic eruption empties an underwater chamber of liquid lava, the resulting cavity may be too fragile to withstand the surrounding water pressure. The collapse of the roof of such a chamber would also collapse any structure above it. See *blue hole, volcano.*

subsurface current Deep water current which is most often slower than a surface current. Its direction is not determined by that of the surface flow. See *Antarctic Bottom Water, deep water.*

Subtropical Convergence A not-quite-continuous line that divides subantarctic from subtropical waters. It hovers at 40 to 50° South Latitude, about 10° north of the Antarctic Convergence. The temperature of the water north of the line is about 10°C warmer in winter than that to the south. The difference in summer is 12 to 18°C. The water that is north of

Name	Site of Operation	Depth	Year
Bathysphere—Barton-Beebe	Bermuda	1000 m (3228 feet)	1934
Kuroshio and Oshoro Maru(mothership)	Japan	205 m (680 feet)	1951–1957
Galeazzi Diving Chamber—Elie Monnier	Mediterranean	300 m (1000 feet)	1955–1956
FNRS—3—Piccard	Dakar	4050 m (13,400 feet)	1954
Trieste—Piccard	Tyrhennian	3700 m (12,200 feet)	1956
Diving Saucer and Calypso(ship)—J.-Y. Cousteau	Mediterranean	300 m (1000 feet)	1960
Archimede—French Naval Vessel	Mid-Atlantic Ridge		1961
Cyana — " " "	" " "		
Alvin—{U.S. (Woods Hole, MA) {Lulu (tender)	Pacific	3000 m (10,000 feet)	1964

the Convergence is also saltier than that just to the south. See *Antarctic water*.

sugar A simple chemical compound composed of carbon, hydrogen, and oxygen, also known as a carbohydrate. There are hundreds of known natural sugars; one of the simplest is glucose

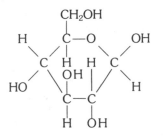

a structure that is one of the possible arrangements (configurations) of the molecular formula $C_6H_{12}O_6$. Glucose is important because it is a fundamental molecule in the metabolism of many sea and land organisms. Glucose can react with other molecules of glucose to form chains of varying length. Maltose, for example, is a disaccharide or a two-glucose unit. Larger chains of glucose (also called polysaccharides) are the glycogens, and still larger polymers are the starches and celluloses. Starch is actually a mixture of two polymers, one a straight-chain polymer of glucose called amylose, the other a branched polymer called amylopectin. Among other polysaccharides are algin (present in brown seaweeds) and chitin (the hard, outer covering, or carapace, of lobsters, crabs, shrimp, and insects).

Ribose is a five-carbon sugar ($C_5H_{10}O_5$) which is a vital part of the adenosine triphosphate (ATP) molecule—the so-called energy-storing compound in many metabolic systems. Ribose is the sugar that is built into the RNA molecule. It forms the polysaccharide skeleton. Ribose with one less $-OH$ group, known as deoxyribose, is a constituent of DNA. See *carbohydrates, polymer*.

Sulawesi Sea Also known as the Celebes Sea; a body of water that separates Sulawesi (Celebes) from the Philippine islands. It is one of several ocean deeps with known volcanic activity. The flat bottom, which has a mean depth of about 4,000 m (13,000 feet), has deep, cold, Pacific bottom water entering the area by spilling over the sill near Mindanao. See *bottom water, Pacific Ocean, sill*.

sulfur A nonmetallic element that occurs in nature either in its free form or combined with oxygen and/or other elements. Sulfur occurs in the sea both in offshore deposits in salt domes, and dissolved in water. The average concentration of sulfur in seawater is about 900mg/liter. It is the third most available element dissolved in seawater. Sulfur is an essential element in proteins, and is therefore present in all organisms. See *element, sulfur bacteria, vent communities*.

sulfur bacteria Filamentous autotrophic (manufacturing their own food) bacteria that derive energy by oxidizing sulfides and building up carbohydrates from carbon dioxide. They occur in several marine habitats from shallow water to hadal depths. Large concentrations of sulfur bacteria were recently discovered living near hydrothermal vents in the seafloor where sulfurous gases escape from the submerged volcanic sites known as "black smokers." The bacteria thrive near the vents, where the water is at temperatures in excess of 100°C (212°F) and under very high pressure. Their unique metabolic system operates on the basis of the chemical equation:

$$ATP + SO_4^{2-} \rightleftharpoons APS + energy$$

In ordinary cells, ATP (adenosine triphosphate) is an energy-storing compound. The phosphate bond of ATP breaks to form adenosine diphosphate and releases energy needed for the cell's metabolic functioning:

$$ATP + O_2 \rightleftharpoons ADP + energy$$

In the sulfur bacteria, the sulfur compound functions in a manner analogous, the ATP reacts with the sulfur compound instead of oxygen, and produces a sulfur-containing adenosine compound with the release of energy. Since the bacteria utilize the energy in the bonds of these sulfur compounds, they function quite well in anerobic (oxygen-free) conditions. Sulfur bacteria exist wherever there are active hydrothermal vents and, subsequent to their discovery in deep seas, they had been found in other areas, including waters off the California coast near La Jolla, on the East Pacific Rise, and in the Bahamas. In addition to the sulfur bacteria, attendant vent colonies have been located near all black smokers. See *autotroph, respiration, vent communities.*

Sunda Shelf   The largest shelf area in the ocean, almost 2 million km² (or 0.6 million square miles) in size. It underlies part of the Java Sea. The entire area, including Malaya, Sumatra, Java, and Kalimantan (Borneo) is a vast, drowned river system. The North Sunda River flowed northeast into the South China Sea from headlands in Malaya and Kalimantan. The South Sunda River also moved from Sumatra east into what is now the Makassar Strait between Borneo and Sulawesi (Celebes). The entire subsided river system is sometimes referred to as the Molengraaff in honor of the principal explorer of the region, G. A. F. Molengraaff, Dutch geophysicist.

The Sunda Strait, which separates Sumatra from Java, is very shallow and probably the result of very recent seismic activity. Krakatoa, in the Sunda Strait, is the most important volcano in the area. See *continental shelf, Krakatau, subsidence.*

surface tension   An intermolecular force that acts in liquids to hold the molecules at the surface more closely to each other than to the molecules within the body of the liquid. Surface tension minimizes the area of the surface. It also causes the curvature of the surface of a full glass of water, and the individual droplets or beads of water that form on a greasy surface. Soaps and detergents are called wetting agents, surface-active

agents, or surfactants, and they reduce surface tension because they interfere with the intermolecular forces holding the water molecules tightly together at the surface.

Surtsey   An island that formed off the coast of Iceland in 1963 as the result of a volcanic eruption. It quickly became both a geological laboratory and a tourist attraction. See *Mid-Atlantic Ridge, mid-ocean ridges.*

suspension feeders   Organisms that strain food from the surrounding water. They can range in complexity from sponges and corals to the baleen whales. They live on plants, animals, or debris that drifts past them, or they may actually seek out small floating or swimming organisms. The feeding mechanisms of suspension feeders vary; some have large mouths that process large quantities of water, from which they extract their food. Such animals are sponges and baleen whales. Others use some organ to screen large quantities of water and pass on the entrapped food to small mouths. Clams, for example, screen the water by using their gills as strainers, and pass the food particles they screen out into their relatively small mouths. See *baleen, Porifera.*

swell   A long and massive wave or succession of long-period waves moving away from the cause of their generation. See *wave.*

swim bladder   An organ regulating the buoyancy of most teleost (bony) fish. It originates as part of the digestive system. This gas-filled sac is used by the fish to equilibrate their internal pressure with that in the surrounding water as they rise or sink.

Some fish use the swim bladder as a sound-generating structure. In some primitive species it is part of the respiratory system and acts as an air (or oxygen) reservoir.

Fish that live very deep in the sea do not have swim bladders, nor do the car-

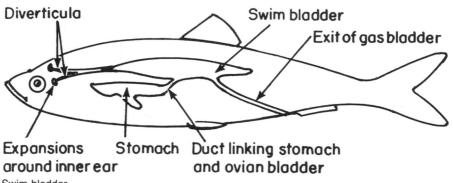

Diverticula   Swim bladder
Exit of gas bladder
Expansions   Stomach   Duct linking stomach
around inner ear   and ovian bladder
Swim bladder

tilaginous fish. These organisms also experience great changes in the pressure of
their surrounding environment, but the
mechanism by which they are able to
adjust to these pressure changes safely is
still largely unknown. See *fish, grunt.*

swordfish   A large oceanic game and
food fish noted for its long, flat, sharp
"sword," which projects forward for at
least 60 cm (25 inches) from its head
and is used to slash prey. Swordfish are
impressive for their size: the average adult
is longer than 4 meters (13 feet), and
weighs between 400 and 500 kg (900 to
1100 pounds).

These warm-water inhabitants are of
the family Xiphidae in the order Perciformes. The most common species is
Xiphias gladius. They are dark blue to
blue-black on their upper surface and
silver on the underside. Their body is
long, scaleless, and slender; swordfish
have neither pelvic fins nor teeth. The
swordfish feeds mostly on other fish, particularly schooling fish. See *fish.*

symbiosis   An interrelationship between two organisms. Mutualism is a term
applied to the relationship when both of
the organisms derive some benefit from
the arrangement. An example of mutualism is the presence of the tiny parrot
fish that live in sea anemones and lure
other fish into the mass of tentacles of
the anemone, after which the parrot fish
dine on the crumbs left by the anemone.

A more usual arrangement is a commensal one, in which organisms live together with only one of them having
some advantages, although the arrangement is not detrimental to the other. Thus,
the fungal colonies that reside on spider
crabs camouflage the crabs, to their advantage, but the only evident benefit to
the fungi is that in being transported from
one locale to another, they may have a
better chance at a food supply. Parasitism
occurs when one species benefits at the
expense of the other without giving anything in return. Many fish have both internal and external parasites. See *commensal relationships.*

T

tablemount See *coral, guyot, island, seamount.*

tack A rope for controlling certain sails, or the sails themselves. When used as a verb, "tack" means resetting the sails in order to take advantage of a wind coming at an angle to the side of the ship (from off-side). The direction in which the ship then moves, and the distance traversed while it is on a tacking course, is also called the tack. By tacking, or setting a series of zig-zag courses, it is possible to sail against a prevailing wind. See *sail.*

Tanganyika, Lake A lake in East Africa in the Great Rift Valley between Zaire and Tanzania. The longest and second deepest freshwater lake in the world. Only Lake Baikal, in the U.S.S.R., is deeper. Lake Tanganyika was first seen by Europeans in 1858, when it was found by Sir Richard Burton and John Speke who were on an expedition to find the source of the Nile. See *rift valley.*

Tasman, Abel Janszoon (1603?–1659) A Dutch explorer who sailed for the Dutch East India Company and after whom Tasmania (an island south of Australia) and the Tasman Sea are named. Van Diemen, the governor-general of the Dutch East Indies, sent Tasman to explore Australia in 1642. Several Dutch explorers had previously stopped at various points on the Australian coast, but Tasman circumnavigated the entire island and also explored Tasmania, which he named Van Diemen's Land for his patron. He sighted the north island of New Zealand—he called it Staten Landt— and thought it was part of South America. He mapped the Tongas, part of the Fiji archipelago, part of the New Hebrides and Solomon islands, and New Guinea. Tasman then returned to Batavia (now Jakarta) after almost a year at sea. On a later expedition (1644) he mapped the Gulf of Carpentaria on the northern coast of Australia, south of New Guinea.

Tasman successfully led a commercial fleet from Batavia to Siam (now Thailand) in 1647. Later he was involved in a military action against the Spanish in the Philippines. He finally retired, rich and famous, to Batavia. See *Australia.*

Tasman Sea An arm of the southwestern Pacific Ocean between Australia and New Zealand. It is named for Abel Tasman, the Dutch navigator who crossed it in 1642. He had previously landed on what he called Van Diemen's Land (Tasmania), believing it to be the Australian mainland.

The Tasman Sea roughly corresponds to a deep basin, with depths in excess of 4000 m, that is known as the Tasman, or Thomson, or East Australian Basin. Its southern reach is Macquarie Island.

The East Australian Current dominates the circulation of surface water of the Tasman Sea. The water has a fairly high salt content near the surface and correspondingly less salt with increasing depth. The deep water contains traces of North Atlantic deep water.

The bottom sediments of the sea are calcareous, with considerable red clay. The Tasman Basin is yet another area assumed by some to have continental origin. According to one theory proposed around 1900 there was once a continent called Tasmantis, a Pacific Ocean analog of Atlantis. But while there are rises in the sea bottom such as the Lord Howe and Chatham, and many seamounts, the bottom of the Tasman Sea is oceanic in nature rather than continental. See *Australia, Australian Plate, bottom water, evolution of oceans, Lord Howe Rise, Pacific Ocean.*

taxonomy The systematic arrangement and naming of organisms. Although Aristotle (384–322 B.C.) and his pupil Theophrastus, (c.371–287 B.C.) did botanical classifications, and there was no explanation for the variation in obviously related living things. Vesalius (1514–1564), and then John Ray (1627–1705) an English botanist, linked organisms by a species concept. The Linnean (1707–1778) innovation was a binomial nomenclature based on Aristotelean logic. The impact of this new classification scheme on Darwin's (1809–1882) work was incredible. It meant that he could explain the rise of new variations that were still related but had eventually diverged. The families of finches were all still finches, but had undergone modifications in their morphology and feeding habits—the various species had changed, but the genus remained constant.

The scheme still used to classify organisms—both those alive today and fossil finds—is a systematic designation of the organism in question by kingdom, phylum, class, order, family, genus and spe-

Taxon	-	Dogfish
Kingdom	-	Animalia
Phylum	-	Chordata
Subphylum	-	Craniata
Class	-	Pisces
Subclass	-	Chondrichthyes
Order	-	Selachii
Family	-	Scyliorhinidae
Genus	-	*Scyliorhinus*
Species	-	*S. caniculus*

Example of an hierarchical classification scheme

cies. All of these categories can have super- and sub-variations. It is important to remember that these designations are manmade, and are therefore subject to change or addition.

Present methods of classification also examine the protein sequences in organisms, since related species have certain biochemical features in common. Related organisms have similar enzyme systems and similar proteins that have the same amino-acid sequences. Thus, differences in the hemoglobins from one mammal to another can be minute, and the enzyme systems in humans and great apes are very similar, while those in humans and mice are less so. These changes occur with taxonomic divergence. See *Aristotle; Linnaeus, Carolus; Rondelet, Guillaume.*

tectonics See *ocean floor, plate tectonics.*

teeth Modifications of the dermal armor plates of primitive fish. The teeth in some fish and reptiles are the same three-layer structures of pulp, dentine, and enamel that are associated with mammalian teeth. They are, however, homodent teeth, meaning that they are all the same and are replaced throughout the life of the animal if they become worn away or are shed. Most mammals have heterodont teeth, consisting of incisors, canines, and molars in a number sequence that is species-specific.

Some teeth are highly specialized: the saw of the sawfish, for example, has the only known set of external but real teeth. The tusks of the walrus are for display and for the defense of the territory of a large bull. The common notion that the walrus tusk is a digging tool is untrue. See: *fish, narwhal, walrus.*

teleost Any of the bony fish. They are also the most advanced in terms of evolution and the largest group of fish. Almost all sport, commercial, and ornamental fish are in this classification, which is not a taxonomic one. The one obvious uniform characteristic of the teleost fish, besides their calcified internal skeleton, is their tail, in which the upper and lower halves are of about equal size, whereas in the cartilaginous fish the tail has two lobes of unequal size.

The range of the teleosts is worldwide, and from habitats in tidal pools to hadal depths. The size of adult telosts may range from 1 to 2 cm (less than 0.5 inches) to longer than 3 m (10 feet). Their reproductive systems vary from species that have distinct male and female members to others that consist of functional hermaphrodites (some of which can self-fertilize), while still others can change sex during their adult lifespan. Some teleosts are ovoviviparous, others viviparous. Most species fertilize eggs in the water, and these hatch to produce planktonic young.

Some teleosts move in schools, others are bottom-dwellers; most are efficient swimmers because of their torpedo-shaped body and powerful tail. See *fish, hermaphrodism, ovoviviparity.*

temperature The degree of hotness or coldness of a body or an environment. Temperature was one of the properties of seawater that Robert Boyle—one of the original members of the Royal Society of London—thought should be studied. He made this suggestion as a part of a master plan for the advancement of science in 1666. Robert Hooke, a friend and later rival of Isaac Newton, designed several devices for the purpose of exploring Boyle's scientific program. Among these was a self-registering thermometer.

Marcet, in 1819, reported on the differences in temperature between bodies of water at the same degree of latitude, and on the difference in density of seawater as compared to fresh water.

Later in the 19th century, despite Humboldt's reports of a cold South American current in equatorial regions, the idea of a uniform temperature in the depths of the sea was accepted as truth. It was held to be almost axiomatic that deeps in the ocean were all at 4°C. Differences in the temperature of different bodies of water had to be disproved repeatedly by Lenz and others.

The temperature of any marine area depends on surface radiation, evaporation, and the movement of water. These are then subdivided according to seasonal variations and tidal oscillations. See *insolation, currents, Humboldt, Friedrich Heinrich Alexander, weather, specific bodies of water.*

temperature of animals Animals that maintain a constant body temperature are called homoiotherms (homeotherms) while animals that adjust their temperature to the ambient temperature are called poikilotherms. The designation of homoiotherm or poikilotherm is somewhat arbitrary. Both forms of animals have a variety of mechanisms for maintaining a temperature that allows them to function at maximum efficiency and to achieve a balance between calorie requirements and temperature. The colder it is, the greater the need for food but the smaller the chance of finding it. Also as poikilotherms cool they become sluggish, and their hunting ability diminishes. The solution to maintaining a constant temperature for many organisms, both homoiotherms and poikilotherms, is hibernation or migration. Mammals and fish face the same temperature-control problem because some of their activities require a particular temperature. For example, swordfish have large fat deposits in and around their brains and eyes, and it is thought that this insulation makes it possible for them to dive deeply into much colder water and continue to function. See *homoiotherm, poikilotherm.*

teredo The shipworm, a highly modified bivalve (mollusk) of the Teredinidae family that bores into wood. The free-swimming larvae settle in ships' bottoms or wharf pilings and continue to enlarge the burrows that they channel for their entire adult lives. Adult teredos may be 30 cm (1 foot) long or more. In these animals the characteristic shell of the bivalve is reduced to two digging scoops, which they bury in the wood they bore through. Only the very long siphons protrude from the burrow opening as evidence of the animal's presence.

See *Mollusca*.

tern   Also known as the sea swallow, any of a group of about 40 species of gull-like birds of the order Charadriiformes, family Laridae, subfamily Sterninae, found almost everywhere in the world.

Terns are related to gulls, look similar, and also have webbed feet. They are generally thinner than gulls, however, and have longer, narrower wings, shorter legs, and forked tails. These birds are agile in the air and frequently migrate over very long distances to breed. Terns feed mostly on small fish and crustaceans, diving into the water while in flight. The lifespan of terns varies, but most are long-lived, and the arctic tern normally lives more than twenty years. Like many other long-lived animals, terns tend to be monogamous. The sound made by the tern is a characteristic "heh, heh", while gulls make a cackling sound.

Some representative species of tern are:

Tertiary Period   The period of the Cenozoic Era that followed the Cretaceous Period and preceded the Quarternary. It started about 65 million years ago and lasted about 55 million years.

The Tertiary was the time when land bridges became submerged, cutting the connections between Siberia and North America, and those between the Indonesian islands. The Himalayas, Alps, Rockies, and Atlas mountains formed in the Tertiary.

The end of the Cretaceous was marked by the extinction of many species in a relatively short time. This left unfilled ecological niches to be filled by other species: snails, bony fish, and mollusks all proliferated. The climate at the start of this period was warm, then underwent a series of oscillations, and finally turned sharply colder. See *Cretaceous Period, Quaternary Period*.

test   The rigid covering of a number of widely divergent organisms. It is a product of the secretion of the animal that lives in it; it may also include compacted nonliving material such as sand grains.

Name	Habitat	Color	Length
Atlantic fairy or white noddy (*Gygis alba*) or (*Anous albus*)	Tropical islands	White	30 cm
Pacific fairy (*Sterna nereis*)	Australia, New Zealand	White	20 cm
Inca (*Larosterna inca*)	Peru	Gray, white	10 cm
Black (*Sterna nigra* or *Chlidonias niger*) [this bird will fly from North to South America, annually]	Europe, Asia in summer; Africa in winter	Black head, white underside, gray wings	25 cm
Common (*S. hirundo*)	North temperate to south coastal zone	White, black head, red legs and bill	35
Arctic (*S. paradisea*) [migrates longest distance]	North temperate to south temp.	White and black body, red legs and gray wings	38
Sooty (*S. fuscata*)	Island colonies	White-fronted black body, gray wings	20
Little (*S. albifrons*)	Europe, Asia, Australia *not* South America	White body with black wings	20

See *bird, gull*.

Among animals that live within tests as opposed to shells are some members of the Sarcodina, the sea urchins, and the tunicates. See *Sarcodina, sea urchin, Tunicata.*

Tethys Sea   The great east-to-west waterway of the Mesozoic Era that separated Laurasia from Gondwanaland. The shift of continental plates produced deformations and the uplifting of sea bed sediments into the European Alps, the Caucasus, and the Himalayas. The remnants of the Tethys Sea are the landlocked Caspian and Aral Seas. These were once linked with the Mediterranean, which also is a remnant of the once larger Tethys. See *Aral, Caspian, Gondwanaland, Mediterranean, Mesozoic Era, Pangaea.*

Thera   Also known as Santorini, a Cycladean (Greek) island north of Crete. It is all that remains of a major volcanic island that exploded around 1500 B.C. The island is half of a volcanic cone composed of lava and pumice (a very light, porous stone resulting from frothy lava). Archeological work has uncovered a Bronze Age culture on Thera with strong ties to that of Crete. This island is thought to be the basis of the Atlantis stories of Greek mythology and the biblical account of the drying-up of the bed of the Red Sea in Exodus. See *Cyclades, tsunami, volcano.*

thermal equator   An irregular and moving line that links temperature maxima. It is most often between 12° North Latitude and 12° South Latitude. It bends northward or southward over continents because land and the air above land heat up faster than does water or the air above water. Since there is more landmass in the Northern Hemisphere than in the Southern, the thermal equator is more northerly than southerly. See *temperature.*

thermal spring   A spring that produces water that has a higher temperature than the temperature of the surroundings. See *vent communities.*

thermocline   A temperature differential in the water. A decrease in water temperature with increasing depth is normal and expected. A thermocline exists where the decrease suddenly becomes sharper than in neighboring areas. This discontinuity at a specific point is either permanent or seasonal.

The permanent thermocline is the result of the descent of uniformly cold polar water. This moves toward the equator, sinking below warmer layers and preserving its own character for thousands of kilometers.

The seasonal thermocline is the result of the rapid warming of surface water in summer. The underlying water stays colder and does not warm up until later in the season. In some temperate regions the thermocline disappears in autumn. This is why coastal water is more agreeable for swimming in the later summer and early autumn. See *Antarctic water.*

Thomson, Charles Wyville (1830–1882)   An English naturalist; organizer and leader of the *Challenger* Expedition, the first oceanographic research voyage. Thomson's early interest in natural history led him away from a medical career at the University of Edinburgh to a lectureship in botany. He was first professor of natural history, then of zoology at several British and Irish universities.

By the 1850s Thomson was an established marine biologist, having studied fossil crinoids, trilobites, and coelenterates. On seeing Michael Sars' collection of organisms that were bagged at depths below the 300-fathom mark (500 m or 1,800 feet) in the Forbes' so-called "azoic zone," he was led to a sequence of dredging expeditions to see what really did live at great depths. This culminated in the organization of the *Challenger* Expedition.

In 1868 Thomson and W.B. Carpenter, a vice-president of the Royal Society, obtained Admiralty support for the project to dredge in the North Atlantic. They found an interesting collection of fauna. In addition, they observed that the temperature of the water at similar depths varied from one location to another. This led to an exhaustive analysis of the composition of the seawater. An extension of these projects in the North Atlantic led to an expedition of global scope.

The organization, writing, and editing of the results of the *Challenger* voyage took years. Thomson, the originator of this seminal project, was knighted in 1876 for his immense contribution to science. Unhappily, he did not live long enough to do more than just begin to arrange the *Challenger* notebooks. They were published by John Murray. See *H.M.S. Challenger; Murray, John; Sars, Michael.*

thresher shark   A shark of the genus *Alopias,* with a typically very long, scythelike tail that is about half the length of the animal, and is the origin of its common name. *A. vulpinus* is a representative species of thresher. It is about 6 m (20 feet) long.

Thresher sharks are found in tropical and warm temperate waters of the Atlantic and eastern Pacific Oceans. Their principal food is cephalopods and schooling fish. It is thought that they use their long tail as a whip with which to stun prey, but not much is known about their feeding behavior. These sharks are not considered maneaters. See *shark.*

tidal bore   See *river bore.*

tidal currents   Local ocean currents that are important in the shallow waters off continental and island shores. The direction of these currents changes throughout the tidal cycle, making a clockwise gyre in the Northern Hemisphere and a counterclockwise one in the Southern. In open ocean the tidal currents become attenuated to the point of being undetected. See *current.*

tidal period   The time between one high water mark and the next. See *tide.*

tidal range   The difference in water height between high tide and the next low tide. See *tide.*

tide   The regular movement of water vertically with respect to some point on land. Tides are classified as either diurnal, with one high tide occurring daily, or semidiurnal, with two high tides each day. There are measurable tides not only in the ocean but on enclosed bodies of water such as bays and large lakes.

The direct cause of the tide is the acceleration, as described by Newtonian mechanics, of the Earth as a response to the gravitational pull of the Sun and the Moon. Thus, the movement of water on Earth is a function of these celestial bodies, of the depth and shape of the coast, and of the Coriolis effect.

Tides are mapped for given areas; a chart that shows all points at which high tide occurs simultaneously is called a *cotidal* chart; another variety of chart connects points at which the high tide comes up to the same height above sea level. At some point in a body of water the lines of radiating tides—the cotidal lines—meet. This is the amphidromic or "no-tide" point. There may be more than one such point in a given area, particularly if that area is large, such as the Caribbean Sea.

The tide at any given location can be diurnal, semidiurnal, or unequal. The last means that the two high tides are not at the same height on any day. This unequal state is an intermediate state between diurnal and semidiurnal. The prediction of the time and height of a tide for a particular place can be done empirically on the basis of careful records of past tidal performance. Thus, the Moon has a specific effect on the tide, as does the Sun. This effect varies over the course of the lunar month. The greatest effect results when the Sun and Moon reinforce each other's gravitational effect. This occurs when the Sun and Moon are in line, on the same or opposite sides of the Earth and the resulting tides are called *spring tides.* At the quadrature of the Moon, the Sun and Moon are out of alignment, with the position of the Sun-Earth line perpendicular to the Earth-Moon line. The result is called *neap tide.* Thus, there is a cycle each month ranging from high high tides (spring) to low high tides (neap). Any study of long-term tidal effects must take into account the variation in the Sun-Moon relationship over a 19-year period. This is called a saronic cycle.

The tide at any given point on the shore can therefore be predicted on the basis of past tidal activity. This point is extremely localized: the tide will exhibit variation from one spot to another on the shore of a single bay. Complicating factors are the movement with time of the amphidromic point, and inertia. It takes energy to move water, and as a result there is a distinct lag between the lunar high, which is the crossing of the meridian of the Moon, and high tide. This is the *intertidal interval,* which also depends on the arrangement of land and water bodies.

It is now possible to model tides with some success. Computer models are used for this and are a navigational asset, as are the older and still extremely useful tide tables. A good model must take into account all factors, and appropriately weight each one. The calculations involved are complex and impossible without computers. See *amphidromic point, Coriolis effect, diurnal tide, lunar tide, Moon, semi-diurnal tide, solar tide, spring tide.*

tiger shark   A large, banded, gray, brown, or black shark, *Galeocerdo cuvieri,* of the family Carcharhinidae found mainly in the Indian Ocean. These animals are about 5.5 m (18 feet) long. They live in warm tropical seas and near coasts. They are omnivorous carrion eaters, feeding on dead and dying animals, and are also known to tear fishing nets to get at the catch inside. While they are reputed to be maneaters, the usual diet of tiger sharks is other sharks, fish, birds, turtles, mollusks, and garbage. See *shark.*

Tigris and Euphrates rivers   Two major rivers of the Middle East (known in Arabic as Nahr Dijlah and Nahr al Furat, respectively) . Both the Tigris and Euphrates rise in Turkey and flow southeast into the Persian Gulf. Almost all of Iraq consists of the Mesopotamian Plain built of the alluvium carried downstream by these two rivers. Baghdad is on the Tigris, the easternmost of the two. The area between the two rivers has been intensively irrigated for millennia. In this extremely hot, dry climate the rivers are the only water sources and every civilization that has lived in the area has used them for irrigation. The canals built outward from the Tigris and Euphrates—the distribution system for the precious water—were the engineering triumphs of their day, and a good deal of the governmental bureaucracy dealt with their maintenence. They were finally abandoned after the devastation resulting from the Mongol invasion. See *delta, Indian Ocean.*

tilefish   A large, temperate-water marine food fish, *Lopholatilus chamaeleonticeps,* of the family Branchiostegidae, order Perciformes. Tilefish live off North America on the continental slope, where the continental shelf drops down toward the ocean floor. They are large, being over 1m (3.3 feet) in length and weighing 15 kg (3 pounds) or more.

These fish are blue-green above with white undersides; their back has irregular yellow spots, while the head is frequently red or has red spots. Tilefish are burrowers: some live in solitary burrows, others in communal burrows that may be 3m (10 feet) long. They hollow out areas in the continental slope and live in the hollows. There has been a noticeable change in the seascape of the continental edge as a result of tilefish activity. Their burrow-making has caused several collapses of the slope structure. See *fish.*

tissue   A functionally specialized group of cells of an organism. Muscle tissue, for example, performs the function of moving an animal's body epidermal (skin) parts, while tissue protects the organism. In highly developed organisms, tissues are assembled into organs.

tombolo   A sand or gravel bar connecting an island to another island or to the mainland. When larger, it may be a spit or a causeway. See *barrier island.*

Tonga Trench   A deep depression in the floor of the southwestern Pacific Ocean. Its average depth is almost 11,000 m (33,000 feet), and it is about 1,400 km (900 miles) long. This fairly narrow, steep-walled break in the Earth's crust lies south-southwest of Samoa. A belt of volcanic and seismic activity is west of the trench.

The Tonga Trench and the Kermadec Trench, with which it is almost continuous, form a line which, together with the Great Alpine Fault of New Zealand, creates a long slip-strike fault that is analogous to the fault in the Caribbean. See *crust, Kermadec Trench, plate tectonics, Puerto Rico Trench, trench.*

toothed whales   Whales of the suborder Odontoceti, with many conical teeth. This group includes the dolphins, sperm whales, norwhal, beluga, porpoise, and killer whale. The toothed whales differ from the baleen whales, which feed by straining tiny organisms from the water. See *individual species, whales.*

torpedo fish   A group of cartilàginous fish of the class Chondrichthyes, subclass Elasmobranchii. The torpedo fish are relatives of the sharks and rays. They are flattened fish whose habitat ranges from shallows to great depths, and from tropical to temperate waters. Their outstanding characteristic is the presence of electric organs on or near their head. These light-emitting fish form significant populations in abyssal ecosystems. They range in size from about 25 to 180 cm (10 to 70 inches). See *Elasmobranchii, light organs.*

toxin   A poisonous substance produced by a spectrum of organisms, some of them marine. Bacteria, dinoflagellates, and algae produce poisonous chemicals, as do fungi (mycotoxins) and animals.

Some toxins will pass through one species (the intermediary) to another without affecting the former. Thus, mollusks may not be poisoned by the "red tides," but fish, birds, and humans who feed on the tainted shellfish will become ill.

Animals that produce highly venomous toxins are usually tropical. They include cone shells, sea snakes, and several fish such as the lion fish and the stone fish. For most of these toxins there are no antidotes.

trade winds   Systems of tropical winds that range from about 30° North Latitude to a similar point south of the equator. They are northeasterlies, following a northeast-to-southwest direction in the Northern Hemisphere, and southeasterlies, with a southeast-to-northwest direction in the Southern Hemisphere. The trade winds meet at either the equator or at the Intertropical Convergence. In the western Pacific and the Indian Ocean, this pattern is overshadowed by the monsoons. See *Intertropical Convergence, monsoon, wind.*

transform fault   A break in the crust of the Earth. Transform faults are characteristic of the mid-ocean ridges, running perpendicular to the line of the ridge. The line of the ridge is displaced (offset) by the fault. See *crust, mid-ocean ridges.*

Transpolar Drift   An arctic current that moves very cold fresh water from north of the Laptev Sea across the Eurasian Basin to the Fram Strait and into the Atlantic. The Transpolar Drift enters the Atlantic by way of the East Greenland Current. See *Arctic Ocean, East Greenland Current.*

trawling   The collection of material from the sea bottom by means of a trawler or gathering vessel, which pulls a conical net that entraps the target population in it. Trawling is used by both commercial fishermen and marine biologists, the former for gathering shrimp or bottom-dwelling fish such as flatfish or cod, the latter—using a net of much smaller mesh—for sampling specific depths for the population at that depth. See *fishing, nets.*

trench   A deep depression in the Earth's crust, marking the boundary between plates. At a trench, one crustal plate dives beneath an adjacent one (subduction). The whole area is one of seismic instability, which is manifested by earthquakes. Examples of trenches are the Aleutian, Puerto Rico, and Tonga trenches. See *crust, individual trenches, subduction.*

Triassic Period   Segment of the Mesozoic Era that follows the Permian and precedes the Jurassic Period. It lasted from about 225 million to 190 million years ago and was named for the three subdivisions of the period as recorded in rock formations: Lower, Middle, Upper. The Middle corresponds to the notably marine part of the Triassic. The rock is sandstone and mudstone; it is found on all continents and is the residue of sedimentation in a hot climate. The landmass during the Triassic was a single supercontinent, Pangaea. The Atlantic Ocean opened at the end of the Triassic.

The biota of the period included the first amphibians and gymnosperms (cone-bearing plants). Algal and coral reef builders and mollusks were present. On the whole, the Triassic did not leave an abundant marine record, possibly because the hot climate evaporated water from the seas and rendered them very salty.

trichocyst   A rod-shaped organelle arranged to be ejected out of the body of an organism to stun, paralyze, or poison prey or predators, or to aid in the recovery of prey. Trichocysts are characteristic structures of ciliates. See *Ciliophora.*

trilobite   An extinct marine arthropod. Trilobites were the dominant animals of the Cambrian Period, and finally disappeared by the Permian. They are used by paleontologists as index fossils for dating rock structures. These animals had a tripartite body similar to that of the arthropods of today. They may have looked like flattened centipedes.

Trilobites ranged in feeding habits from plankton and detritus feeders to carnivores. Their sizes varied from tiny species less than 1 cm (0.3 inches) long to those 40 to 50 cm (16 to 20 inches) long.

See *Arthropoda, Cambrian Period, fossils.*

trough   An undersea depression that is U-shaped rather than V-shaped like a trench or canyon, which it otherwise resembles; also, an atmospheric low about which winds circulate. See *canyon, trench, wind.*

tsunami   A Japanese word for a wave system following a major sea disturbance such as an earthquake. It is not a tide; therefore the phrase "tidal wave" is an inaccurate one.

The tsunami is a gravity wave, and as such is influenced by the water depth and the shape of the shoreline. Because of these variables, it is difficult to predict the outcome.

In this wave system, energy is transferred between the ocean and the atmosphere. These interchanges include an increased atmospheric pressure, which causes the ocean to become shallower. Thus, waves in the ocean can move faster than they normally would in deep ocean waters. This helps explain the very rapid movement of the water created by the Krakatoa eruption in 1883. See *island, Krakatau, volcano, waves.*

Tsushima (or Tashima) Current   A western Pacific Ocean current which flows out of the Yellow Sea through the Korea Strait and into the Sea of Japan. It is a warm, low-salt surface current that branches into the East Korea Current and then again into the Tsugaru and Soya Currents, all of which are warm-water flows.

The Tsushima Current is a spring and

summer current; it disappears in the winter. See *current, Sea of Japan.*

tube worm   See *Pogonophora, vent communities.*

tuna   A large food and sport fish, of the family Scombridae, order Perciformes, that is related to the mackerel. Tuna are long, chunky fish characterized by a thin, shapely tail and a series of small fins on their dorsal and ventral surfaces, anterior to the tail. Like other long-distance swimmers, they have temperature-regulatory structures under their skin. These are energy conservers which can raise the temperature of the fish if it is in very cold water.

Tuna are generally dark blue or black on their upper surface, light or silvery below the lateral line, and frequently iridescent. They move in dense schools, a habit fully exploited by commercial fishermen. Tuna are found in temperate to tropical water worldwide, and are a food fish almost everywhere. They are carnivores and feed on herring. See *fish, schooling.*

Tunicata   A subphylum of marine chordates including the sessile (stationary) sea squirts. There are over 2,000 classified species of tunicates. Their range is worldwide and they are found at every depth. Their notable characteristics are a flexible protective cover—the test or tunic;

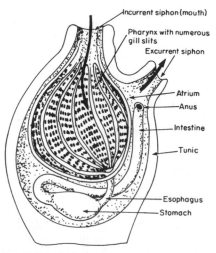

An adult tunicate

and a notochord, which is a feature of the larval state only. The lifestyles of tunicates vary from the pelagic to sessile, and they may vary in size from 1 to 2 mm (0.03 inches) to over 30 cm (1 foot). Colonies such as the Thaliaceae may be larger, some exceeding 1 m (40 inches) in diameter.

The tunicates are classified as Ascidiacea—sea squirts—which are sessile and either solitary animals or colonial species; the Thaliacea, which are pelagic and have a complex reproductive system that includes both sexual and asexual phases; and the Larvacea, which are tiny pelagic animals.

Tunicates have hearts and a network

Common Name	Scientific Name	Length	Weight	Habitat
Bluefin	(*Thunnus thynnus*)	4–5 m	800 kg	Mediterranean or warm Atlantic
Albacore	(*T. alalunga*) This fish migrates over great distances	1 m	30 kg	Warm Pacific
Yellowfin	(*T. albacares*) The name is derived from the yellow pectoral fins and gillcovers	2.5 m	200+ kg	Tropical
Big eye	(*T. obesus*)	2.5 m	200+ kg	Tropical
Skipjack	(*Katsuwonus pelamus*) This fish has stripes, not spots.	85–95 cm	22–24 kg	North and Baltic Sea, Pacific

See *fish, schooling*

of blood vessels and nerves. The distributive function is, however, performed by cells called macrophages, which move within the animal. These cells are also the medium for gas exchange.

The question of the notochord has been debated for years. Tunicates were once described as "degenerate" organisms. While this view is no longer held, there is still the problem of why this animal has a supporting skeletal unit as a larva, only to lose it in the adult form. Tunicates also use both iron and vanadium in their metabolic processes, and both metals are concentrated in their tissues. The need for either or both is unexplained. See *Larvacea, pelagic environment, sea squirt, sessile.*

Turbellaria   A class of free-living flatworms. Most of these animals are aquatic. They are usually small, and many are barely visible to the unaided eye. Some are larger, and a few range in size up to about 60 cm (2 feet).

Many turbellarians are brilliantly colored. However, the best-known, the freshwater planarians are black. Turbellarians move by undulating the muscular layer that lies beneath their epidermis. The respiratory function is performed by the epidermis, through which gases move in and out by diffusion. The excretory system is very simple, and as with respiration, some of its function is undertaken by the epidermis. There is a "head" end and a distinct mouth on the ventral surface near the "head."

Turbellarians are hermaphrodites. They produce both eggs and sperm, but can also reproduce by splitting laterally (down the middle from head to tail).

These organisms are frequently commensals or symbionts. They live on or near other marine invertebrates, and sometimes harbor protists or nematodes. See *commensal relationships, flatworms, symbiosis.*

turbidity currents   Currents in which there is suspended sediment. These currents occur on the continental slopes. Sediment in the water moves to a point where, due to gravity, it is deposited. Since most of the sediment in the ocean is the result of river transport, this material is deposited in unconsolidated patterns at river outflows.

Turbidity currents arise quickly as the result of a flood, earthquake, or rapid deposit of river sediment. These trigger rapid water movements, which are the undersea equivalents of atmospheric storm surges. The rapid water flow will break up the unconsolidated deposits of sediment, causing rapid changes in the undersea landscape—usually landslides.

Turbidity currents are thought to be the cause of submarine cable breaks. They are also thought to be the mechanism by which undersea canyons are originally carved or maintained free of sediment. See *sediment, submarine canyon.*

Turkish Plate   A small crustal plate that is moving between the African and Hellenic plates. It is headed for the Eurasian Plate. The Caucasus Mountains are the seam of subduction along which this plate dips down under the Eurasian Plate. The buildup of pressure in this area makes it one of frequent seismic activity. See *plate tectonics, subduction.*

turtle   See *sea turtle.*

Tyrrhenian Sea   A roughly triangular part of the Mediterranean Sea between the Italian mainland, Sicily, and Sardinia. It is named for the mythical founder of the Etruscan kingdom. The deepest point, which is almost 3,600 m (12,000 feet) below sea level, is between the Gulf of Salerno and Sardinia. Atlantic water enters the Tyrrhenian Sea south of Sardinia and circulates counterclockwise. See *Mediterranean.*

U

ultraplankton   See *plankton.*

United Nations Conventions   See *Law of the Sea.*

upwelling   The rise of subsurface cold, dense water. It is most noticeable on the western coasts of continents. The deep water coming up does so to replace wind-displaced surface water. If the rising water is colder than the ambient air, fog results. The California coast near San Francisco is a prime example of a site where such colder water wells up. If the deep water is nutrient-rich, the result is a bloom of plankton at the surface. This in turn feeds a large fish and bird population, as evidenced on Antarctica and along the Chile-Peru coast of South America. See *bottom water, nutrients, Peru Current, plankton.*

V

veliger The larva of a mollusk after it has developed its velum (swimming membrane).

Venezuelan Basin The second largest portion of the Caribbean; it lies north of South America, west of Grenada, east of the Colombia Basin, and south of Santo Domingo and Puerto Rico. The average depth of the bottom of the Venezuelan Basin is about 4,400 m (14,000 feet). The Beata Rise separates the Venezuelan from the Colombia Basin.

The bottom topography of the Basin is interesting, since it gives two reflections of sound. One explanation for this double echo is based on the theory of formation of the Caribbean Sea. If this Sea was always oceanic, and then sank further still, it would have a different bottom composition than if it was once terrestrial and then sank. If the latter is the correct scenario, then an eroded landscape would have been overlain by sediment carried to it by wind and waves. See *Beata Ridge, Caribbean Sea.*

vent communities Recently discovered ecosystems, not based on photosynthesis, which occur at volcanic sites and at rifts where mineral-containing water emerges from breaks in the ocean floor. The ecosystem of the vent community is extraordinary because it exists in areas where the average depth is 2,500 m (8,200 feet) and the temperature is above 350°C (about 700°F) inside the vents and about 250°C (480°F) in the surrounding water. These communities are associated with black smokers. When its vent ceases to emit mineral oxides and sulfides, the vent community changes in character.

Ocean water, sulfur, manganese, and iron support the sulfur bacteria of the vent community, which depend on metabolic systems that utilize sulfur instead of oxygen. They also thrive in a high-temperature, high-pressure environment. The bacteria in turn support populations of tube worms, huge crabs, large oysters and clams, fish, and other biota that live in the constant dark of deep ocean at these seemingly poisonous sites.

Vent communities were first discovered in the 1970s, and are thought to be models of life on the primordial Earth. As such they are being intensively studied. So far, these habitats have been found in the Pacific Ocean at the East Pacific Rise, at the Galapagos Rift, near the Gulf of California, and on the Juan de Fuca Ridge. They have also been located in the Caribbean. See *photosynthesis, Pogonophora, sulfur bacteria.*

ventral The lower or underside of an animal. See *dorsal.*

Venus comb See *Murex.*

Venus's flower basket The skeleton of a tropical sponge. The "basket" is a graceful, J-shaped, tapered lattice about 25 cm (10 inches) long and tubular. The species *Euplectella* (class Calcispongiae, phylum Porifera), which give rise to the Venus's flower basket, attach themselves to sandy bottoms in tropical seas. They are most often found near the Philippines, and range as far north as Japan. They have also been collected in the Western Pacific and in the Indian Ocean. These sponges are prized by collectors. See *Porifera, sponge.*

Venus's girdle A member of the phylum Ctenophora. This beautiful undulating, luminous, transparent ribbon is actually a pelagic (open-sea) animal. See *Ctenophora*.

Verrazzano, Giovanni da (1485–1528)

Florentine navigator and explorer. He sailed to Newfoundland in 1508, and apparently made voyages to Carthage, Damascus, and the Eastern Mediterranean for the French. He met Magellan in 1517 in Seville, where both were looking for royal sponsors for voyages of discovery.

Verrazzano was financed and given command of a voyage of discovery by François I of France. The Florentine bankers in Rouen also contributed to this, and were given shares of the possible profits. Four ships set sail for the New World, but only the largest, *La Dauphine* and *La Normande,* crossed the Atlantic. After a storm, the party made landfall on Cape Fear, in what is today North Carolina. When Verrazzano sailed around the Outer Banks he thought he had found the Pacific. He then sailed north, looking for the passage to India, and entered what is now called New York Bay on April 17, 1524. He continued up the coast to Rhode Island, Maine, and Newfoundland before turning east and back to France.

Verrazzano made careful calculations of latitude for each of his anchorages. He kept logs of distance, and estimated longitude with reasonable success. Although he erred in some calculations, he realized that the landmass he found was nowhere near Asia, and that it was no fringe of islands, but that a large continent was interposed between the Atlantic and the Pacific oceans.

On a second voyage, again sponsored by France, Verrazzano sailed to Brazil for wood that was in demand by the dyers. The profits made on this trip encouraged his backers, who then financed another attempt to find the strait to the Indies by sailing south. Verrazzano was captured and killed by the inhabitants of what is now Guadaloupe in 1528, and his brother, Girolamo, brought the ships back to France.

vertebrate   An animal with a backbone. These include the fish, amphibians, reptiles, birds, and mammals. As a group, the vertebrates have an internal skeleton with a separate head; the head has hinged jaws. The vertebral column is a series of bones, whose number is significant. Vertebrates also have two pairs of appendages that have been highly modified, and in some cases lost.

The bony vertebral column that protects the spinal cord—the great trunk of nervous tissue that extends from the brain down the back of vertebrates and gives off nerve branches to the body—is preceded in embryonic development by a notochord which develops ventral to the nerve cord. In many primitive animals this remains throughout life. The animals known as tunicates have a notochord as juveniles and then develop an adult body without the support of a bony or cartilaginous spinal column and skeleton. In more advanced forms it is lost and is replaced by a calcareous or cartilaginous skeleton.

The vertebrae of different species of vertebrates are of different shape, size, and number, depending on the animal. Fish have trunk and caudal vertebrae, snakes have the greatest number, but most of the bones of the spinal column are the same size and shape; in turtles several bones are fused together to form the carapace. Some of the vertebrae of birds also have been fused together for better aerodynamics. See *bird, fish, mammal, notochord, Reptile, Tunicata.*

vertical migration   See *migration.*

Vespucci, Amerigo (1454–1512)

An explorer born in Florence to a family with Medici connections. As a member of this extended family he worked in their banking house and then in an affiliated bank and ship chandlery in Seville, Spain, where he outfitted the third of Columbus's voyages in 1498. In 1499 Vespucci sailed to the New World with Ojeda but left the voyage in Hispaniola, returning on his own. Based on that trip, he broadcast his own reputation as a mariner, which led to an invitation from King Manuel of Portugal to join Coelho's expedition. Vespucci accepted the royal invitation to accompany and chronicle the voyage which lasted from 1502 to 1504. His accounts, entitled *Lettera* and *Mundus Novus,* were published by a friend in Florence (1504–1506). This led to the association of Vespucci's name with that of the new continent he explored, when Martin Waldseemuller, a publisher, produced a new edition of Ptolemy's Geography that showed the "new world" discovered by Americus (Amerigo), and the name was gradually adopted for the entire landmass, not only South America. Waldseemuller's map showed another ocean between America and India, because Vespucci—a keen observer of the geography, scenery, flora, fauna and indigenous population of the new lands— was one of the first to realize that the newly discovered lands were not a part of the Asian coast.

Victoria Nyanza, Lake The largest lake in Africa and the chief reservoir of the Nile. It occupies the plateau between the East and West Rift valleys in East Africa. Lake Victoria is the second largest lake in the world, with a surface area of 68,500 km² (26,500 square miles). Only Lake Superior in North America exceeds it. The Lake was first seen by Europeans when John Speke, searching for the headwaters of the Nile, sighted it in 1858.

Vitiaz A Russian vessel that charted the Pacific Ocean trenches during the International Geophysical Year (IGY) of 1956–7. See *IGY.*

volcano A mound associated with a vent in the crust of the Earth, on land, or under water, from which molten or hot rock and steam are ejected. Undersea volcanic activity forms the Earth's undersea ridges. In an oceanic volcano, as in a terrestrial one, magma, the liquid rock from the Earth's interior, oozes out through

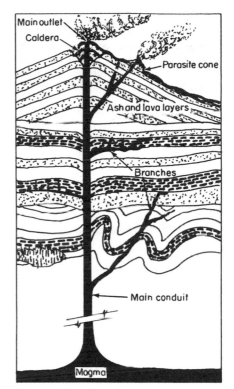

Cross section of a volcano

long fissures or rises in a central channel, although the former is the more likely phenomenon at the crustal boundaries.

Fissure-flow volcanoes produce large areas of lava called pillows. These are similar to the gently sloped Hawaiian volcanoes, and resemble gentle, rolling hills. Pillows are found in the Atlantic, and typical of slow, spreading ridges. The sheet flows of such areas as the East Pacific Rise are formed by very fluid lava with a glassy crust.

The volcano that has a well-defined central channel will form a cone whether it is on land or undersea. Seamounts have their origin in volcanoes of this central-channel type. If there is a depression in the center of a seamount, it is called a crater if it is small and a caldera if it is large. Seamounts are usually bounded by fault lines; these sites of seismic and volcanic activity produce a fairly constant supply of molten rock. If it leaks into the rock at the top of a volcanic cone, this rock will eventually fill it, and the cone will become leveled off. This leveled structure is then called a guyot.

Volcanoes also occur within plates, at points of geologic stress. Such volcanoes tend to be in chains located obliquely in relation to the mid-ocean ridges. See *crust, earthquake, guyot, hot spot, lava, plate, seamount.*

W

Wagenaer, Lucas Janszoon (1534?–1605) A Dutch sailor and cartographer who became a pilot and in 1582 began a compilation of pilot's charts for western and northern Europe. This collection eventually became *Spieghel des Zeevaerdt* (Mirror of Seafaring), which was published in 1584–1585. It was a useful navigational aid, and faithfully mapped the region from Cadiz, Spain, to the north of Holland. The Lord High Admiral of England was so impressed with it that he had it translated. The translation appeared in 1588, with the title *The Mariner's Mirror*.

Wallace, Alfred Russell (1823–1913) A British naturalist who discovered natural selection independently of Charles Darwin (1809–1882). Wallace began his career as a surveyor and then became a self-taught botanist. He explored the Amazon River and survived a shipwreck in 1852. After a dramatic rescue, he transferred his interests to the Malay Archipelago, where he did intensive exploration (1854–1862), and finally returned to England as an established naturalist. His book on the natural history of the East Indies, *The Malay Archipelago,* was first published in 1869. It was a continual best seller, and was reissued in several subsequent editions.

Wallace's *Species Notebook* was a scheme for grouping animals in terms of their geographical location and related species. This formed the basis of a scientific paper that appeared in 1855. The explanation given by Wallace for natural selection was formulated in 1858. He sent it—with a long explanation—to Darwin, who was already an authority on the subject. Darwin presented both his and Wallace's work simultaneously in 1858 (Darwin's evolutionary views dated to the late 1830s, but he had not yet publicized them).

Wallace continued his research on evolution and the study of mimicry in both related and unrelated species. Toward the end of his life his ideas on the evolution of humans became entangled with his adherence to spiritualism, then a very popular pastime in England. *See Darwin, Charles; evolution; Wallace Line.*

Wallace Line A zoogeographical concept introduced by Alfred Wallace on the basis of his observations of the flora and fauna of South America and Malaysia. A "Wallace" line is said to exist between different but related species that are separated by a natural barrier such as a large river.

Wallace published a work on the zoogeography of Southeast Asia in 1860. He introduced the concept of the line in

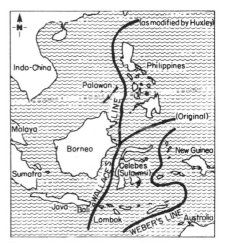

Division of zoogeographic regions

1863, and with it he divided Asian from Australian species. He drew the line between Bali and Lombok in the south, between Borneo and Sulawesi in the north, and in the east around the Philippines. As more information became available, the line shifted.

In 1902, Webber proposed a modification and expansion of the line. According to him, it would run from the Indian Ocean through Timor, then north through Molucca and into the Philippine Sea. See *evolution; evolution of oceans; migration; Wallace, Alfred Russell.*

walrus   A large, tusked, seal-like mammal of arctic waters, of the family Odobenidae. The walrus has many of the characteristics and the body plan of the seal, and no external ears. It has short red hairs on its wrinkled hide and a comical, brushlike moustache and whiskers.

The male is considerably larger than the female, with an average length in excess of 3 m (10 feet); walruses frequently weigh more than 1200 kg (2500 pounds). The tusks on the male are also large; these enlarged canine teeth are not digging tools, but are used for both courtship displays and territorial defense.

The walrus is a colonial animal. It lives in large groups dominated by enormous harem-collecting males. The most usual habitat is near a stretch of reasonably shallow water. The walrus feeds almost exclusively on clams. The only serious danger to this fat and lazy animal comes from humans. See *pinniped.*

Walvis Ridge   An undersea elevation near the western coast of Africa, which extends about 3,000 km (1,900 miles) into the southern Atlantic Ocean. It makes a rough line from the Tristan da Cunha island group on the Mid-Atlantic Ridge, northeast to the coast of Namibia. This aseismic ridge is very narrow and steep-sided, with drops of more than 1,000 m (3,300 feet) on its southern side. The expeditions that have mapped turbidity currents have shown that there is great current activity in this area. Rapid currents scour the deeps near the ridge, and it is sediment-free.

On some maps the Walvis Ridge is called the Walfisch Ridge. See *aseismic ridge, Mid-Atlantic Ridge, turbidity currents.*

warm-blooded animals   Animals that maintain a constant body temperature; also called homoiotherms (homoiotherms). See *homeotherms, poikilotherms, temperature in animals.*

water   See *seawater.*

water flea   A small freshwater crustacean, of the genus *Daphnia,* that moves about on the surface of the water with sudden leaps.

waterspout   A spinning, funnel-shaped column of water or spray pulled up by a whirling wind from the ocean surface. This rising wind and water funnel looks like a tornado. Waterspouts are formed in the same humid, hot, tropical air that builds thunderclouds. While they are most often tropical, they are sometimes found in middle latitudes, particularly on hot summer days. Waterspouts dissipate very rapidly if they move over land. See *cyclonic storm, hurricane.*

wave   Crest of water moving along the surface of the water in lines at right angle to the direction of motion. Waves are divided into two large groups: small waves, called ripples, with crests of equal size less than 2 cm apart, which result from

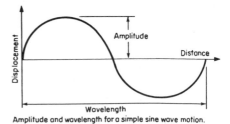

Amplitude and wavelength for a simple sine wave motion.

Amplitude and wavelength for a simple sine wave motion

surface tension, and longer waves, called gravity waves.

Gravity waves are then subdivided into shallow-water waves and deep-water waves, depending on the depth of the water through which they move. In shallow-water waves the water particles move forward and backward in a horizontal plane. The crests represent crowds of water particles; the troughs are sparse populations. A tidal or river bore is a classic shallow-water wave. A solitary wave is one that occurs when a mass of water moves in one direction and an equal mass of water moves in the reverse direction. The water both in front of and behind the wave looks like it is standing still, but in fact is moving forward.

When one observes a shoreline, one sees a train of waves. They will decrease in velocity in shallow water. Thus, since the positive direction (i.e., the direction in which the wave train is going) moves faster than the negative bore, crests pile up and produce breakers. Breakers are the foaming delivery of water onto the beach. If the beach is one of gentle slope the breakers spill water onto it; if the beach has a steep slope, the breakers plunge onto it.

Deep-water waves are generated by storms, and can have crests miles apart. Wind generated waves start out short. There are conflicting theories about how these short waves become long. The velocity of fast, storm-generated waves is generally less than that of the generating wind—at most about 80% of it. Thus, long waves travelling from a storm center move more rapidly than the short surface waves. This is the "groundswell" that will precede and predict a coming storm.

By clocking the time between crests, it is also possible to determine where the storm is and where it was when the first waves arrived. This is extremely important in coastal areas, since violent storms at sea, or seismic activity undersea, will produce deep-water waves that will travel very rapidly and only slightly alter the condition of the surface. Enormous quantities of energy are expended in lateral motion in deep water; in shallower water this energy is expended by smashing against the shore. The result of this impact with the shore is an instant and dramatic rise in sea level. Winds whip up the water, aggravating the situation. The effect on land is flooding of coastal areas to a variable degree depending on the height of the water and on the coastal area. See *internal wave, rip current, river bore, tsunami.*

wave height The vertical distance between the top or crest of a wave and the bottom of the preceding (or succeeding) trough. See *wave.*

wave period The time between the passage of one wave crest and the next crest through a specific point. See *wave.*

weather The state of the atmosphere. This results from the combined effect of clouds, precipitation, temperature, humidity, visibility, wind velocity, and barometric pressure. Weather is treated on two bases: geographic (global, regional, or local) and chronological (long- or short-term).

Much of the equipment of the meteorologist consists of the products of 17th- and 18th-century science. The 20th century contributed radar, dating techniques, very sensitive photography, and computer modeling, among other techniques. Today it is possible to track storms by radar and to use satellites to photograph developing weather patterns.

The study of seemingly local meteorologic events in terms of global weather is now an unfolding area of research. The phenomenon of El Niño in the southern Pacific has a profound effect—a chilling one—on the weather in North America. This in turn affects the Atlantic Ocean, with an increase in iceberg formation as a direct result of the colder winter that comes with the Niño phenomenon. Another result of this weather disturbance is the increase in heavy rain in equatorial regions, which diminishes the velocity of the prevailing westerly winds.

Accurate meteorologic record-keeping and model-building by simulation both depend on large computers with extensive memory and calculating capability. The increasing accuracy of weather prediction depends on both such records and such models. See *atmosphere, El Niño, meteorology, ocean-atmosphere, wind.*

Weddell, James (1787–1834)

An English sea captain in both the Royal Navy and the maritime trade. Weddell sailed in antarctic waters, hunting seals in the post-Napoleonic period (after 1815). He visited the South Shetland Islands and named the South Orkney Islands in 1822. In 1822–1824 he charted the waters near the South Shetlands and explored the sea named for him. His record mark for the most southerly point reached was 74° 15' South Latitude before his path was blocked by ice. His memoir, *A Voyage Toward the South Pole,* was a popular success. It was published in 1822 and initiated efforts in several countries to explore the Earth's polar regions. See *Dumont D'Urville, Jules-Sebastien César; Wilkes, Charles; Ross, John.*

Weddell Current

An antarctic coastal flow of water at the ocean surface. It moves in an easterly direction from the Weddell Sea, and is outside the West Wind Drift. See *Antarctic Ocean, current, West Wind Drift.*

Weddell Sea

The southernmost part of the Atlantic Ocean; it is usually ice-filled. The area was explored by James Weddell in 1823. Coastal fossil-bearing limestone was found along the sea in 1902–1904. The continental shelf of Antarctica runs under the sea for about 150 km (90 miles); the continental slope is relatively steep, and descends to depths of about 4,500 m (15,000 feet). This depth may be the result of the weight of the ice shelf. There is controversy about the origin of the Weddell Sea Basin.

The Weddell Sea is generally accepted as the source of most Antarctic bottom water. It is a cold sea, its water temperature generally being below −1°C, and is fairly salty for polar water at a salinity of about 3.41%, with higher percentages in some areas. There is little wind, which means that the huge icebergs produced in the sea tend to stay and fill the basin.

A recently discovered marsupial fossil located on the Antarctic Peninsula, western boundary of the Weddell Sea, is further evidence of the definite connections between South America, Australia, and Antarctica. The fossil was of South American origin; the link between this continent and the Antarctic mainland may therefore have continued into the Eocene. See *Antarctic water, Eocene Epoch, fossils.*

Wegener, Alfred Lothar (1880–1930)

A German meteorologist and geophysicist best remembered for his theory of continental drift. He published this work in 1915, but was certainly not the first person to have noticed the coastal congruencies of the continents. That observation had been made early in the 17th century.

Wegener's theory was based not only on the similarities of the coastlines of various land masses, particularly in the Southern Hemisphere, but on similar fossil records on now widely separated landmasses as evidence of ancient climate conditions. This evidence had been rather inadequately explained away by the evocation of convenient "land bridges" linking these landmasses at some previous time. Wegener called his primordial, warm, Permian continent Pangaea, from Gaea, the mythological Greek "earth-mother." In 1937, Du Toit, a French paleogeophysicist, divided this single, then hypothetical continent into two. He called the northern portion Laurasia and the southern portion Gondwana.

One of the great flaws of Wegener's hypothesis was that he could not adequately explain the mechanisms by which the fragments of his single continent had separated, even though he made continuous and determined efforts to do so, using paleoclimatology, balloon observations, and geodetic measurements.

The theory received considerable attention until 1928, when a panel of fourteen geologists voted on it. Five were in favor of it, and suggested future surveying to attempt to prove it; two were uncertain, and seven turned it down. From that time until the work of Fred Vine, Drummond Matthews, and others on paleomagnetism in the 1950s, the Wegener theory was ignored.

Wegener was lost in a blizzard in central Greenland. He was there as a member of a meteorological survey. See continental drift, magnetic anomaly, plate tectonics.

West Australian Current A portion of the West wind drift, the current generated in antarctic waters, which turns north along the western coast of Australia and continues into tropical regions as a coastal current. The West Australian Current is a part of the Indian Ocean gyre. This surface stream is strongest in the summer. See gyre, Indian Ocean, West Wind Drift.

westerlies Winds found in both the Northern and Southern hemispheres, between 30° and 60° of latitude. These winds move from west to east. The westerlies arise in subtropical high-pressure zones. The westerlies are very extensive, and may reach up 15 km (9 miles) into the stratosphere. They have velocities of about 10 km (6 miles) per hour at the Earth's surface. Since wind velocity increases with altitude, this means that at jet-stream heights of over 10 km (10,000 m, 33,000 feet), the wind is moving at velocities greater than 300 km per hour. See monsoon, trade winds, wind.

west-flowing currents Surface currents that are propelled by the prevailing westerly winds. Most are at about the 45° parallel in both hemispheres, and they occur in all oceans. The west-flowing current in the North Atlantic is the Gulf Stream; in the South Atlantic, the Brazil and Cape Horn currents move west. In the Pacific Ocean, the northern west-flowing current is the Kuroshio, while the Equatorial and East Australian currents are in the Southern Hemisphere. In the Indian Ocean there is no northern current; the landmasses are in the way. The Agulhas Current in the Southern Hemisphere is the end-result of the west-flowing Equatorial Current. See individual currents.

West Greenland Current An Atlantic Ocean current; a branch of the East Greenland Current. It moves north to the Davis Strait. The coastal water is cold and of a low salt content, ranging from 3.1 to 3.4%, which is similar to that of the East Greenland Current. The water farther out to sea on the western side of Greenland—the North Atlantic gyre—is warmer and saltier than that of the West Greenland Current, with an average salt level of about 3.5%. The West Greenland Current eventually loses its particular character by dividing and merging with the Labrador Current as the latter moves southward; the rest disappears into Baffin Bay. See Atlantic Ocean, currents, East Greenland Current, gyre.

West Spitsbergen Current A current that brings North Atlantic water through the Fram Strait between eastern Greenland and Svalbard (Spitsbergen). It makes a sweep around Jan Mayen Land before disappearing into the East Greenland Current. See Arctic Ocean, East Greenland Current, Fram Strait.

West Wind Drift The common name of the Antarctic Circumpolar Current. This is the tremendous volume of surface water that is moved by the prevailing westerly winds. See Antarctic Ocean.

whale oil Oil obtained by steam-cooking the blubber or fat layers of captured whales. This was why whalers went to sea. The increasing rarity of whales, as a result of overhunting, has made them an unreliable source of oil. The problem of overhunting was first noted by Alexander Agassiz and underscored by Prince Albert of Monaco, two prominent, early 20th century oceanographers. Some species may have been hunted to extinction. While some whale oils are quite specialized in use, such as spermaceti, the uses of whale oil could be filled by other means. See *Agassiz, Alexander; Albert I; extinction; marine oils; whale oil.*

whale shark   A worldwide, tropical, and harmless giant shark. It is the largest living fish. Specimens of more than 9 m (30 feet) in length have been documented, and sightings of animals three times that size have been reported but not verified. The whale shark is gray or gray-brown on its upper surface, white below, and marked with small dark spots. This fish is a slow swimmer that ingests plankton and small fish in its path. See *plankton, shark.*

whales   Large marine mammals of the order Cetacea, with highly modified forelimbs that have become flippers, no hindlimbs, a vestigial pelvis, and a flattened tail with flukes that act as stabilizers. There are ten fossil forms and nine existing groups of whales, some of which are endangered.

The toothed whales are homodonts, each tooth looking like every other, and have a single nostril at the top of their head. This group includes river dolphins; beaked, sperm, and beluga whales; narwhals; porpoises; dolphins; and killer whales.

The baleen whales have a flanged lower jaw that acts as a giant strainer of the tiny animals that these whales feed on. The head is disproportionately large for the body. These animals, which include the gray, right and humpback whales, have paired nostrils.

Most of the toothed whales school. The baleen whales tend to live in small groups called pods, or are solitary creatures. Schools of whales are hierarchically arranged in the same way that elephant herds arrange themselves. Dominant males lead, others protect the herd's females and children, and dominant females organize the other females and juveniles. Play behavior among juveniles and adults is common, as is cooperative action and assistance. The latter may cross species lines—whales of one species will assist those of another.

Both the sight and hearing of whales are well developed. Most cetaceans make complicated and intelligible (to other cetaceans) sounds. The mechanism for the rapid swimming of whales is obvious. The body is extremely streamlined, sleek, and equipped with a massive and powerful tail. The explanation for the mechanism of whales diving rapidly to great depths and returning just as rapidly is not established. It is not known how a mammal can accomplish this without suffering severe brain damage, hypothermia, or the bends.

Biologists tend to correlate the longevity of whales (which tend to live as long as or longer than humans) with their long gestations of a year or more, and long infancy of at least 18 months. Whales are born large and grow very rapidly on a diet of high-fat, high-protein milk.

Most whale species are well known for long migrations from feeding to breeding grounds. Some, however, don't migrate at all. Thus, lifestyle varies from the almost stationary to the baleens, which may travel 5,000 km (3125 miles) to calving areas in warm water.

Toothed whales feed by capturing prey such as squid. The dentition of these whales varies; some have teeth in both jaws, others, such as the sperm whale, only on the lower jaw. The upper jaw of the baleen whales has plates suspended from it that are a curtain of bristling fringes of horn. This keratin material is chemically the same as hair or human fingernails. Baleens either skim the water or swim openmouthed, straining the water as they move forward, or gulp. Gulpers take mouthfuls of water and then expel it past the baleen. They then swallow whatever remains caught on the baleen.

Paleontologists believe that whales made the gradual transition from land to the sea in the early Eocene epoch, 55 million years ago. Fossil remains in what was the eastern Tethys Sea, and is now in Pakistan, indicate a move to a shallow-water, intermediate form of these animals. The evidence for this theory is a particular ear development of the whales. See *baleen, dolphin, echolocation, extinction, krill, porpose, blue, sperm, killer, right and humpback whales.*

whelk   A large, univalvular marine snail of the family Buccinidae, subclass Prosobronchia, class Gastropoda that resembles the conch. Whelks feed on carrion but also eat other mollusks and fish, which they kill with their long proboscis. These animals have a worldwide range, living on sandy bottoms. The common North Atlantic species is *Buccinum undatum,* the waved whelk. Its shell is about 8 cm (under 3 inches) long, pale yellow to cream colored, and heavy for its size. The conchs are more often tropical or Pacific Ocean creatures. *Murex* or oyster drills are rock whelks. See *Gastropoda, Murex.*

White Sea   Also known as the Byela More; an embayment of the Barents Sea south of the Kola peninsula in the U.S.S.R. It is fairly shallow, with a maximum depth of less than 350 m (1,000 feet), and most of it is between 100 and 200 m (330 to 660 feet) deep. A sill prevents bottom water from mixing efficiently with the water of the Barents. The White Sea's water has a very low salt content, of 2.4 to 2.6%. Even the bottom water is of low salinity, at about 3.0 to 3.05%. This sea, like the Baltic, was an ice lake in the Pleistocene epoch, about 4 million years ago. See *Arctic Ocean; Barents Sea; Deryugin, Konstantin.*

white shark   A dangerous, carnivorous shark, *Carcharodon carcharias,* of the family Isuridae, found in tropical to warm temperate waters worldwide. This is the shark most likely to be a maneater. It is a handsome animal, gray-blue or gray-brown on its upper surface and creamy white on the underside. Old individuals tend to become whiter as they

age. The reported maximum length of the white shark is greater than 10 m (33 feet), but the average individual is about 5 to 6 m (16 to 20 feet) long.

The white shark literally feeds on whatever it can catch. It will attack fish, birds, turtles, garbage, porpoises, and much else. Its voracious appetite and varied diet enhances its fearsome reputation.

In some classification schemes, the white shark is grouped with the mackerel sharks. See *shark.*

whiting   A marine food fish and commercially important relative of the cod. The *Gadus* or *Merlangus merlangus* is the North Sea species of whiting. It is most often about 70 cm (28–29 inches) long, and silvery-white with a black spot near the pectoral fins as well as the characteristic chin barbel of the codfish. The American species of whiting is *Menticirrhus.*

Other related species are the kingfish *(Menticirrhus saxatilis),* the rock whitings *(Odacidae)* of Australia and New Zealand, the *Sillaginidae* of the Indian and Pacific oceans, and the lake whitefish of North America. See *cod.*

Wiechert-Gutenberg discontinuity
The layer that separates the Earth's mantle from the core. Its depth is estimated at about 2,900 km (1,800 miles). See *crust, mantle, Moho.*

Wilkes, Charles (1798–1877)   An American admiral and explorer who joined the U.S. Navy after service in the merchant marine. He worked with Ferdinand Hassler, the founder of the U.S. Coast and Geodetic Survey, and sailed on two surveying voyages between 1826 and 1833. In 1834 Wilkes headed the Navy's Department of Maps and Charts.

After much lobbying, Congress finally authorized and funded an expedition to the antarctic. After buying instruments in Europe, Wilkes, in command of six ships, set sail in 1838. The first landfall on the voyage was Samoa. After refitting, the expedition sailed south and sighted land in January 1840. Taking advantage of the antarctic summer, Wilkes' group sailed along the coast of this land for 2,400 km (1,500 miles), and Wilkes named the territory Antarctica. The expedition then sailed north along the coast of the American continent, and finally ended its voyage in New York in June 1842. Their cargo of artifacts eventually formed the basis of the Smithsonian Institution's collection of Pacific Islands ethnology in Washington, D.C.

Willoughby, Hugh (?–1554) An English explorer remembered for his attempt to find the Northeast Passage. Willoughby had a military career of some note, but no sailing experience before he accepted an appointment as captain-general of a party of three ships which sailed from London in May 1553. The ships were separated by a storm in the far north. Two ships, the *Bona Esperanza* and the *Bona Confidentia,* were kept at sea by bad weather, and their crews, unprepared for the arctic winter, died of exposure. The last entry in Willoughby's log was in early 1554. Willoughby's body and possessions were reclaimed by his pilot, Chancellor. The log of the expedition was included by Hakluyt in his series. See *Chancellor, Richard; Hakluyt, Richard; Northeast Passage.*

wind A movement of air relative to the land or water surface. Winds are local or general. General winds move over great distances and have reasonably fixed locations and directions on a global scale. They are the Polar Easterlies and West Winds, and the Trade Winds.

Winds are the result of several components; a principal one is that air heats more rapidly over land than over water. This warmed air is then moved toward the closest pole by convection currents in the atmosphere. As it moves it is deflected by the rotatory motion of the Earth.

Local winds include coastal winds and winds moving down mountains or slopes. The coastal winds are subdivided into several categories, such as a sea breeze, which is a wind from cooler water moving

onshore, and the Bora, a cold mountain wind blowing down on a warm coastal area and the adjacent sea. The bora is a common phenomenon along the Mediterranean coast in the winter. See *ocean-atmosphere, roaring forties, trade winds, westerlies.*

windward Facing the wind, or moving in the direction from which the wind originates. See *leeward.*

Windward Islands See *Caribbean Sea.*

Woods Hole Oceanographic Institution A private institution, founded in 1930, whose charge is the investigation of all aspects of oceanography. Interrelated research in physiology, microbiology, ecology, and animal and plant behavior is carried out by the permanent staff and visiting fellows. The Institution maintains the Marine Biological Laboratory, also at Woods Hole, and several research vessels. One of them, *Alvin,* has been used on a number of deep dives in the Pacific to explore vent communities, and in the Atlantic to explore rift valleys and the wreck of the *Titanic.* See *Alvin, submersibles, vent communities.*

worm A term loosely applied to invertebrates that are longer than their cross sections. There are several phyla and many individuals of marine worms. They are Acanthocephala, Annelida, Aschelminthes, Chaetognatha, Echiuroidea, Nematoda, Nemertea, Pogonophora, Sipuncula, and Platyhelminthes. These include free living and parasitic species. See *individual phyla.*

wrack See *eelgrass.*

Wrangell, Baron Ferdinand Petrovich von (1796–1870) A Russian diplomat-explorer who sailed to the Arctic on several voyages between 1820 and 1824, searching for islands north of Siberia. He failed to find what was eventually charted as the island group around 176° East Longitude, the largest member

BEAUFORT SCALE OF WIND FORCE

BEAUFORT NUMBER	GENERAL DESCRIPTION	MANIFESTATIONS	KILOMETERS PER HOUR	MILES PER HOUR	KNOTS
0	Calm	Smoke rises vertically.	Less than 1	Less than 1	Less than 1
1	Light air	Wind direction shown by smoke but not by weathervanes.	2 - 6	1 - 3	1 - 3
2	Slight breeze	Wind felt on face; leaves rustle; ordinary vane moved by wind	7 - 12	4 - 7	4 - 6
3	Gentle breeze	Leaves and small twigs in constant motion; wind extends light flags	13 - 18	8 - 11	7 - 10
4	Moderate breeze	Raises dust and loose paper; small branches are moved	19 - 26	12 - 16	11 - 14
5	Fresh breeze	Small trees in leaf begin to sway; crested wavelets form on inland waters.	27 - 35	17 - 22	15 - 19
6	Strong breeze	Large branches in motion; whistling heard in wires; umbrellas used with difficulty.	36 - 44	23 - 27	20 - 24
7	Moderate gale (high wind)	Whole trees in motion; difficulty in walking against wind.	45 - 55	28 - 34	25 - 30
8	Fresh gale	Twigs broken off trees; progress generally impeded.	56 - 66	35 - 41	31 - 35
9	Strong gale	Slight structural damage occurs; shingles and slate removed.	67 - 77	42 - 48	36 - 42
10	Whole gale	Trees uprooted, considerable structural damage; seldom experienced inland.	78 - 90	49 - 56	43 - 49
11	Storm	Very rarely experienced; widespread damage.	91 - 104	57 - 67	50 - 56
12	Hurricane	Air at sea filled with foam and spray.	Above 105	Above 68	Above 56

of which is now called Wrangell Island. Between 1825 and 1827 Wrangell led a Russian circumnavigational expedition. From 1829–35 Wrangell was the governor of Russian America—today known as Alaska—and opposed to its sale to the United States.

wrasse   An elongate, prickly-spined fish that is thin for its length and characterized by thick lips and sometimes protruding teeth. There are over 300 species of these members of the family Labridae, order Perciformes. They are usually inhabitants of tropical or warm temperate waters, and range in size from 5 cm to 2 m (2 inches to 6.5 feet).

Wrasses are frequently part of the biota of coral reefs, where they establish commensal or mutually beneficial relationships with other fish. For example, they clean the external parasites from groupers, eels, and snappers.

Most wrasses are brilliantly colored. The group also includes a number of species that change color with age. Wrasses can also change their sex as they change color. Interestingly, the sex change can occur more than once in the lifetime of an individual fish. See *commensal relationships, coral reefs, symbiosis.*

wrecks   See *marine archeology, salvage.*

Wüst, Georg Adolf Otto (1890–1977)   A leading German oceanographer whose name is inexorably linked to the *Meteor* Expedition. After the death of his mentor, Alfred Merz, Wüst was precipitated into the task of chief oceanographer of the German Atlantic Expedition of 1925–1927. This was the first study of all aspects of an entire ocean, and from the careful and extremely detailed measurements that were taken, an understanding of circulation, deep currents, temperature, and salinity strata was eventually achieved.

After World War II, Wüst reestablished the *Institut für Meereskunde* (Institute for Oceanography) at Kiel, Germany, and was its director. See *Atlantic Ocean; Ekman, Vagn Walfrid; ocean currents.*

Y

Yangtze River Also known as the Ch'ang Chiang, Change Jiang, or Blue River; a river flowing east through central China from Tibet to the East China Sea. It is Asia's longest river, at over 5,000 km (3,100 miles) long, and in Chinese its name means "long river." The English name Blue River is peculiar since the water is most often a yellow-brown.

The Yangtze rises in Tibet and runs southeast, dropping in a series of waterfalls, then turns to the northeast, traveling through the North China Plain to the East China Sea and Yellow seas. While it is not as muddy as the Huang Ho (Yellow River), it still carries a great volume of alluvium, and extends its delta by about 20 m (about 65 feet) per year. Extensive irrigation and terracing work that has gone on for centuries has made efficient use of the great volume of the Yangtze's water. The valley of the River is an extremely important area agriculturally.

The outlet of the Yangtze is at Shanghai, which remains a principal port of China. The River is navigable for almost 2,000 km (1,250 miles) of its length by vessels of significant commercial size. See *East China Sea, Yellow Sea.*

yawl A fore-and-aft rigged sailing vessel with two masts. The main mast is much larger than the mizzen. See *sail.*

Yellow River See *Huang Ho.*

Yellow Sea Also known as Huang Hai; a marginal arm of the Pacific Ocean that lies between China north of the mouth of the Yangtze River and Korea. Its average depth is about 45 m (145 feet) because a great deal of the bottom is very shallow, particularly in the area of the Gulf of Chihli. The Sea is postglacial in origin and only about 11,000 years old. Excepting for the Gulf of Chihli, the average depth of the Yellow Sea is 60 to 80 m (200 to 260 feet).

Both the Huang Ho (Yellow River) and the Yangtze empty into the Yellow Sea, accounting for its color. Its deep water is both cold and salty; the surface water, because of the great volume of river influx, is about 3.0 to 3.3% salt. This may be lower still in the summer.

The tidal range of the Sea is high, at about 8 m or 26 feet. There are strong tidal currents in some areas.

The Tsushima Current flows through the Yellow Sea, bringing warm and more saline water into it. The climate is characterized by cold and dry weather in winter, warm and humid weather in summer. See *monsoon, tidal range, Tsushima Current.*

Yoldia Sea See *Baltic Sea, Finland, Gulf of.*

Yucatan Current A surface stream in the Caribbean Sea, flowing from Honduras through the Yucatan Channel to the east of the Gulf of Mexico. In summer its maximum velocity is about 2 m per second (4 mph). See *Caribbean Sea, Gulf of Mexico, Gulf Stream.*

Yukon River A river in the Canadian northwest and in Alaska. It is one of the longest of North American rivers, at 3,185 km (1,979 miles), and is named Yukon below the confluence of the Pelly and Lewes rivers. The merged stream flows from British Columbia (Canada) northward and then northwest. The White, Stewart, and then the Klondike rivers flow into the Yukon. This large volume of water from the northwest of Canada turns to enter Alaska at Fort Yukon. Only part of the Yukon is navigable; the channel is a shifting one. This factor, and the cold weather, make it unreliable for commercial shipping. The largest river town on the Yukon is Dawson in the Canadian Northwest Territories. The River flows through Alaska and empties into the Bering Sea. See *Bering Sea*.

Z

Zoanthidea Colonial polyps that re-semble sea anemones. They are antho-zoans of the phylum Cnidaria, and are related to the corals. They inhabit sandy bottoms in tropical to temperate water, an ecological niche similar to that of the sponges.

The Zoanthidea collect prey by means of a mucus coating that they generate and to which plankton adhere. They tend to be small; the largest species are less than 20 cm (8 inches) in size. See *Anthozoa, polyp.*

Zoochlorellae Green algae that are symbionts or commensals with the coral or coralline algae. See *commensal relationships, symbiosis, Zooxanthellae.*

zoogeography A branch of biogeo-graphy concerned with the distribution of living animals in ecological space. Any species in any area is in an equilibrium with immigrants (i.e., new arrivals from other places), and with other species that are on the edge of extinction. Both of these processes are normal, but both are influenced by the activity of humans.

Continental drift has explained numer-ous finds of animal fossils in places where such animals could not exist today be-cause of climate. It also accounts for the appearance of the same or similar species in areas that are now widely separated geographically but were once in close proximity. See *continental drift, ecology, habitat.*

zooplankton See *plankton.*

Zooxanthellae Symbiotic protozoans or dinoflagellates. They derive their name from their flagellae and yellow pigments. These organisms conduct photosynthesis and live within other protists such as foraminiferans and radiolarians. They uti-lize the carbon dioxide produced meta-bolically by these hosts to manufacture sugars. These relationships are mutual or commensal, since the Zooxanthellae can leave and establish an independent ex-istence.

The presence of Zooxanthellae in coral is a stimulus to reef growth, since these protists increase the food supply available to the reef-building coral and coralline algae. See *coral reef, dinoflagellate, For-aminifera, flagella, Protozoa, Radiolaria.*

Zostera See *eelgrass.*

Appendix I

GEOLOGIC TIME SCALE

Cenozoic Era
    Quaternary Period
        Holocene (Recent): 0 to 1.5 million years ago
        Pleistocene 1.5 to 4 million years ago.
            The ice ages occurred in this period. In western Europe, the Alps were uplifted and the Mediterranean receded. In the east, the collision of India with Asia closed the Tethys Sea.
    Tertiary Period
        Pliocene about 10 million years ago
        Miocene   25 million years ago
        Oligocene   40   "   "   "
        Eocene   55   "   "   "
        Paleocene   65   "   "
Mesozoic Era
    Cretaceous, Upper 100 million years ago
        _____, Lower 135   "   "   "
            The primitive continent of Gondwanaland broke into the pieces now recognized as Africa, South America, and India. The western Alps were uplifted. Many species disappeared during this period, an event referred to as the Cretaceous extinctions.
    Jurassic, Upper 160 million years ago
        _____, Middle 170   "   "   "
        _____, Lower 190   "   "   "
            The Tethys Sea extends to the Pyrenees. There are shallow seas east of the Rockies in North America.
    Triassic, Upper 205 million years ago
        _____, Middle 215   "   "   "
        _____, Lower 225   "   "   "
The Tethys Sea, which at its greatest extension reached from the Atlas Mountains in North Africa to the Himalayas, opens eastward. Similar plants and animals exist in Africa, South America, and India.
Paleozoic Era
    Permian, Upper 240 million years ago
            The Paleozoic and particularly the Permian-Triassic transition is marked by extinctions of organisms, most noticeably the sessile ones, with the notable exception of the bivalves. The Permian epoch was particularly devastating, especially for organisms living near the surface of the sea. In general, the fish

populations followed those of the corals, ammonites, and crinoids, with many genera disappearing at the Permo–Triassic boundary.

_____, Lower 280 million years ago

Carboniferous, Upper (Pennsylvanian) 315 million years ago

_____, Lower (Mississippian) 350 million years ago

Major plant growth on land, the basis of coal deposits. The ferns were the dominant plants.

Devonian, Upper 360 million years ago

_____, Middle 370   "   "   "

_____, Lower 400   "   "   "

The folding of land rocks, such as in Australia. Fish were the dominant animals.

Silurian, 430 to 440 million years ago

The late Silurian is marked by the development of plants with cambium.

Ordovician, Upper 445 million years ago

_____, Lower 500 million years ago

The major diversification of invertebrates occurred in the mid-Ordovician. Calcium carbonate shells appear on mollusks. Corals, bryozoans, brachiopods, crinoids and nautiloids develop. Calcareous deposits are laid down over the Baltic Shield in Europe. Bone begins to form in some animals.

Cambrian, Upper 515 million years ago

_____, Middle 540   "   "   "

_____, Lower 570   "   "   "

At the division of the Cambrian and Precambrian, or somewhat later, the Caledonian geosyncline divides North America from Europe. Fossils and landforms that predate this event are identical. Terranes, inclusions of "foreign" territory, have been found in the Blue Ridge Mountains and the Piedmont Plateau in the southeastern United States.

Precambrian 4 to 1 billion years ago.

Sponges with horny or calcareous spicules appeared, carbonate plankton appeared about 1 billion years ago, filamentous green algae about 2 billion years ago. The first living system is very tentatively dated at about 4 billion years ago.

Appendix II

MARINE HISTORY: A CHRONOLOGY OF SIGNIFICANT EVENTS

Year	Person	Event
7th C BC	Kaleus	sailed through the Pillars of Hercules (Gibraltar)
609	Necho II	commissions Phoenicians to circumnavigate Africa
6th C	Necho II	starts canal to connect Red Sea to the Nile
600		Greek settlement at Marseille
5th C	Anaximander	first idea of evolution. He thought all land animals came from amphibians
4th C	Hanno of Carthage	sailed down the west African coast
350–300	Alexander the Great	commissioned Admiral Nearchus to explore the Persian Gulf and the Indian Ocean
344	Aristotle	on Lesbos studying marine biology
110		mariculture (oysters) near Naples
81	Sujin	Japanese emperor encourages shipbuilding and the fishing industry
40 AD	Hippalus	Greek merchant sailor goes from Berenice (Egyptian Red Sea port) to India and back using monsoon direction change
79	Pliny	dies in Vesuvian eruption
140	Ptolemy	publishes map of the world—longitude
2nd C		Chinese mariners sail to India
274	Ojin	Japanese long galley
450?	Hawaii Loa	Polynesian sailed from Raiatea to Hawaiian Islands

Year	Person	Event
7th C	Brendan	sails the North Atlantic
700?		Arab sailors in Spice Islands (Moluccas)
850		astrolabe in common use
874		Iceland discovered
900		Greenland discovered
982		Greenland colonized
998	Biarni Hierulfson	sights North America
1000	Leif Ericsson	colonized North America
1224	Abdullah ibn Ruml	geographical encyclopedia
1292		Venetian long galley developed
1312		Canary Islands rediscovered
1337	William Merlee	at Oxford. Attempted scientific weather forecasting
1405	Yung Lo	sends a fleet of 63 junks on the first of six expeditions. Cheung Lo, admiral. They go to Philippines, Aden, East Africa
1415		Dutch fishing boats using drift nets
1419		Portuguese land on Madeiras
1421	Henry the Navigator	establishes a scientific observatory at Sangres
1432	Goncàlo Cabral	discovers the Azores, for Portugal
1434	João Dias	Portuguese, sails around Cape Bojador (Africa)
1455	Alvise da Cadamosto	sailing for Prince Henry discovers Cape Verde Islands
1470		Portuguese sail to the Gold Coast (Africa)
1486		Portuguese sail to Angola
1487	Bartolomeu Dias	sails around Cape of Good Hope
1492	Christopher Columbus	sails for Spain to the Spice Islands

Year	Person	Event
1496	John Cabot	sails for Henry VII of England to North America
1497	Vasco da Gama	Portuguese, sails around Africa to India
1498	Columbus	discovers the Orinoco River
1500	Pedro Cabral	sails to South America and claims it for Portugal
1501	Amerigo Vespucci	sails to Brazil for Portugal. He realizes it is on a continent nowhere near Asia
1507	Martin Waldseemuller	map of the world with America as the name of the New World
1511		Portuguese reach Amboyna and Moluccas
1513	Juan Ponce de Leon	sights Florida
1517	Jorge Alvarez	Portuguese, sails to Canton
1519	Fernao Magellan	sails for Spain to the Orient by a westward route
1521	Francisco de Gordillo	sails up the coast of North America to the Carolinas—for Spain
1524	Giovanni da Verrazzano	sails into New York Bay
1534	Jacques Cartier	sails to Labrador for France
1538	Gerardus Mercator	uses the name America to mean all of the lands of the western hemisphere
1539	Olares Magnus	map of the world
1542	Antonio da Mata	first European in Japan
1551	Pierre Belon	*Histoire Naturelle des Etranges Poissons*
1553	Richard Chancellor	travels to Moscow via Archangel for England
1554	Guillaume Rondelet	*Livre des Poissons Marines*
1567	John Hawkins and Francis Drake	sail to the West Indies to harass the Spanish fleet
1569	Mercator	publishes his atlas, cosmographic and navigational maps of the world

Year	Person	Event
1570	Abraham Ortelius	first modern atlas
1576	Martin Frobisher	discovers Baffin Island and Frobisher Bay in the Canadian Arctic
1577	Francis Drake	begins his circumnavigation of the earth
1581	William Borough	*Discourse on the Variation of the Compass*
1582	Richard Hakluyt	*Diverse Voyages Touching on the Discovery of America*
1584	Walter Raleigh	discovers and annexes Virginia (England)
1585	John Davis	sails into Davis Strait
	Lucas Waggenart, or Wagenar or Wagenaer	*Spiegel de Zeewart*
1586	Thomas Cavendish	leaves on voyage of circumnavigation
1587	Konrad Gesner	posthumous publication of *History of Animals*
1591	James Lancaster	first voyage to the East Indies
1592	Juan de Fuca	discovers British Columbia (Canada)
1595		Dutch expedition to the Orient
1596	Willem Barents	Dutch explorer to north, finds Spitsbergen and Barents Sea
1598	Olivier van Noort	Dutch, circumnavigation
	Visunsin	Korea, first ironclad ship
1599	Richard Hakluyt	map of North America
1600	William Gilbert	*De Magnete,* theoretical discussion on magnetism
1602	Galileo Galilei	laws of oscillation and gravitation
1603	Samuel de Champlain	explores and names New France (Canada)
1606	Luis Vaes de Torres	sails between Australia and New Guinea
1609	Henry Hudson	explores Delaware Bay and Hudson River
1616	William Baffin	explores Baffin Bay

Year	Person	Event
	Willem Schouten and Jakob Le Maire	sail around the tip of South America and name the cape for Hoorn (Netherlands)
1617	Willibrod Snellius	uses trigonometric triangulation as a surveying technique
1622	Edmund Gunter	variation in the declination of the compass
1623		New Netherlands established
1626		Jardin des Plantes established in Paris
1627	Johannes Kepler	Rudolphine Tables of Fixed Stars
1642	Abel Tasman	discovers Tasmania and New Zealand
1643	Evangelista Torricelli	designs the barometer
1661	Robert Boyle	*Skeptikal Chymist*
1662	Charles II	charters the Royal Society
1676	John Flamsteed	Royal Observatory (Greenwich) Astronomer
1683	Isaac Newton	mathematical theory of tides as response to gravitation of Sun and Moon
1699	William Dampier	sails to the Pacific for the Admiralty
1701	William Kidd	the pirate, hanged
1714	Gabriel Fahrenheit	mercury thermometer designed
		British government establishes a prize for an accurate measure of longitude
1728	Vitus Bering	sails through Bering Strait
	John Harrison	spring driven clock wins Admiralty prize for longitude measure
1735	Carolus Linnaeus	*Systema Naturae*
1738	Daniel Bernoulli	*Hydrodynamica* explained pressure and velocity of fluids
1741	Alexei Cherikov	lands on California coast
1766	Louis de Bougainville	sails to the Pacific, discovers Tahiti

Year	Person	Event
1767		American whaling ships in Antarctica
1768	James Cook	sails to establish a scientific station in Tahiti
1769	Benjamin Franklin	charts the Gulf Stream
1771	Bougainville	circumnavigation
1772	Joseph Priestley	discovers nitrogen
1777	Antoine Lavoisier	composition of air determined
1785	Comte de La Pérouse	voyage to the Pacific
1788	James Hutton	*New Theory of the Earth*
1789		mutiny on the HMS *Bounty*
1791	George Vancouver	explores the western coast of North America
1798	Baron Cuvier	publishes *Elementary Survey of Natural History of Animals*
1800	Alexander von Humboldt	explores the Orinoco
1801		metric system becomes the only system of weights and measures in France
1802	John Dalton	publishes his work on the atomic theory
	Nathaniel Bowditch	*Practical Navigator*
1804	John Stevens	of Hoboken, New Jersey, launches a steamship. The *Little Juliana* is a technical success and a financial failure
1805	Francis Beaufort	scale of wind velocities
1807	Robert Fulton	commercial success of a steamboat, the *Clermont*
1809	Chevalier Lamarck	*Système des Animaux sans Vertèbres*
1812	Baron Cuvier	*Recherche sur les Ossements Fossiles des Quadrupeds*
1815		volcanic explosion of Tamboro (Indonesia) and resultant tsunami
1818	John Ross	Northwest Passage attempt. He finds cliffs in the Arctic near Baffin Bay

Year	Person	Event
1822	Chevalier Lamarck	*Histoire naturelle des Animaux*
1830	Charles Lyell	described geologic system of earth's formations
1831	Charles Darwin	sails on the *Beagle*
	James Clark Ross	determines the magnetic North Pole
1832	Nils Nordenskjold	sails the entirety of the Northeast Passage
1839	James Clark Ross	sails to the Antarctic with the *Erebus* and *Terror*
1840	Louis Agassiz	*Etudes sur Glaciers*
	Charles Darwin	*Zoology of the Voyage of the Beagle*
1842	Matthew Maury	begins his research on sounding of the ocean bottom
1846		Smithsonian Institution founded
1858	C. Darwin and A. Wallace	in a joint paper at the Linnaean Society present the Theory of the Survival of the Fittest
1859	Ferdinand de Lesseps	begins the Suez Canal
1865	Antonio Snider-Pelligrini	another proposal that the continents were once a single entity
1873	Julius Payer and Karl Weyprecht	discover Franz Josef Land—name it for the Austrian emperor
1876		H.M.S. *Challenger* returns to England
1904	Robert Scott	to the Antarctic
1906	F. Nansen	publishes his account of the Norwegian North Pole Expedition
1909	Robert Peary	first man at the North Pole
1911	Roald Amundsen	first man at the South Pole
1912	R. F. Scott	attempts to reach the South Pole, dies
1916	Paul Langevin	designs submarine detection device
1919	Ernest Shackleton	publishes *South*, the account of the 1914–1917 expedition to Antarctica

Year	Person	Event
1923	K. M. Deryugin	USSR polar research station
1925		Meteor Expedition to study the Atlantic
1926	Amundsen, Ellesworth and Nobile	fly over the North Pole in an airship
	Richard Byrd and Floyd Bennett	fly from Spitsbergen to the North Pole
		Scott Polar Research Institute established in Cambridge (England)
1935	William Ewing	seismic study of the sea floor
1947	R. E. Byrd	Little America, polar research station in the Antarctic
1957		International Geophysical Year (IGY) begins
1961	Robert Dietz	Continent and Ocean Basin Evolution
1977		discovery of vent communities

Appendix III

TAXONOMIC CLASSIFICATION OF LIVING ORGANISMS

Kingdom	Phylum	Class	Subclass
Monera No nuclear membrane	Schizophyta (bacteria) 1,500 species		
	Cyanophyta (blue-green algae) 200 species		
Protista Nuclear membrane	Chlorophyta (green algae) 7,000 species		
	Chrysophyta (yellow-brown algae) 6,000 species		
	Pyrrophyta (flagellated algae) 1,100 species		
	Phaeophyta (or Xanthophyta) (brown algae) 1,500 species		
	Rhodophyta (red algae) 4,000 species		
	Protozoa (no chlorophyll) 27,400 species	Mastigophora (dinoflagellates) 5,200 species	
		Sarcodina (foraminiferans) 11,500 species	
		Ciliophora 6,000 species	
Metaphyta Plants	Tracheophyta 287,200 species	Angiospermae (flowering plants) 275,000 species	

Kingdom	Phylum	Class	Subclass
Metazoa Animals	Porifera (sponges) 10,000 species	Calcarea (calcium sponges) 50 species	
		Desmospongiae 9,500 species	
		Sclerospongiae (coralline) 7 species	
		Hexactinellida (glass sponges) 450 species	
	Cnidaria (or Coelenterata) 10,000 species	Hydrozoa (polyps) 3,000 species	
		Scyphozoa (jellyfish) 250 species	
		Anthozoa (corals, sea anemones) 6,500 species	
		Cubozoa	
	Ctentophora (comb jellies) 80 species		
	Platyhelminthes (flatworms) 25,000 species	Turbellaria (free-living flatworms)	
		Monogenea	
		Trematoda (flukes)	
		Cestoda (tapeworms)	
	Nemertea (ribbon worms) 800 species		
	Aschelminthes (roundworms) (phylum is questionable. All the classes may be separate phyla.)	Nematoda 5,000 marine species	
		Rotifera	
		Gastrotricha	
		Kinorhyncha	
		Priapulida	
	Acanthocephala		

Kingdom	Phylum	Class	Subclass
	Ectoprocta (bryozoans) (mosses) 4,500 species Entoprocta 60 species		
	Phoronida (horseshoe worms)		
	Brachiopoda (lampshells) 300 species		
	Echiuroida		
	Sipuncula (peanut worms) 325 species		
	Chaetognatha (arrow worms) 50 species		
	Pogonophora (tube worms, or beard worms) 150 species		
	Mollusca (the classification of these animals is being revised) 80,000–100,000 species	Monoplacophora 10 species Amphineura (or polyplacophora; chitons) 600 species	
		Gastropoda (snails) 64,500 species	
		Lamellibranchia (or Pelecypoda) 7,500 species	
		Scaphopoda (tusk shells) 350 species	
		Cephalopoda (octopus) 600 species	
	Annelida (segmented worms) 10,000 marine species	Polychaeta (marine segmented worms)	
		Arachiannelida	
		Myzostomaria	
		Hirudinea (leeches)	
		Oligochaeta (earthworms)	

Kingdom	Phylum	Class	Subclass
	Arthropoda (segmented legs) 30,000 marine species	Tardigrada (water bears)	
		Chelicerata (or Xiphosurida) 4 species (horseshoe crab)	
		Pycnogonida (sea spider)	
		Crustacea (lobster, crab) 26,000 species	
			Ostracoda
			Copepoda
			Branchiura
			Cirripedia (barnacle)
			Malacostraca
		Insecta	
	Echinodermata 6,000 species	Crinoidea (sea lilies) 630 species	
		Asteroidea (starfish) 1,600 species	
		Ophiuroidea (brittle stars) 2,000 species	
		Echinoidea (sea urchins) 860 species	
		Holothuroidea (sea cucumbers) 900 species	
	Hemichordata 90 species	Enteropneusta (acorn worms)	
	Tunicata 1,375 species	Ascidiacea (sea squirts)	
		Thaliacea (salps)	
		Larvacea	
	Arcania 25 species	(Lancelets)	
	Chordata 52,500 species	Pisces (fish) Agnatha (jawless fish) 50 species	
		Chondrichthyes (sharks) 625 species	

Kingdom	Phylum	Class	Subclass
		Osteichthyes (bony fish) 30,000 species	
		Amphibia (frogs and salamanders) 2,600 species (almost none marine)	
		Reptilia (snakes, turtles, alligators) 6,500 species	
		Aves (birds) 8,600 species	
		Mammalia (seals, bears, whales) 4,100 species	

Appendix IV

MARINE SCIENCE RESEARCH PROJECTS

Large scientific oceanographic studies are presently international affairs. The days of the private amateur scientists—Agassiz was one—are long gone. Single nations are similarly finding it difficult to initiate and run expensive oceanic research efforts. Some of the best-known international efforts are:

Agency	Program
UNESCO	Intergovernmental Oceanographic Commission. Office of Oceanography. This group sponsored the International Indian Ocean Expedition and the International Tropical Atlantic Expedition
IHB	International Hydrographic Bureau, based in Monaco. It standardizes survey data and names
WMO	World Meteorological Organization
IIP	International Ice Patrol
ICSU	International Council of Scientific Unions
IAPSO	International Association for Physical Science of the Ocean
WDC	World Data Center. It disseminates material from the International Geophysical Year
DSDP	Deep Sea Drilling Project. This started in 1968 and its charge is to sample the Earth's crust below the sea, from depths of 6000 m and below
BOMEX	Barbados Oceanographic and Meteorological Experiment. This is sponsored by the United States, Canada, and Barbados
IDOE	International Decade of Ocean Exploration (1970–1980)

The Facts On File Dictionary of ARCHAEOLOGY: The best single volume for finding a variety of terms used in archaeological studies.
ISBN 0-8160-1893-6.

The Facts On File Dictionary of ASTRONOMY (Revised and Expanded Edition): Listing more than 2,300 entries, and using more than 85 diagrams, illustrations and tables, this is the most up-to-date reference to the field of astronomy.
ISBN 0-8160-1892-8.

The Facts On File Dictionary of BIOLOGY (Revised and Expanded Edition): An up-to-date book encompassing both the pure and applied aspects of plant science revised and expanded to include every aspect of biological studies.
ISBN 0-8160-2368-9.

The Facts On File Dictionary of CHEMISTRY (Revised and Expanded Edition): Clear definitions of the most important and most commonly used chemical terms.
ISBN 0-8160-2367-0.

The Facts On File Dictionary of GEOLOGY AND GEOPHYSICS: More than 3,000 clear and concise entries defining the entire range of geological phenomena.
ISBN 0-8160-1929-0.

The Facts On File Dictionary of MATHEMATICS (Revised and Expanded Edition): The most important and commonly used mathematical terms, including ones recently coined in fields such as computer graphics, artificial intelligence and robotics.
ISBN 0-8160-2365-4.

The Facts On File Dictionary of PHYSICS (Revised and Expanded Edition): 3,000 clear and concise definitions covering almost everything in physics from absolute temperature to Zener diode.
ISBN 0-8160-2366-2.